A LEVEL BIOLOGY

A G Toole
Vice Principal, Pendleton College, Salford

S M Toole
Head of Biology, Hulme Grammar School for Girls, Oldham

First published 1982
Revised 1984, 1986, 1988, 1991, 1993, 1995
Reprinted 1991, 1992, 1994, 1996

Letts Educational
Aldine House
Aldine Place
London W12 8AW
Tel: 0181 740 2266

British Library Cataloguing in Publication Data
A CIP record for this book is available from the British Library

ISBN 1 85758 335 3

All photographs reproduced by permission of Science Photo Library and Biophoto Associates.

Printed and bound in Great Britain by Ashford Colour Press

Letts Educational is the trading name of BPP (Letts Educational) Ltd

PREFACE

As the authors of this book we fundamentally rewrote the text in 1993 in order to take account of modern advances in Biology and the changed methods of learning and assessment now current in education. More of the book was given over to subject content, thus providing a more detailed coverage of all the major topics required by the examination boards. A greater number of diagrams was included to improve clarity. All nomenclature has been revised to take account of the recommendations of the Institute of Biology and the Association for Science Education. The inclusion of many new issues, especially those of a social and biotechnological nature, increased both the size and scope of the book making it appropriate for A- and AS-level students of both Biology and Human and Social Biology.

In this latest revision we have maintained that approach. The introduction of new syllabuses by all boards has meant that all topics have been revised in line with these changes; some have disappeared altogether to be replaced by new ones, such as biotechnology. Many diagrams have been updated and the total number has been increased to aid clarity and understanding.

We trust you will find this book informative and helpful and that it will stimulate further interest in Biology. We hope it will support your studies and so contribute to a satisfactory outcome at the end of your course. We wish you well in your efforts to achieve A-level success.

Susan and Glenn Toole 1995

CONTENTS

SECTION 1: STARTING POINTS

SECTION 2: BIOLOGY TOPICS

SECTION 3: TEST RUN

STARTING POINTS

In this section:

How to use this book

The structure of the book

Using your syllabus checklist

Syllabus checklists and paper analysis

Examination board addresses

Studying and revising Biology

The difference between GCSE and A/AS-level

Modular courses

Study strategy and techniques

Practical assessment and coursework

Revision techniques

The examination

Question styles

Examination techniques

Taking modular tests

Final preparation

HOW TO USE THIS BOOK

THE STRUCTURE OF THIS BOOK

The key aim of this book is to guide you in the way you tackle A-level Biology. It should serve as a study guide, work book and revision aid throughout any A-level/AS-level Biology course, no matter what syllabus you are following. It is not intended to be a complete guide to the subject and should be used as a companion to your textbooks, which it is designed to complement rather than duplicate.

We have divided the book into three sections. **Section One, Starting Points,** contains study tips and syllabus information – all the material you need to get started on your A-level study – plus advice on planning your revision and tips on how to tackle the exam itself. Use the **Syllabus Checklists** to find out exactly where you can find the study units which are relevant to your particular syllabus.

Section Two, the main body of the text, contains the core of A-level Biology. It has been devised to make study as easy – and enjoyable – as possible, and has been divided into chapters which cover the themes you will encounter on your syllabus. The chapters are split into units, each covering a topic of study.

The **Chapter Objectives** direct you towards the key points of the chapter you are about to read. The **Chapter Roundup** at the end of the chapter gives a summary of the text just covered, brings its topics into focus and links them to other themes of study. To reinforce what you have just read and learned, there are **Illustrative Questions and Worked Answers** at the end of each chapter. All questions are similar to those recently set by the examination boards (including Scottish Higher). The tutorial notes and suggested answers give you practical guidance on how to answer A-level questions, and provide additional information relevant to that particular topic. There is also a **Question Bank**, with further examples of A-level exam questions for you to attempt.

In **Section Three, Test Run,** we turn our attention to the examination you will face at the end of your course. First, you can assess your progress using the **Test Your Knowledge Quiz** and analysis chart. Then, as a final test, you should attempt the **Mock Exam,** under timed conditions. This will give you invaluable examination practice and, together with the specimen answers specially written by the authors, will help you to judge how close you are to achieving your A-level pass.

USING YOUR SYLLABUS CHECKLIST

Whether you are using this book to work step-by-step through the syllabus or to structure your revision campaign, you will find it useful to use our checklist to record what you have covered – and how far you still have to go. Keep the checklist at hand when you are doing your revision – it will remind you of the chapters you have revised, and those still to be done.

The checklist for each examination – A, AS or Higher Grade – is in two parts. First there is a list of topics covered by this book which are part of the syllabus. Although the checklists are detailed, it is not possible to print entire syllabuses. **You are therefore strongly recommended to obtain an official copy of the syllabus for your examination and consult it when the need arises.** The examination board addresses are given after the syllabus checklists.

When you have revised a topic make a tick in the column provided and, if there are questions elsewhere in the book, try to answer them.

The second part of the checklist gives you information about the examination, providing useful details about the time allocated for each paper and the weighting of the questions on each paper. The different types of questions which may be set are explained in detail later in this section under the heading The Examination.

SYLLABUS CHECKLISTS AND PAPER ANALYSIS

ASSOCIATED EXAMINING BOARD
A level (terminal) 0607

Syllabus topic	Covered in Unit No	✓
Cell ultrastructure and microscopy	1.2	
Molecules of biological importance	1.1	
Enzymes	5.1	
Cellular respiration and ATP	5.5	
Photosynthesis	5.2	
DNA and the genetic code	3.1	
Mitosis and meiosis	3.2	
Variation	3.4	
Natural selection and evolution	3.4	
Ecosystems	6.2	
Distribution of organisms	6.1	
Populations	6.1	
Energy flows in a community	6.2	
Recycling of nutrients	6.2	
Human activities and conservation	6.3	
Classification	2.1, 2.4	
Digestion	5.1, 5.3	
Saprobionts and parasites	5.4	

Syllabus topic	Covered in Unit No	✓
Gas exchange	7.1	
Transport in flowering plants	7.3	
Transport in animals	7.2	
Immunity	7.2	
Homeostasis	8.1	
Liver	8.1	
Nerve impulse	8.4	
Receptors	8.4	
Hormones	8.2, 8.3	
Behaviour	8.4	
Support	1.3, 7.3, 8.5	
Muscular contraction	8.5	
Asexual reproduction	4.1	
Sexual reproduction	4.2, 4.3	
Coordination of reproduction	4.2, 4.3	
Development	4.4	
Adaptations to extreme conditions	8.1	

Paper analysis

Number of written papers: 2, plus practical coursework assessment

Paper 1	*2 hours*	20 short-answer questions Drawn from whole syllabus All compulsory Testing recall, application and interpretation 35% of total marks
Paper 2	*3 hours*	Essay, comprehension, interpretation and analysis questions 5 questions to be answered Choice of essays but other questions compulsory 45% of total marks
Coursework		Teacher assessment of practical coursework 2 pieces of practical work to be planned, carried out, observations made, evaluated and conclusions drawn.

ASSOCIATED EXAMINING BOARD
A level 0601 (modular)

Syllabus topic	Covered in Unit No	✓
MODULE 1 THE ORGANISATION OF LIVING ORGANISMS		
Cell ultrastructure and microscopy	1.2	
Molecules of biological importance	1.1	
Enzymes	5.1	
Cellular respiration and ATP	5.5	
Photosynthesis	5.2	
Homeostasis	8.1	
MODULE 2 VARIATION AND THE MECHANISMS OF INHERITANCE AND EVOLUTION		
DNA and the genetic code	3.1	
Mitosis and meiosis	3.2	
Variation	3.4	
Natural selection and evolution	3.4	
Ecosystems	6.2	
Distribution of organisms	6.1	
Populations	6.1	
Energy flows in a community	6.2	
Recycling of nutrients	6.2	
Human activities and conservation	6.3	

Syllabus topic	Covered in Unit No	✓
MODULE 3 SUPPLY AND DEMAND IN LIVING ORGANISMS		
Classification	2.1, 2.4	
Photosynthesis	5.2	
Digestion	5.1, 5.3	
Saprobionts and parasites	5.4	
Gas exchange	7.1	
Transport in flowering plants	7.3	
Transport in animals	7.2	
Immunity	7.2	
Homeostasis (liver and kidney)	7.4, 8.1	
Nerve impulse	8.4	
Receptors	8.4	
Hormones	8.2, 8.3	
Behaviour	8.4	
Support	1.3, 7.3, 8.5	
Muscular contraction	8.5	
Asexual reproduction	4.1	
Sexual reproduction	4.2, 4.3	
Coordination of reproduction	4.2, 4.3	
Development	4.4	
Adaptations to extreme conditions	8.1	

Paper analysis

4 compulsory modules

4 × 1¼ hour tests	4 × 20% total marks
Teacher-assessed practical	20% total marks

AS level 0970 (modular)

Content: Modules 1 and 2 of the A-level modular syllabus

Paper analysis

2 × compulsory modules

2 × 1¼ hour tests	2 × 40% total marks
Teacher assessed practical	20% total marks

ASSOCIATED EXAMINING BOARD
AS level (terminal) 0979

Syllabus topic	Covered in Unit No	✓
Cell ultrastructure and microscopy	1.2	
Molecules of biological importance	1.1	
Enzymes	5.1	
Cellular respiration and ATP	5.5	
Photosynthesis	5.2	
DNA and the genetic code	3.1	
Mitosis and meiosis	3.2	
Variation, natural selection and evolution	3.4	
Ecosystems	6.2	
Populations	6.1	

Syllabus topic	Covered in Unit No	✓
Energy flow and recycling of nutrients	6.2	
Human activities and conservation	5.3	
Classification	2.1, 2.4	
Digestion	5.1, 5.3	
Gas exchange	7.1	
Transport	7.2, 7.3	
Homeostasis	8.1	
Nerve impulse and receptors	8.4	
Hormones	8.2, 8.3	
Behaviour	8.4	
Reproduction	4.1	

Paper analysis

Number of written papers: 1, plus coursework

Paper 1	*2½ hours*	Short-answer, structured and essay questions Approx. 10 short-answer questions, 3 structured questions and 3 essay questions Short-answer and structured questions are compulsory; 1 essay chosen from 3 80% of total marks
Coursework		Teacher assessment of practical coursework 20% of total marks

UNIVERSITY OF CAMBRIDGE LOCAL EXAMINATIONS SYNDICATE
A level (terminal)

Syllabus topic	Covered in Unit No	✓
Cell structure	1.2	
Biological molecules	1.1	
Enzymes	5.1	
Cell and nuclear division	3.2	
Genetic control and inheritance	3.1	

Syllabus topic	Covered in Unit No	✓
Inherited change and evolution	3.3, 3.4	
Energetics	5.2, 5.5	
Ecology	6.2, 6.3	
Transport	7.2, 7.3	
Regulation and control	7.4, 8.1, 8.4	

Two options chosen from: Biodiversity; Applied plant and animal science; Application of genetics; Growth, development and reproduction

Paper analysis

Number of written papers: 3, plus assignment and coursework

Paper 1 *2½ hours* Structured and free-response questions on option topics
All compulsory
35% of total marks

Paper 2 *1 hour* Multiple-choice questions on core topics
40 questions
All compulsory
15% of total marks

Paper 3 *1½ hours* 2 sections on core topics
Section A: consists of compulsory short-answer questions
Section B: 3 free-response questions from which **one** is chosen
30% of total marks

Assignment Teacher-assessed investigative assignment
10% of total marks

Coursework Teacher assessment of practical coursework
10% of total marks

UNIVERSITY OF CAMBRIDGE LOCAL EXAMINATIONS SYNDICATE
AS level (terminal)

Syllabus topic	Covered in Unit No	✓
Cell structure	1.2	
Biological molecules	1.1	
Enzymes	5.1	
Cell and nuclear division	3.2	
Genetic control and inheritance	3.1	

Syllabus topic	Covered in Unit No	✓
Inherited change and evolution	3.3, 3.4	
Energetics	5.2, 5.5	
Ecology	6.2, 6.3	
Control and communication	8.3, 8.4	

One option chosen from: Biodiversity or Human health and disease

Paper analysis

Number of written papers: 2, plus assignment and coursework

Paper 1 *1½ hours* 2 sections on core topics
Section A: compulsory short-answer questions.
Section B: 3 free-response questions, from which **one** is chosen
55% of total marks

Paper 2 *¾ hour* Questions on the option
25% of total marks

Paper 3 Teacher assessment of extended investigation
20% of total marks

UNIVERSITY OF CAMBRIDGE LOCAL EXAMINATIONS SYNDICATE
AS and A level (modular) 8534, 9534

Syllabus topic	Covered in Unit No	✓
MODULE 4801: FOUNDATION*		
Cell structure	1.2	
Cell and nuclear division	3.2	
Biological molecules	1.1	
Enzymes	5.1	
Cell membranes	1.2	
Control and communication	8.1, 8.3, 8.4	
MODULE 4802: CENTRAL CONCEPTS IN BIOLOGY*		
Energy and photosynthesis	5.1, 5.2	
Respiration	5.5	
Energy and ecosystems	6.2	
Genetic control of development	3.1	
Passage of information to offspring	3.3	
Selection and evolution	3.4	

Syllabus topic	Covered in Unit No	✓
OPTIONAL MODULES		
MODULE 4803: ECOLOGY AND CONSERVATION	5.4, 6.1, 6.2, 6.3	
MODULE 4804: TRANSPORT, REGULATION AND CONTROL	7.2, 7.3, 7.4, 8.1	
MODULE 4805: GROWTH, DEVELOPMENT AND REPRODUCTION	4.1, 4.2, 4.3, 4.4, 8.2	
MODULE 4806: MICROBIOLOGY AND BIOTECHNOLOGY	2.2, 2.3	
MODULE 4807: APPLICATIONS OF GENETICS	3.1, 3.3, 3.4	
MODULE 4808: HUMAN HEALTH AND DISEASE	5.3, 7.1, 7.2	
MODULE 4845: TRANSPORT†	7.2, 7.3	

* compulsory module † AS syllabus only

Paper analysis

At A level, students study five content-based modules and one Experimental Skills module. All modules have an equal weighting of 16⅔%.

Content-based modules

2 compulsory modules plus 3 of the optional modules. Candidates may choose as an optional module not more than one of the following: 4841 Instrumentation electronics, 4842 Scientific communication, 4843 Food technology.

Each is assessed by an end of module paper of 1½ hours. This consists of two sections:
Section A: compulsory short-answer/comprehension questions
Section B: one free-response question chosen from 2 options

Coursework/practical examination

Teacher assessment of an extended investigation and either teacher assessment of coursework or a practical examination.

At AS level, students study the two compulsory, content-based modules, plus module 4845. All modules have an equal weighting of 33⅓%.

Content-based modules

Modules 4801 and 4802. As for A level, each module is assessed by an end-of-module paper of 1½ hours.

AS-module

Module 4845 is a half-module of subject content and a half-module of experimental skills assessment.

Module 4845 is assessed by:
Written paper of ¾ hour: compulsory short-answer/comprehension questions
and either, separate assessment of 3 skills during the course:
A planning
B implementing
C interpreting and concluding
Or, an extended investigation.

UNIVERSITY OF LONDON EXAMINATIONS AND ASSESSMENT COUNCIL
AS and A level (modular) 8041, 9041

Syllabus topic	Covered in Unit No	✓
CELL BIOLOGY AND GENETICS*†		
Cells and organelles	1.2	
Molecules	1.1	
Enzymes	5.1	
Metabolic pathways	5.5	
Chromosomes and the genetic code	3.1, 3.2	
Patterns of inheritance	3.3	
Sources of new inherited variation	3.1, 3.4	
THE ORGANISM AND THE ENVIRONMENT*†		
Autotrophic nutrition	1.1, 5.2	
Heterotrophic nutrition	5.1, 5.3, 5.4	
Regulation of internal environment	7.3, 7.4, 8.1	
Adaptations to the environment	2.1, 2.3, 2.4, 6.1	
Ecosystems	6.1, 6.2	

Syllabus topic	Covered in Unit No	✓
Environmental change and evolution	3.4	
Human influences on ecosystems	6.3	
SYSTEMS AND THEIR MAINTENANCE*		
Exchanges with the environment	7.1, 7.4	
Transport systems	7.2, 7.3	
Sensory receptors	8.4	
Chemical coordination	8.2, 8.3	
Nervous coordination in mammals	8.4	
Support and movement	1.3, 7.3, 8.5	
Reproduction	4.1, 4.2, 4.3	
MICROORGANISMS AND BIOTECHNOLOGY	2.2	
APPLIED PLANT BIOLOGY	1.3, 3.4, 4,1, 4.3, 5.2, 8.2	
APPLIED ANIMAL BIOLOGY	3.4, 4.3	
FOOD AND HEALTH	2.2, 3.4, 5.3, 7.2	

* compulsory module (A level) † component modules of AS syllabus

Paper analysis

Theory

For A level there are 3 compulsory modules plus any 1 other module.
For AS there are 2 compulsory modules. Each module is assessed by a modular test, which may be taken in January or June during or at the end of the course. Each *module test* is a paper of 1 hour 20 minutes, which consists of 3 sections.

- Short structured questions:
  5 compulsory questions of 4–7 marks
- Longer structured questions:
  2 compulsory questions of 11–12 marks
- Free-prose question:
  1 compulsory question of 10 marks

Each module test comprises 30% of total mark for AS and 15% for A level

Synoptic paper *2 hours* (A level)

- Short structured questions:
  2 compulsory questions of 7–8 marks
- Longer structured questions:
  3 compulsory questions of 14–15 marks
- Essay: 1 question of 20 marks

Synoptic paper *1 hour 20 minutes* (AS)

- Longer structured questions:
  2 compulsory questions of 12–13 marks
- Essay: 1 question of 15 marks

Each synoptic paper comprises 20% of total mark

Coursework

Teacher assessment of coursework, including individual studies. 20% of total marks

NORTHERN EXAMINATIONS AND ASSESSMENT BOARD
A level (terminal) 4161

Syllabus topic	Covered in Unit No	✓
Biological molecules	1.1	
Cells	1.2, 5.1	
Energy relationships in living organisms	5.2, 5.5	
Uptake of materials	5.1, 5.3, 7.1, 7.3, 8.1, 8.2, 8.3	
Transport and waste removal	7.2, 7.3, 7.4	
Coordination and homeostasis	8.1, 8.2, 8.3, 8.4, 8.5	
Variation	3.4	

Syllabus topic	Covered in Unit No	✓
Genetic material	3.1	
Cell division and continuity	3.2	
Mechanisms of inheritance	3.3, 3.4	
Natural selection	2.1, 3.4	
Ecosystems	6.1	
Energy transfer and nutrient cycles	6.2	
Adaptation and diversity	4.3, 6.1, 6.2, 7.3, 7.4	
Human activity and the environment	6.3	

One option chosen from: Microorganisms and biotechnology; Biology of food production; Health and disease; Social biology

Paper analysis

Number of written papers: 2, plus coursework

Paper 1 *3 hours* Compulsory short-answer and structured questions
45% of total marks

Paper 2 *2½ hours* 1 essay question chosen from 3; 1 compulsory comprehension question and 1 hour optional topic test of short-answer and structured questions
37% of total marks

Coursework Teacher assessment of practical coursework
18% of total marks

NORTHERN EXAMINATIONS AND ASSESSMENT BOARD
A level (modular) 4164, AS level (modular) 3164, AS level (terminal) 3161

Syllabus topic	Covered in Unit No	✓
PROCESSES OF LIFE*		
Biological molecules	1.1	
Cells	1.2	
Enzymes	5.1	
Photosynthesis	5.2	
Respiration	5.5	
Energetics	6.2	
Nutrient cycles	6.2	
Response to external environment	8.3, 8.4	
Homeostasis	8.1	
CONTINUITY OF LIFE*		
Variation	3.4	
Genetic information	3.1	
Cell division	3.2	
Mutations	3.4	
Patterns of inheritance	3.3	

Syllabus topic	Covered in Unit No	✓
Natural selection and diversity	2.1, 3.4	
Interactions between organisms	6.1	
Human influences on environment	6.3	
PHYSIOLOGY	5.3, 7.1, 7.2, 7.3, 7.4	
BIOLOGICAL BASIS OF BEHAVIOUR	8.2, 8.3, 8.4, 8.5	
ECOLOGY	6.1, 6.2, 6.3, 7.3, 7.4	
MICROORGANISMS AND BIOTECHNOLOGY	2.2, 5.1	
BIOLOGY OF FOOD PRODUCTION	3.1, 3.4, 4.1, 5.2, 6.3, 8.2, 8.3	
HEALTH AND DISEASE	2.2, 3.4, 5.3, 7.1, 7.2, 8.1	
SOCIAL BIOLOGY	3.4, 4.2, 6.3, 7.2, 8.3	

* compulsory modules (A and AS level)

Modular A level comprises two compulsory modules and either four other Biology modules or three Biology modules and one other taken from a related discipline within the modular suite. AS comprises two compulsory modules plus one other.
The AS level course can be assessed either in modules or by an end-of-course assessment.

Paper analysis

Theory *3 hours*

End of module test *($1\frac{1}{2}$ hours)*
- Section A: short-answer questions (*1 hour*)
- Section B: structured questions that may require continuous prose *($\frac{1}{2}$ hour)*

All questions are compulsory
Each module test:
- $13\frac{2}{3}$% of total marks (A level)
- $27\frac{1}{3}$% of total marks (AS level)

Coursework

Teacher assessed coursework within each module
- 3% of total marks (A level)
- 6% of total marks (AS level)

NORTHERN IRELAND COUNCIL FOR THE CURRICULUM EXAMINATIONS AND ASSESSMENT
A level

Syllabus topic	Covered in Unit No	✓
Molecules of biological importance	1.1	
Enzymes	5.1	
Cells and their organelles	1.2	
Mammalian tissues	1.3	
Angiosperm tissues	1.3	
Variety of organisms	2.1, 2.2, 2.3, 2.4	
Populations	6.3	
Mitosis	3.2	
Meiosis	3.2	
Inheritance	3.3	
Protein synthesis	3.1	
Variation	3.4	
Evolution	3.4	
Asexual reproduction	4.1	
Reproduction in flowering plants	4.3	
Mammalian reproduction	4.2	
Autotrophic nutrition (photosynthesis)	5.2	

Syllabus topic	Covered in Unit No	✓
Heterotrophic nutrition	5.3	
Respiration	5.5	
Economic role of microorganisms	2.2	
Energy flow through the ecosystem	6.2	
Carbon cycle	6.2	
Nitrogen cycle	6.2	
Plant hormones	8.2	
Thermoregulation	8.1	
Osmoregulation	7.4	
Receptors	8.4	
Nervous coordination	8.4	
Support in plants	1.3, 7.3	
Support in animals	1.3, 8.5	
Uptake and transport in plants	7.3	
Mammalian circulatory system	7.2	
Gas exchange	7.1	
Nitrogenous excretion	7.4	

Paper analysis

Number of written papers: 3, plus coursework

Paper 1 *$1\frac{1}{2}$ hours*

Short-answer questions
- Number of questions varies
- All compulsory
- 20% of total marks

Paper 2 *2½ hours* Between 5 and 10 structured questions
All compulsory
30% of total marks

Paper 3 *2½ hours* Various styles but including an element of free response
3 questions to be answered
Limited either/or choice
30% of total marks

Coursework Teacher assessment of practical coursework
20% of total marks

NUFFIELD (NORTHERN EXAMINATIONS AND ASSESSMENT BOARD) A level

Syllabus topic	Covered in Unit No	✓
Gas exchange	7.1, 7.2	
Circulation and transport	7.2, 7.3	
Cells, organisms and chemical reactions	1.1, 1.2, 5.1, 5.3, 5.5, 8.1	
Heterotrophic nutrition	5.3, 8.1	
Photosynthesis	5.2	
Water relations of cells and organisms	7.3, 7.4	
Excretion	7.4	
Control systems	8.1, 8.3, 8.4, 8.5	
Responses to stimuli	8.2, 8.4	
Behaviour	8.4	
Reproduction and life cycles	4.1, 4.2, 4.3	

Syllabus topic	Covered in Unit No	✓
Development	4.4, 8.2	
Cell division	3.2, 2.2, 3.4	
Mendelian genetics	3.3	
Structure and function of nucleic acids	3.1	
Gene action	3.1, 3.3	
Population genetics and evolution	3.4	
Ecology	6.1, 6.2, 6.3	
Systematics and classification	2.1	
Immunology	7.2	
Human health and medical technology	4.2, 7.2, 7.2, 7.4	
Biotechnology	2.2, 3.1	

Paper analysis

Number of written papers: 2, plus coursework

Paper I *3 hours* Section A: 7 compulsory structured questions (*2 hours*)
Section B: compulsory questions testing understanding of a written passage (*1 hour*)
40% of total marks

Paper II *3 hours* Section A: 30 compulsory objective test questions (*1 hour*)
Section B: an answer in continuous prose (*½ hour*)
Section C: 3 questions out of 8 (*1½ hours*)
40% of total marks

Practical assessment A written project
20% of total marks

UNIVERSITY OF OXFORD DELEGACY OF LOCAL EXAMINATIONS
A and AS level (modular)

Syllabus topic	Covered in Unit No	✓
MOLECULES AND LIFE*		
Molecules	1.1	
Enzymes	5.1	
Photosynthesis	5.2	
Energetics	6.2	
Cycling of nutrients	6.2	
Respiration	5.5	
CELLS AND STRUCTURES*		
Viruses	2.2	
Cell membranes and organelles	1.2	
Tissues	1.3	
Homeostasis	8.1	
Cell division	3.2	

Syllabus topic	Covered in Unit No	✓
GENETICS AND EVOLUTION†		
DNA and genes	3.1	
Inheritance	3.3	
Variation	3.4	
Evolution	3.4	
Classification	2.1, 2.3, 2.4	
BIOTECHNOLOGY	2.2, 3.1, 5.1	
HUMAN HEALTH AND DISEASE	2.2, 3.3, 3.4, 5.4, 7.1, 7.2, 7.4	
ENVIRONMENTAL BIOLOGY	5.4, 6.1, 6.2, 6.3	
PLANT STUDIES	1.1, 3.1, 4.1, 4.3, 5.2, 5.5, 6.2, 7.3, 8.2	
ANIMAL STUDIES	4.2, 5.3, 7.1, 7.2, 7.4, 8.1, 8.3, 8.4, 8.5	

* compulsory module for AS and A level

† compulsory for A level; part compulsory for AS level

A level comprises the **three** compulsory modules and **two** others, plus a coursework module.

Modules may be assessed together at the end of the course, or during the course, in December, March and June.

AS level comprises two compulsory modules, plus module 3, which is half of the Genetics and Evolution module and coursework.

Paper analysis

Each module test	*(1½ hours)*	All questions compulsory and short-answer or structured. Each module comprises 16% of the total marks at A level, 32% at AS level.
AS module 3 test	*(¾ hour)*	One written paper on part of the Genetics and Evolution module 16% of the total marks.
Coursework		Teacher-assessed coursework, including a longer investigation 20% of total marks

OXFORD AND CAMBRIDGE SCHOOLS EXAMINATION BOARD
A and AS level (terminal) 9672, 8400

Syllabus topic	Covered in Unit No	✓
CELLULAR ACTIVITIES*		
Cells	1.2	
Biological molecules	1.1	
Respiration	5.5	
Enzymes	5.1	
GENETICS AND EVOLUTION*		
Nucleic acids	3.1	
Genetic code	3.1	
Cell division	3.2	
Genetics	3.3	
Mutations	3.4	
Evolution	3.4	
Classification	2.1	
ECOLOGY AND ENVIRONMENTAL PHYSIOLOGY*		
Photosynthesis	5.2	
Energetics	6.2	
Cycling of nutrients	6.2	

Syllabus topic	Covered in Unit No	✓
Populations	6.1	
Effect of human activity	6.3	
Response to changes in environment	8.2, 8.3, 8.4	
PHYSIOLOGY OF ANIMALS AND PLANTS	4.2, 5.3, 7.1, 7.2, 7.3, 7.4, 8.1, 8.4, 8.5	
DIVERSITY OF ORGANISMS	2.1, 2.2, 2.3, 2.4, 3.4, 4.3, 7.1	
COLONISATION OF LAND	2.3, 4.2, 4.3, 5.2, 7.1, 7.3, 7.4, 8.5	
HUMAN HEALTH AND DISEASE	5.3, 7.1, 7.2, 8.1, 8.5	
ANIMAL BEHAVIOUR	4.2, 8.3, 8.4	
CONSERVATION	3.1, 6.3	
BIOTECHNOLOGY	2.2, 3.1, 5.1, 6.3, 7.2	

* Compulsory units for AS and A level. In addition, A-level candidates study three optional units.

Paper analysis

Paper 1 (A and AS) *3 hours*	Short-answer, structured and extended answers. All compulsory 40% of total marks for A level 80% of total marks for AS level
Papers 2–8 (A level only)	3 1-hour papers on optional units Short and structured questions 40% of total marks
Coursework	Teacher assessment of practical work 20% of total marks

OXFORD AND CAMBRIDGE SCHOOLS EXAMINATION BOARD

Structured science scheme (Biology components)

Within the scheme

A level 9683 – 6 modules
AS level 8383 – 3 modules
Double award – 12 modules

AS-Biology modules – B1, B2, B3 (all compulsory)
A-level Biology – B1, B2, B3 (compulsory) and 3 chosen from B4, B5 or B6, B7, B8, B9, B10, S1

Syllabus topic	Covered in Unit No	✓
MODULE B1: CELLULAR ACTIVITIES		
Structure of plant and animal cells	1.2	
Tissues and organs	1.3	
Molecules	1.1	
Membrane structure	1.2	
Energy transformations	5.5	
Properties of enzymes	5.1	
MODULE B2: GENETICS AND EVOLUTION		
Nucleic acids	3.1	
Protein synthesis	3.1	
Mitosis	3.2	
Meiosis	3.2	
Mendelian genetics	3.3	
Evolution	3.4	
Genetic engineering	3.1	
Taxonomy and classification	2.1	
MODULE B3: ECOLOGY AND ENVIRONMENTAL PHYSIOLOGY		
Photosynthesis	5.2	
Energy and ecosystems	6.2	
Populations	6.1	
Nutrient cycles	6.2	
Man and the environment	6.3	
Response to the environment	4.2, 8.1, 8.2, 8.3, 8.4	
MODULE B4: PHYSIOLOGY OF ANIMALS AND PLANTS	4.2, 5.3, 7.1, 7.2, 7.3, 7.4, 8.1, 8.4, 8.5	
MODULE B5: DIVERSITY OF ORGANISMS	2.1, 2.2, 2.3, 2.4, 3.4	
MODULE B6: COLONISATION OF LAND	2.3, 4.2, 4.4, 5.2, 7.1, 7.3, 7.4, 8.5	
MODULE B7: HUMAN HEALTH AND DISEASE	5.3, 7.1, 7.2, 8.5	
MODULE B8: ANIMAL BEHAVIOUR	4.2, 8.3, 8.4	
MODULE B9: CONSERVATION	6.4	
MODULE B10: BIOTECHNOLOGY	2.2, 3.1, 5.1	
MODULE S1: SCIENCE AND TECHNOLOGY IN SOCIETY	2.2, 3.4, 4.2, 6.3	

Scheme of assessment

Each module is examined by means of a test of 1½ hours, comprising compulsory questions. In total these make up 80% of the total mark.

Teacher-assessed experimental and investigation work comprise the remaining 20%

SCOTTISH EXAMINATION BOARD

Higher grade

Syllabus topic	Covered in Unit No	✓
Variety of cells	1.2, 1.3	
Cell membrane	1.2	
Photosynthesis	5.2	
Energy release	1.2, 5.5	
Proteins	1.1, 3.1	
Viruses	2.2	
Cellular defence	7.2	
Meiosis	3.2	
Inheritance	3.3	
Mutation	3.4	
Selection	3.1, 3.4	
Growth	4.4	
Genetic control	3.1	
Animal hormones	8.3	
Plant hormones	8.2	
Mineral requirements	1.1, 5.1, 5.3	
Homeostasis	7.4, 8.1	
Population dynamics	6.1	
Water balance	7.3, 7.4	
Obtaining food	6.1	

Paper analysis

Number of written papers: 2

Paper 1 *(1¼ hours)* 40 multiple choice questions
33.33% of total marks

Paper 2 *(2½ hours)* Section A: structured questions
Section B: 3 questions, 2 based on data interpretation and 1 assessing aspects of compulsory practical activities
Section C: 2 essays out of 4
66.66% of total marks

WELSH JOINT EDUCATION COMMITTEE
A and AS level (terminal)

Syllabus topic	Covered in Unit No	✓
SCIENTIFIC BASIS OF LIFE†		
Biological compounds	1.1	
Cell organisation	1.2, 1.3	
Enzymes	5.1	
Role of ATP	5.5	
Photosynthesis	5.2	
Respiration	5.5	
INTERACTIONS WITH THE ENVIRONMENT†		
Energy transfer	6.1, 6.2	
Response to environment	7.4, 8.1, 8.2, 8.3, 8.4	
Populations	2.1, 3.4, 6.1	
Variation	3.1, 3.3, 3.4	
Inheritance and evolution	3.3, 3.4	
MICROORGANISMS AND MAN†		
Eukaryotes and prokaryotes	1.2, 2.2	
Microorganisms	2.2	
Bioengineering	2.2, 3.1	
Immunity	7.2	

Syllabus topic	Covered in Unit No	✓
BIOLOGICAL RESOURCES AND THEIR PRODUCTION		
Populations	6.1	
Exploitation by humans	6.3	
Plant reproduction	4.3, 4.4	
Plant hormones	8.2	
Animal reproduction	4.2, 8.3	
Cloning	3.1, 4.1	
Selection	3.4	
Pest control	6.3	
EXCHANGE AND TRANSPORT		
Gas exchange	7.1	
Photosynthesis	5.2	
Heterotrophic nutrition	5.3, 5.4	
Transport in plants	7.3	
Blood system	7.2	
INFORMATION PROCESSING		
Response to the environment	8.2, 8.3, 8.4	
Nervous system	8.4	
Movement	8.5	
Plant hormones	8.2	

For A level all topics are studied.
For AS level only those topics marked with a dagger (†) are studied.

Paper analysis

A level

Paper A1: *(2½ hours)* 4 essays from a choice of 7
40% of total marks

Paper A2 *(2½ hours)* Variety of compulsory structured questions
40% of total marks

AS level:
One written paper *(2½ hours)* Examination will be on the units marked with a dagger (†)
80% of total marks
Section A: compulsory structured questions
Section B: 2 essays out of 4

Coursework: For both A and AS level there will be internal assessment of investigational skills; 20% of total marks.

WELSH JOINT EDUCATION COMMITTEE
A and AS level (modular)

The A-level modular course is first available for award in 1997 (1996 for the modular AS course).

Syllabus topic	Covered in Unit No	✓
MODULE 1: ORGANISATION OF LIVING SYSTEMS		
Molecular structure	1.1	
Biological organisation	1.2, 1.3	
Molecular uptake	1.2	
Enzymes	5.1	
Energy	5.1	
Photosynthesis and energy	5.2	
Respiration releases energy	5.1	
Energy transfer	6.2	
Responses to internal and external stimuli	7.4, 8.1, 8.2, 8.3, 8.4	
MODULE 2: INTERACTIONS BETWEEN THE ORGANISM, ENVIRONMENT AND MAN		
Genetic variation/transcription	3.1, 3.2	
Genetic principles	3.1, 3.3	
Evolution/classification	2.1, 2.2, 2.3, 2.4, 3.4	
Structure: prokaryotic and eukaryotic organisms	1.2, 2.2	
Laboratory culture of microorganisms	2.2	
Population growth	2.2	
Applications for microorganisms	2.2, 6.3	
Genetic engineering	3.1	
Harmful microorganisms	2.2	
Deterioration of food	2.2	
Control of pathogens	2.2, 7.2	
Food preservation	2.2	

Syllabus topic	Covered in Unit No	✓
MODULE 3: BIOLOGICAL RESOURCES AND THEIR PRODUCTION		
Population control	6.1	
Management of ecosystems	6.3	
Instability of managed ecosystems	6.3	
Sexual reproduction in plants	4.3	
Plant hormones: economic implications	8.2	
Sexual reproduction in animals	4.2	
Animal hormones in food production	4.2	
Genetic cloning	4.1	
Artificial selection	3.4	
Control of pests and diseases	6.3	
Autotrophic nutrient supply	5.2, 7.3	
Heterotrophic nutrition/digestion	5.3	
MODULE 4: TRANSPORT AND COORDINATION		
Exchange and transport systems	7,1, 7.2, 7.3	
Gas exchange in plants and animals	7.1	
Transport systems in plants and animals	7.2, 7.3	
Responses to stimuli by plants and animals	8.2, 8.3, 8.4	
Sensory receptors in mammals	8.4	
Structure of mammalian nervous system	8.4	
Muscle structure and function	8.5	
Hormonal responses in plants	8.2	
Behaviour	8.4	

Paper analysis

A level

Each A-level examination will consist of:
Modules 1–4 (compulsory)
4 × 1 hour 20 min tests 4 × 20% of total marks
Teacher-assessed practical
20% of total marks

AS syllabus

Each AS examination will consist of:
Modules 1 and 2 (compulsory)
2 × 1 hour 20 min tests 2 × 40% of total marks
Teacher-assessed practical
20% of total marks

Compensatory AS

A-level candidates who fail to achieve a final grade across the four modules can be awarded a compensatory AS if they reach an appropriate standard.

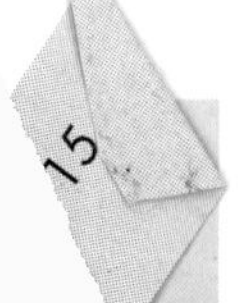

EXAMINATION BOARD ADDRESSES

AEB	The Associated Examining Board Stag Hill House, Guildford, Surrey GU2 5XJ Tel: 01483 506506
Cambridge	University of Cambridge Local Examinations Syndicate Syndicate Buildings, 1 Hills Road, Cambridge CB1 2EU Tel: 01223 553311
NEAB (including Nuffield)	Northern Examinations and Assessment Board 12 Harter Street, Manchester M1 6HL Tel: 0161 953 1180
NICCEA	Northern Ireland Council for the Curriculum Examinations and Assessment Beechill House, 42 Beechill Road, Belfast BT8 4RS Tel: 01232 704666
Oxford	University of Oxford Delegacy of Local Examinations Ewert House, Ewert Place, Summertown, Oxford OX2 7BZ Tel: 01865 54291
Oxford and Cambridge	Oxford and Cambridge Schools Examination Board (a) Purbeck House, Purbeck Road, Cambridge CB2 2PU Tel: 01223 411211 (b) Elsfield Way, Oxford OX2 8EP Tel: 01865 54421
SEB	Scottish Examination Board Ironmills Road, Dalkeith, Midlothian EH22 1LE Tel: 0131 663 6601
ULEAC	University of London Examinations and Assessment Council Stewart House, 32 Russell Square, London WC1B 5DN Tel: 0171 331 4000
WJEC	Welsh Joint Education Committee 245 Western Avenue, Cardiff CF5 2YX Tel: 01222 265000

STUDYING AND REVISING BIOLOGY

THE DIFFERENCE BETWEEN GCSE AND A/AS LEVEL

If you have studied GCSE Biology, you will probably find that the major differences between GCSE and A/AS level lie not so much in subject content as in the depth of and approach to study. For you the topics studied at A level will be very familiar, but you will explore them in much more detail. Some, like biochemistry and cell ultrastructure, may be new, but on the whole you will cover the same areas as at GCSE but to a depth that enables you to have a much greater understanding of the material.

If you have studied Human Biology, Environmental Biology or associated syllabuses and are now taking A-level Biology, the differences in content will be greater, but the principle of studying in more depth and detail still applies. Those of you embarking on A-level Human or Social Biology after GCSE Human Biology will not find the differences in subject content so marked.

Most of you, however, will be starting an AS or A-level course in a biological science not having studied Biology as a separate entity at GCSE, but rather as part of a single or dual award science course. For you the distinctions between Biology, Physics and Chemistry may not be as clear, but this should not prove to be any disadvantage. Indeed your knowledge of the other two main branches of science will be of much benefit, especially over those who have studied a biological GCSE as their only science. Whatever your background you should not be daunted by the prospect of studying a biological science at A level – interest, commitment and hard work are the essential ingredients for success and count for more than any particular combination of GCSEs.

Few students find the differences in subject content at A level a problem. The same cannot always be said of the study methods and skills required at this level; it may take you a little while to adjust to the different style. Most importantly, you will be following fewer subjects – probably just three or four compared to around double that number at GCSE. This means you will be spending at least twice as long studying each subject at school or college. What many students fail to appreciate is that the same is true of time spent outside the classroom. This too must increase for each subject, 5 hours per week being a basic minimum for each A level and half this for each AS level.

More so than at GCSE, you will be required to appreciate the underlying principles of each topic and the concepts and ideas behind them, as well as the factual content associated with them. An even greater emphasis is placed on the ability to apply general biological knowledge and principles to novel situations, rather than simply being able to recount a previously learned set of facts. The ability to interpret data in a variety of forms is required and this is often deliberately obscure data which you will never have previously encountered. In this way examiners are able to test understanding as opposed to rote learning. Compared to GCSE you are expected to know a much wider range of species and be able to use these to illustrate specific points. More open-ended questions will appear and you may not be required to give specific answers, but rather to argue a case for and/or against a particular view. At A level the ability to analyse experiments, data and information critically and to evaluate the accuracy of results and theories is an extension of the basic training given in these skills at GCSE. You will need to maintain a balanced view and argue logically. In order to develop these attributes you will need to read widely on all aspects of Biology. It is this, above all, that is essential if you wish to obtain a high grade at A level where a greater range of sources of information must be used. Scientific journals, for example, are invaluable as a means of keeping up to date in a subject that is developing as rapidly as Biology. Such reading should be done continuously throughout the course and you should be familiar with the Biology section of your school, college, local and central libraries. At this level essays are not so much a way of testing your learning of a topic as a way of getting you to expand your knowledge and incorporate information from other sources.

One major difference from GCSE is the large body of factual information which you will need to learn. This must be done as you proceed; there is simply too much to be mugged up on in a last minute programme of revision. You will need to learn quickly the skill of taking notes during class lectures, or from films, videos, TV or books (discussed later in this section). The ability to isolate the main points from what is being communicated is essential. These skills are not evident at GCSE where lecturing and note taking are less used.

There is a much greater onus on you as an A-level student to organise your own work, use your initiative and develop your own pattern of study. There is greater freedom to pursue and develop your interests. However, all this brings the need for greater self-discipline and personal sacrifice in order to work diligently over at least two years and not just in the run-up to the examination. Probably the major cause of failing to attain the desired grade at A level is deluding yourself that provided adequate revision is done in the weeks preceding the examination success is assured. A-level study can be stimulating and rewarding but it is always demanding and time consuming. Above all you will need to maintain a keen interest in the subject, the motivation to do well and the self-discipline to sustain your efforts throughout the course.

Practical skills are assessed at A level and at GCSE. While the skills you will require at A level are more extensive and more refined, they are not fundamentally different. You may, however, find that a longer term project, often encompassing up to 30 hours of study over many weeks or even months, is part of your syllabus. This demands many skills some of which may be new: these are discussed in detail later in this section.

There is an increasing trend towards the use of mathematical methods, especially statistics, in Biology; these demand greater mathematical skill than needed at GCSE.

Another major difference with many A-level examining boards is the pattern of a core syllabus, plus one or more optional topics. You may have some say in which options are chosen, although often the resources and expertise available constrain schools and colleges from providing a choice.

The AS level

Like A levels, the Advanced Supplementary (AS) syllabuses are based upon the interboard common core in Biology agreed by the GCE boards in 1983. These courses have been designed with three groups of students in mind:

1. those doing A-level subjects unrelated to Biology, e.g. arts subjects, in order to broaden their curriculum and give them some grounding in a science subject;
2. those studying complementary A levels, e.g. Physics, Chemistry and Maths, in order to act as a supporting subject;
3. those not wishing to follow a full A-level course, but who have a personal or career interest in a biological qualification beyond GCSE, e.g. those hoping to do nursing.

AS syllabuses occupy half the teaching and study time of an A-level subject. While usually extending over two years, you may be able to take AS examinations after a shorter period depending on the resources and policy of your school or college. A levels and AS levels can be taken in any combination, but typically you might study three A levels and one AS level, two A levels and two AS levels, or two A levels and one AS level. The standard of work expected at AS level is the same as for A level; it is not half way between GCSE and A level. Instead the amount, rather than the standard, of the work is reduced to take account of the shorter teaching time available. The grading system will be the same as for A levels and the grading standards will likewise correspond.

MODULAR COURSES

Most examining boards now offer modular courses as a flexible scheme which allows you to follow a variety of different routes in order to achieve an AS or A level. The syllabus is divided into small units (modules), each of which is examined by an end-of-module test. These tests may be taken at appropriate times throughout the course or, if preferred, at the end of the course. The advantages of taking the tests throughout the course are that there is a regular feedback on how well you are doing and there is the possibility of retaking modules in order to improve your grade. Furthermore, the results of modular tests may be 'banked', and then 'cashed-in' at some time within four years to obtain an AS or A level. This allows for a break in a course of study without loss of the credit already obtained. Some modules are compulsory, others may be selected from a list of optional ones. At the end of a modular course there may be a **synoptic assessment**. This is a test made up of questions which are taken from across the compulsory modules.

By varying the number and type of modules studied you can obtain an AS or A level in Biology, Human Biology or, by combining Biology modules with those from Physics and Chemistry, an AS or A level in Science. Currently, an AS level is either two or three modules and an A level four or six modules, depending on the examination board.

STUDY STRATEGY AND TECHNIQUES

We should make it clear at the outset that study techniques are very much an individual matter, dependent on your own character, temperament, resources and environment. There is no single method which guarantees good grades. What exists, and what we attempt to survey in the following, are a number of strategies which have been shown to work for others. It is up to you to try them and decide which best meets your needs. They are not an easy, short-cut route to success – all entail hard work and commitment over a long period, and without these essential ingredients, no amount of advice on study and revision will bring success.

Note taking

A-level Biology entails absorbing information from a variety of sources: lectures, lessons, practical sessions, books, periodicals, films, newspapers, television, radio, etc. Whatever the source, you will need to make notes to remind you of the important points.

Organising notes

Firstly you should get exercise books or preferably loose-leaf files in which notes can be kept. Have them to hand – never rely on scraps of paper for notes. Divide up the files into sections using card dividers. Maintain a clear index of what is in each file. Numbering of pages is not, however, to be advised as additions and deletions will quickly make this redundant. You may need to reorganise notes from time to time. There is rarely any point in maintaining them purely chronologically – it is far better to keep specific topics together.

Divide up the notes on the page into sections as you write. Try from the outset to use a hierachy of headings so that each topic can be divided into relevant subsections. Emphasise important words within the text by using capitals or (freehand) underlining. Make points in the form of a numbered list – this makes revision easier. Organising your notes in this way may seem unnecessarily formal but it will quickly become automatic and you will save considerable time when trying to find information at a later date, or when isolating the main points for revision.

Isolating the main points

When making notes you must determine what is relevant and what is superfluous. It is best to assume that all that is written in a Biology textbook, or is said by a teacher in a lesson, has some importance. When it comes to periodicals, newspapers, radio and TV the same is not always true. Try to pick out only the relevant items. To do this requires a good knowledge of your syllabus – try to have the appropriate section with you or at least to have read it before taking your notes.

Even when considering only relevant information, you will not have time to write everything down. The best idea is to jot down headings, important terms, diagrams, etc. without necessarily joining them in fluent prose. Never attempt to write word for word what is said unless specifically instructed to do so. Try to distinguish between the main points of an argument and those which are peripheral aspects included to reinforce an idea. One excellent practice is to re-read your notes shortly after having written them (say within eight hours). On doing so add some of the extra material you remember or further information to help you understand the topic.

Writing notes quickly

Not everyone writes rapidly. Courses are available on speed writing and you may wish to consider these if note taking proves to be laboriously slow. They are not essential however – a few useful hints can often do the trick.

1. Use standard biological abbreviations, e.g. NAD, CNS, RNA, etc.
2. Use truncated forms of words, e.g. temp.
3. Use accepted shorthand notations, e.g. 1° (= primary), ∴ (= therefore).
4. Make up your own shorthand for commonly occurring words, e.g. P/S (= photosynthesis), Arp (= Arthropod). (NB Do *not* use these in examinations as they are likely to be incomprehensible to others.)
5. Never use rulers or coloured pencils to highlight points at the time of writing – they can be added later. (NB They may be essential in some diagrams, however.)
6. Do not use correction fluid – crossing out neatly saves time.

Adding to your notes

Always leave some space at the end of each section and start new topics on a clean sheet. This enables you to add new material at the relevant point – such updating is essential in a subject which changes as rapidly as Biology.

Reading around the subject

With an increasing emphasis in syllabuses on the technological, social, economic, medical and environmental aspects of Biology, it is very important to keep up with current affairs. Newspapers and magazines can be scanned for items of biological significance, although the information needs to be treated with some caution – such articles do not always give a balanced or scientific view. Try to isolate the factual information for future use, and look critically at any arguments or solutions put forward.

There are some excellent periodicals on the market – *Scientific American*, *New Scientist* and *Nature* are usually available in your school, college or local libraries. One excellent publication designed specifically for students studying A-level biological subjects is the *Biological Sciences Review*. It contains up-to-date articles on a range of relevant topics written by experts in their field. Your school or college may have a group subscription allowing you to buy it at reduced cost. General Biology textbooks and more specific books on individual topics provide a wealth of knowledge relevant to syllabuses.

Read on whatever topic interests you. While it is an added advantage if it has direct relevance to your syllabus, do not make this the only reason for reading. Read for enjoyment on the topics in which you are interested – only in this way will you develop a love of the subject and a desire to know more. Any biological reading will be of benefit as it should improve your English, widen your vocabulary (both biological and nonbiological) and therefore make you a more efficient communicator. Make notes, or add to those you already have, as you go along. Jot down page or book references as these may prove very useful when compiling essays or carrying out projects at a later date.

Presentation of data

Biology involves data which need to be presented in an appropriate way. Many methods are available and the difficulty often lies in choosing the best one for the circumstances.

Tables This is the simplest and probably the most common method. It is often the first stage before further treatment, e.g. drawing of a graph, but in itself may not be very helpful. Remember to clearly define the parameters and the units they are measured in. Provide a title. Table 1 illustrates a typical table of data.

Table 1 *Frequency of heights (measured to the nearest 2 cm) of a sample of humans*

Height/cm	Frequency	Height/cm	Frequency
140	0	166	443
142	1	168	413
144	1	170	264
146	6	172	177
148	23	174	97
150	48	176	63
152	90	178	46
154	175	180	17
156	261	182	7
158	352	184	4
160	393	186	0
162	462	188	1
164	458	190	0

Line graph This comprises two axes each of which measures a variable. The **independent variable** has fixed values which are selected by the experimenter. The **dependent variable** is the parameter being measured. The values of the independent variable are plotted on the horizontal axis (also called the *x*-axis or abscissa) while those of the dependent variable are plotted on the vertical axis (*y*-axis or ordinate). The values plotted on the graph are called the **co-ordinates** and the line drawn between them is referred to as the **curve** (even if it is a straight line!). Fig. 1 is a graph of the data in Table 1.

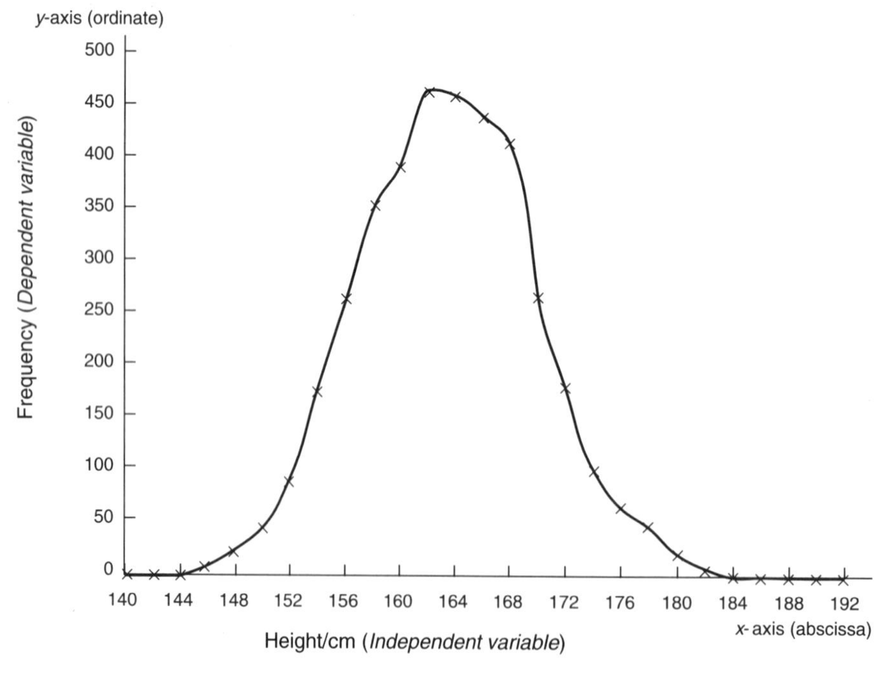

Fig. 1 Graph of frequency against height/cm for a sample of humans

Histograms The axes are drawn in much the same way as for a line graph, but the values for the independent values are normally reduced by grouping them into convenient classes. In Fig. 2 this has been done by grouping the heights in Table 1 into sets of 5 cm (e.g. 141–145, 146–150, etc.). Vertical columns are drawn instead of points.

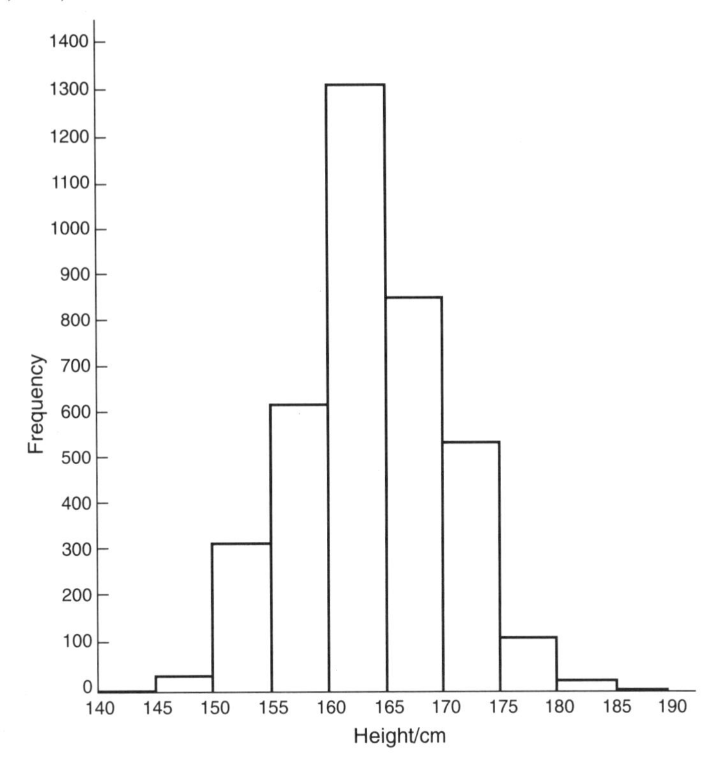

Fig. 2 Histogram of height frequencies in a sample human population

Bar graphs These are like horizontal histograms except that they are used when a nonnumerical grouping is required, e.g. Fig. 3 considers height classes in three racial groupings.

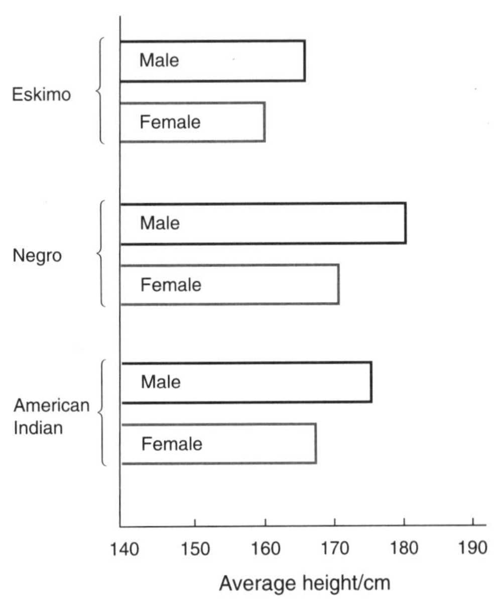

Fig. 3 Bar graph showing average height variation according to racial group and gender

Pie charts These are a simple and very clear means of showing how a sample is divided into specific parts. The circle (pie) represents the whole sample and is subdivided into different segments according to the relevant proportions of each part. It is normal to start with a vertical line from the centre to the upper part of the circle and then to partition off the segments, beginning with the largest, in a clockwise direction. Fig. 4 illustrates this when the data in Table 1 are divided into height groupings.

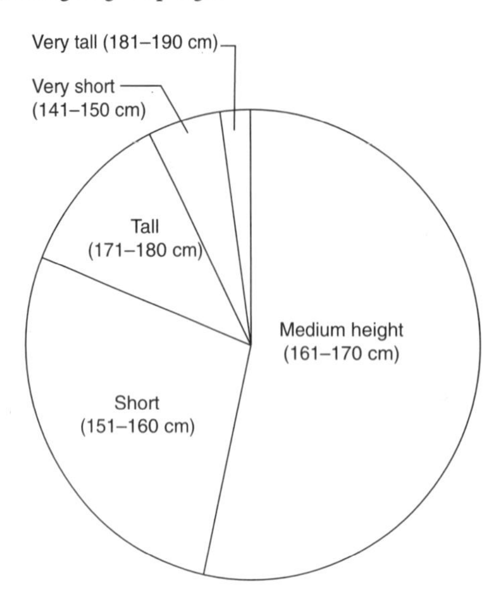

Fig. 4 Pie chart showing the proportions of five height categories in a population

Other methods More complex analyses of data, including the use of standard deviations, χ^2 tests and t tests, are increasingly to be found on examination syllabuses but are beyond the scope of this book.

Diagrams and drawings

Diagrams and drawings are an essential part of practical work, but they also play their part in theory. Some questions specifically ask for drawings, e.g. 'Give an illustrated account...', 'By means of labelled drawings...', 'With the aid of diagrams...', etc. In these cases drawings must be included. Where drawings are not specifically requested it is more difficult to advise on their use. If drawings are included they should be drawn well; rough sketches rarely bring credit. To do well, drawings take time, so ask yourself whether the information can be provided as clearly but more quickly in words. Do not make drawings for the sake of it, only include them where they are entirely relevant to the main theme of a question and they increase the understanding of the reader. Some students draw a diagram whenever they use a particular word, e.g. villus. In a question specifically on absorption of digested food in mammals, a drawing of a villus may be appropriate. In an essay on different mechanisms of feeding in animals it comprises such a tiny proportion of the total answer that it barely deserves a mention, let alone a diagram. If you decide a diagram is appropriate, use the following guidelines:

1. Use a good quality, sharp pencil. The hardness will depend upon the individual but HB is best for most people.
2. Consider the size of what you intend to draw in relation to the size of your paper and leave room for labels and annotations.
3. Make large, clear line drawings without the use of ink, or coloured pencils except possibly when distinguishing oxygenated and deoxygenated blood.
4. Do not shade unless this is essential as a means of distinguishing various parts.
5. Give each drawing a suitable title.
6. State the magnification, scale or actual size of the object drawn.
7. Fully label all drawings and *add appropriate annotations* (notes attached to each label).
8. Do not label too close to the drawing and never on it.
9. Do not cross label lines and, if possible, arrange labels vertically, one beneath the other.

PRACTICAL ASSESSMENT AND COURSEWORK

Practical assessment will comprise anything from 20% to 40% of your total A-level percentage. It may take the form of a practical examination, a project or continual assessment, or any combination of the three.

Projects and continual assessment

Projects and longer term experiments provide an opportunity for you to demonstrate your skills without the time constraints that apply in an examination. Remember, however, that quantity is no substitute for quality and that more time available should not lead to unduly long and rambling accounts.

Be sure at an early stage of your course that you are aware of what practical skills will be assessed and how. Start work early to give yourself practice at mastering techniques. Remember that many practical techniques are tested by questions on theory papers and so you must revise them prior to written examinations.

Practical examinations

The detail of practical work is beyond the scope of this book and practical examination papers take many forms, but most include:

1. microscope work, including the ability to make temporary slides, stain specimens, use the microscope efficiently and make drawings;
2. identification and drawing of specimens including the ability to put specimens into their major groups and make identification keys (details on the drawing of specimens are given in the following);
3. experimental work including the ability to design and carry out biological investigations (further details are given in the following).

Drawing specimens

These specimens may be living or dead, whole or in part, microscopic or macroscopic, real or photographed. Use the following guidelines:

1. Use a good quality, sharp pencil of a hardness that suits you (HB is normally the best).
2. Draw on good quality plain paper which is capable of withstanding rubbing out.
3. Ensure the diagram, along with all labels and annotations, will fit comfortably on the page.
4. Make large, clear line drawings without the use of ink or coloured pencils.
5. Make single pencil lines without sketching or shading.
6. Keep the drawing simple by providing only an outline of all the basic structures.
7. Draw accurately and faithfully what can be seen. Never draw anything you cannot see, even if it is expected to be present. Never copy from books.
8. Draw individual parts of a specimen in strict proportion to each other.
9. Provide suitable headings which clearly indicate the nature of the drawing. For microscope drawings, the section (TS/LS, etc.) should be stated.
10. State the magnification, scale or actual size of each specimen.
11. Label fully all biological features keeping labels away from the diagram, and never label on the actual drawing.
12. Avoid crossing label lines and, if possible, arrange labels vertically, one beneath the other.
13. Use annotations (notes added to labels) if at all possible. In particular, try to relate structure to function.

⓮ With microscope drawings it may be necessary to include two drawings described as follows:

(a) a low-power map indicating the main regions of each cell type, but *without drawing individual cells*;

(b) a high-power drawing of a section or wedge passing through all the major cell areas. Draw a few cells of each type and, in at least one cell, include cellular detail, e.g. nucleus, cytoplasm, storage grains.

⓯ Keep all drawings for assessment carefully, e.g. in a hardback loose-leaf file. Not only are they a permanent record of your work, but they are also invaluable for future reference.

Experimental investigations

Experimental work usually forms an important component of both continual assessment and the practical examination. It tests a number of skills:

❶ Ability to follow instructions

❷ Construction of suitable hypotheses

❸ Planning of experiments

❹ Design and manipulation of apparatus

❺ Making accurate observations and recordings

❻ Making precise measurements

❼ Presenting results in a suitable form

❽ Interpreting results accurately

❾ Making logical deductions from results by applying biological knowledge

❿ Discussing critically the methods used and results obtained

The writing up of experiments must be in accordance with any instructions given by a teacher or the rubric of an examination paper. In general the account will include:

❶ **A title** This should indicate the broad purpose of the experiment, e.g. 'Investigation of the effect of temperature on enzyme activity'.

❷ **The aim of the experiment** This should give the precise aim of the experiment, e.g. 'To determine the rate of starch breakdown by amylase at temperatures in the range 0–100 °C'. Alternatively a hypothesis may be tested, e.g. 'Sodium chloride increases the rate of starch breakdown by amylase'.

❸ **Method** This is a precise, step-by-step account of the procedures carried out. It should be written in the past tense and impersonally, e.g. 'A test tube was taken...'. If properly written another biologist should be able to perfectly imitate the experiment.

❹ **Results** These are a complete account of your recordings and observations. They should be presented in some appropriate form, e.g. descriptive prose, table, graph, histogram.

❺ **Conclusion** This is a *brief* statement of the single main fact determined by the experiment, e.g. 'The optimum temperature for the breakdown of starch by amylase was found to be 45 °C'.

❻ **Discussion** This should be an attempt to relate known biological knowledge to the results to try and explain them. It might also include:

(a) criticisms of the method employed;

(b) possible sources of error in the results;

(c) suggested improvements to the experiment.

For continual assessment, experiments may be spread over several weeks. Commonly these are investigations on genetics, growth or ecological changes, but almost any investigation can be designed to be performed over a long period.

For examination purposes, the choice of experiment is restricted by the need for it to be carried out within about two hours. Examination investigations therefore often involve food tests, enzymes or osmotic experiments. Even within these topics there is a vast number of possible investigations. Provided appropriate instructions are given, unfamiliar materials may be provided.

During the actual examination it is important to keep calm and work carefully. Do not rush; there is normally sufficient time and haste will only lead to mistakes. Do not become flustered if an experiment does not yield the expected results. Your findings are not necessarily wrong and even if they are, much credit can still be achieved if you make logical conclusions based upon the results you obtained.

REVISION TECHNIQUES

Examination success requires adequate preparation through proper revision. As with study methods, revision techniques are very much a personal affair. You should try as many methods as possible so that by the time of your final examination you have perfected one that suits you.

Revision begins on the first day of your A-level course. It is not a process which can be left to the few weeks preceding final examinations – the sheer volume of material to be learned and understood is too vast for this. You should try to develop the habit of reading over the day's notes later that day and, if possible, the whole of the week's notes at the weekend. Treat all interim tests and examinations as the real thing; try out different techniques each time and so discover what works for you.

Whatever methods you choose to employ, a few basic principles should be followed:

1. Don't work for too long at a single session – the length will vary from individual to individual but for most one hour is a reasonable maximum.
2. Take breaks of 10–15 minutes between sessions during which you should enjoy a complete change, e.g. a walk outside, a drink in front of the television.
3. Work where distractions are unlikely – if possible in a separate, silent room, sitting at a desk or table.
4. Be aware of the fact that you may consciously or subconsciously be trying to avoid revision, and fight the urge to give up. Don't jump up to answer the telephone or door – let someone else do it. You can manage an hour without having to get a drink – the thirst you feel is almost certainly an 'escape' mechanism.
5. Bring variety to revision – vary the topics, subjects, times and places you work.
6. Try to get feedback on the effectiveness of your revision. Test yourself (or get others to test you) at regular intervals – preferably at least 12 hours after the revision session.
7. Try past examination questions (such as those in this book) to test your knowledge and to familiarise yourself with question styles and answer techniques.
8. Organise revision programmes before examinations. Start early but be realistic – you are unlikely to be able to work six hours every day for six months. Leave periods of time unaccounted for – you will need these to make up for time lost due to illness, distractions, idleness, etc., or to allow yourself the reward of a break from the routine of revision.
9. Every day we have spare moments when waiting for a meal, travelling on a bus or train, etc. These periods amount to hundreds of hours over a two-year course. Therefore try to perfect the habit of using them to do snippets of revision like learning the photosynthesis equation, a comparison of nervous and endocrine systems, the regions of the alimentary canal, etc.

THE EXAMINATION

QUESTION STYLES

Essay questions

Examination boards use essay questions less than they used to. Nevertheless all boards include them to a greater or lesser degree. They are included because, being open ended, they give considerable scope for you to demonstrate a wide range of knowledge and to show the depth and breadth of your reading in Biology. It is often for this reason that some candidates dislike them – they are unsure what the answer requires and/or they do not possess the breadth and depth of knowledge to answer them effectively. Equally there are some who fare badly on essay questions for the opposite reason – they digress widely from the question, using it as an opportunity to relate the last biological article they read or indulge their opinions on an issue, with no regard to their relevance. The skill lies in achieving a compromise between these two extremes.

Essay questions may be structured to some degree. In a highly structured essay the question comprises a number of different parts which often guide you step by step through the question, making clear at each stage exactly what is required. To take a typical example:

(a) Explain what is meant by the terms:
 (i) enzyme
 (ii) coenzyme (4)
(b) With reference to the lock and key theory and active sites explain how an enzyme works. (8)
(c) What are the factors which influence the rate of enzyme activity?
 Briefly point out why the factors you mention alter the rate. (8)

The figures in brackets represent the marks allocated to each section – a common feature of the structured essay. These marks provide clear guidance on how long should be spent on each part. In this example it is relatively clear what the examiner is looking for at each stage. The answer to this question might not be very different from one entitled 'Write an essay on enzymes' but we suspect you would find the latter more problematic because you cannot be sure what the examiner wants and therefore what to include and what to leave out. The structured essay overcomes these problems to some degree, but you should never neglect careful planning. Under each subsection of the question it is necessary to jot down the essential points to be made before starting your answer. Only in this way can your response be logically argued with well-marshalled ideas and relevant supporting evidence. You should answer each part separately – never attempt to merge them into a general essay. Take care not to use information more than once as you cannot be given credit twice for the same explanation. In the question cited it would be feasible to include the lock and key theory in all three parts. This would be a waste of examination time. It is almost certainly to avoid this problem that the examiners have guided you to include this information in part (b).

An alternative type of essay is the unstructured one. This may be as open ended as 'Write an essay on enzymes/carbohydrates/animal reproduction/photosynthesis/hormones', etc. In practice they are often a little more explicit:

Describe the properties of water and show why it is vital to living organisms.
Describe how animals depend for their survival on the activities of plants.
Argue a case for and against the use of pesticides and fertilisers.

In this style of essay planning is even more crucial. The number of points which could be made is vast and the skill is in limiting them to the most important, i.e. those most likely to be on the examiners' mark scheme. In the 30–40 minutes typically allowed for an essay question, only perhaps 20 or 30 individual facts can be included in sufficient detail to warrant marks. You must ensure that these points cover the full range of the topic being examined. Take, for example, the following essay:

Write an essay on movement in organisms.

There is a danger here that all points made will be biased towards the movement of whole organisms (locomotion) and animals. Good planning (and adequate knowledge) can avoid this. Movement in Biology includes the movement of material within an organism and within and between cells. All living things use movement. Your essay should therefore encompass all types of movement and a wide range of organisms – plant, animal and protoctista. Avoid any bias towards animals, especially mammals. The main areas on your plan should be:

Types of movement	Example
Subatomic movement	electron movement in photosynthesis
Atomic/ionic/molecular	diffusion, osmosis, active transport into and out of cells; transpiration, translocation
Cellular	amoeboid, ciliate, flagellate locomotion; movement of macrophages
Tissue/organ	heart pumping, circulation of blood
Organism	locomotion, swimming, flying, burrowing, etc.
Population	migrations

Include each of the six areas listed but try to link them in a coherent manner. Some examination boards give marks for the manner in which the essay is written, its fluency, clarity and logical arguments.

The wording of essay questions is all important. There are subtle differences between terms such as 'compare', 'describe', 'discuss' and 'distinguish'. Unless you appreciate this you are likely to needlessly throw away marks. The following list is a guide to the appropriate meaning of a number of commonly occurring question terms:

Describe	give an account of the main points, with reference to (visual) observations, if possible
Explain **Account for**	show how and why, give the reasons for – with reference to theory, if possible
Compare	point out the similarities and differences
Distinguish **Contrast**	make distinctions between, point out differences
Discuss	debate, giving the various viewpoints and arguments
Criticise	point out faults and shortcomings
Survey	give a comprehensive and extensive review
Comment on	make remarks and observations on
Illustrated	use figures, drawings, diagrams
Annotate	add notes of explanation
Calculate	a numerical answer is required and working should be shown
Briefly **Concisely**	give a short statement of only the main points
Outline	give the essential points, briefly
List	catalogue, often as a sequence of words one beneath the other
Significance	importance of
State	set down concisely with little or no supporting evidence
Define	give only a formal statement
Suggest	put forward ideas, thoughts, hypotheses
Devise	construct, compose, make up
Estimate	give a reasonable approximation

Short-answer (structured) questions

These forms of question are immensely varied ranging from a single-word response, e.g. 'Which organic base pairs with adenine in DNA?', to ones requiring a detailed explanation, e.g. 'Explain how ultrafiltration takes place in the kidney'. Intermediate forms include filling in blanks in passages, completing tables and labelling diagrams. Data in a variety of forms are popular.

Answers to short-answer questions should be especially concise. It is content rather than style which is being tested and it is less important than in essays to concern yourself with the niceties of sentence construction. Responses must, however, be detailed and comprehensible. The space allocated for an answer is a good guide to the length of the answer required.

Multiple-choice questions

Now used by a number of examining boards in Biology, this form of questioning varies in style, but generally comprises a statement or question followed by four or five alternative answers. Questions of this type, where you are required to select the single correct alternative, are called **single-completion questions**. For example:

In the electron (hydrogen) carrier system of aerobic respiration, which of the following is the last stage?

A formation of ATP
B reduction of oxygen
C production of carbon dioxide
D reduction of cytochrome
E production of H^+ (Answer = B)

In approaching multiple-choice questions you should read through the whole question and *all* the alternatives before making your choice because often more than one will be correct to some degree. It is the best, or most appropriate, answer that is required. On reading again reject those responses you feel are clearly incorrect. Always reject on a sound biological basis and never because 'it doesn't sound right' or because 'it can't be B for the fifth time on the trot'. Should you still be left with more than one answer try to find a good reason why one response is better than another. If you simply cannot decide for valid reasons then, as a last resort, guess. Unless the rubric states otherwise there is no penalty for a wrong answer and so a guess is better than giving no response.

A second type of these questions is the **multiple-completion** one. Here you will be provided with three or four statements, any number of which may be correct. You are required to select those you think are accurate and then respond by giving the correct letter according to a key which is provided. A typical example follows:

For the following question, determine which of the responses that follow are correct. Give the answer A, B, C, D or E according to the key below:
A 1,2 and 3 are correct
B 1 and 3 only are correct
C 2 and 4 only are correct
D 4 alone is correct
E 1 and 4 only are correct

Active transport:

1 proceeds at the same rate for all molecules
2 is unaffected by changes in temperature
3 continues to occur in the presence of cyanide
4 may occur against a concentration gradient (Answer = D)

A further multiple-choice variant is the **matching pairs** type. A list of four or five items is given which are followed by individual questions, the answers to which are one of the items. These items may be words, statements, a series of diagrams or labels on diagrams. You should bear in mind that any item from the list can be used once, more than once or not at all. Many a candidate has misguidedly ensured that their five responses correspond to each of the five listed items. The following is an example:

A sucrose
B amylase
C cellulose
D sucrase (invertase)
E lactose

For each of the statements below select the most appropriate response from the list above.

1 A polysaccharide (Answer = C)
2 An enzyme which hydrolyses starch (Answer = B)
3 A sugar found in milk (Answer = E)
4 A substance commonly transported in phloem (Answer = A)
5 A disaccharide produced by combining glucose and fructose (Answer = A)

EXAMINATION TECHNIQUES

Bad examination technique is often blamed for poor A-level results. It may be that in many cases it is inadequate revision which is the main cause, but nevertheless there are countless marks squandered at every sitting by students who fail to read questions properly or pace themselves sensibly. Below are listed a number of points which, if stuck to, will help avoid marks being needlessly wasted.

1. Read all instructions carefully. Do not assume them to be exactly the same as those you have seen on past papers.
2. Note the number of questions to be answered and keep strictly to this.
3. Act on any guidance given in the general instructions about the use of English, necessity for diagrams, need for orderly presentation, etc.
4. Read all questions with great care. Do not be in a hurry to get started. Be sure you understand what the question requires before answering.
5. Where there is a choice of questions read *all* the questions first before making any selection. Read the paper a second time, making your choices, and finally read the selected questions a third time to ensure you have chosen ones you are able to answer. Answer questions in any order but number them clearly.
6. Often the total marks for the paper are stated. Divide this number into the time provided to find approximately how many minutes are available for each mark, e.g. 100 marks on a three hour paper gives

$$\frac{180 \text{ minutes}}{100 \text{ marks}} = 1.8 \text{ minutes per mark}$$

   Allowing time for reading this gives about 1.5 minutes per mark. You should therefore spend 15 minutes on a questions worth 10 marks. Allocate your time for every question or part of a question in proportion to the marks available.

7. Refer back to the question a number of times during the writing of an answer to ensure you have not strayed from the point. Re-read the whole question once you have completed your answer in case you have omitted any part.
8. Try to isolate the key word or words in a question and answer precisely in accordance with them (see the list of typical key words on p.39).
9. Try to be completely relevant, clear and concise in your answers. Do not ramble.
10. Check during the last quarter of a paper that you have followed all instructions carefully and have answered (or are about to) the required number of questions. Do not leave this until the last five minutes – it will be too late to put right should you have made an error.

Remember, however, that good examination technique commences early in your course. It comprises learning to think for yourself, writing adequate notes, reading widely, learning as you go along, practising questions and revising thoroughly in plenty of time. Only if these matters are attended to will the points made above be of value to you.

TAKING MODULAR TESTS

Preparation for modular tests and the examination techniques you need to apply to them are largely the same as for other written examinations. However, the following points should be remembered when taking modular tests.

1. The modular examinations, even where taken after only one term of study, are marked to the same standard as terminal examinations. No concession is made for your relative inexperience at A-level study.
2. The modular examination usually covers a narrower range of material than other examinations. Do not be lulled into thinking that preparation can be less thorough and/or less urgent.
3. The time interval between practice (mock) test and the actual examination is likely to be short, leaving little time to compensate for deficiencies in knowledge and/or technique.
4. Terminal examinations (which are those taken after 14 February in the final year) must contribute at least 30% for the final mark/grade. These later modules often incorporate skills and understanding which have been acquired in previous modules. Do not assume all previous knowledge can be forgotten.
5. Modular tests are typically 1–1½ hours in duration. Their relative shortness makes the management of your time especially important as there is little opportunity to compensate for lost time. Answers need to be concise and to the point.
6. There is little respite from examinations, no sooner is one complete than the next appears on the horizon. Take care not to become fatigued but to adequately build yourself up mentally for the next module.

FINAL PREPARATION

Provided you have prepared an adequate revision timetable, begun the process early and followed it through faithfully, there should be little left to do on the evening prior to your examination. If you have not revised adequately it is a mistake to imagine that a few hours of cramming can compensate for your earlier omissions.

The best advice we can give for the evening prior to an examination is to try to relax and have a good night's rest. You may find it difficult to follow this advice as it demands great confidence that your revision has been complete. If you simply cannot bring yourself to do no work, we would suggest a quick skim through your notes, textbook or this companion guide, perhaps reading headings and subheadings, to generally absorb ideas and principles. Equally, there may be a few equations, definitions or mnemonics which you might wish to reinforce in your memory. Avoid, however, detailed revision of topics as it is almost certainly too late for this to be of benefit and there is the very real risk of inducing panic as you struggle to come to grips with a difficult concept. This will only create confusion and undermine your confidence, making matters worse than if no work had been done at all.

Our advice is therefore to keep any revision low key, to prepare the necessary examination materials (pens, pencils, calculator, etc.) for the following day, relax, set your alarm and get a good night's sleep.

BIOLOGY TOPICS

In this section:

Each chapter features:

- *Units in this chapter*: a list of the main topic heads to follow.
- *Chapter objectives*: a brief comment on how the topics relate to what has gone before, and to the syllabus. Key ideas and skills which are covered in the chapter are introduced.
- *The main text*: this is divided into numbered topic units for ease of reference.
- *Chapter roundup*: a brief summary of the chapter.
- *Illustrative questions and worked answers*: typical exam questions, with tutorial notes and our suggested answers.
- *Question bank*: further questions, with comments on the pitfalls to avoid and points to include in framing your own answers.

CHAPTER 1

ORGANISATION

Units in this chapter

Chapter objectives

Biology covers a vast field of knowledge across a considerable range of sizes. On the one hand, it includes the movement of electrons during photosynthesis and, on the other, the migrations of individuals around the globe. Size, in part, determines the various levels of organisation, three of which – molecular, cellular and tissue organisation – are discussed in this chapter. The information given is fundamental to a proper understanding of Biology as the ideas and concepts covered are the basic building blocks from which the rest of the subject is constructed.

1.1 MOLECULAR ORGANISATION

ATOMS

An **atom** is the smallest unit of a chemical element which can exist independently. It has a nucleus made up of positively charged particles called **protons** (the number of which is referred to as the **atomic number**) and **neutrons**, particles which have no charge. Around this positively charged nucleus orbit negatively charged particles called **electrons**. As the number of electrons in an atom is equal to the number of protons, there is no overall charge. While protons and neutrons contribute to the mass of an atom, electrons have such a small mass that it can be ignored for all practical purposes. The number of electrons, however, determines the chemical properties of the atom.

In general, the number of protons, neutrons and electrons in an atom is the same but circumstances sometimes arise where the proportions vary. Changes to the number of electrons lead to the formation of an **ion**; where the number of neutrons varies, an **isotope** is the result.

Ions

We have already seen that, if the number of electrons and protons is the same, there is no overall charge on an atom. If, however, an atom loses an electron (a process called **oxidation**), it becomes positively charged due to the excess of protons over electrons. It is now called a **positive ion**. In the same way, the gain of an electron (called **reduction**) produces a negative ion. Ions are depicted by writing the chemical symbol of the atom followed by a superscript indicating the appropriate charge. For example:

hydrogen ion (hydrogen atom which has lost one electron) = H^+
chloride ion (chloride atom which has gained one electron) = Cl^-
calcium ion (calcium atom which has lost two electrons) = Ca^{2+}
sulphate ion (sulphate group which has gained two electrons) = SO_4^{2-}

Isotopes

An atom may sometimes gain a neutron which causes it to become heavier without altering its overall charge; such an atom is termed an isotope. For example, a hydrogen atom (atomic mass = 1) can gain a second neutron to form the isotope deuterium (atomic mass = 2). Isotopes have important applications in Biology.

MOLECULES

We have seen that atoms possess electrons which orbit around the nucleus. These electrons orbit in fixed quantum shells and there is a limit to the number in each shell. The one nearest the nucleus may possess a maximum of two electrons and the next shell a maximum of eight. An atom is most stable when its shells have their full complement of electrons. If any shell has less than the required number of electrons, the atom will tend to combine with another atom which also has an incomplete shell. In so doing they share electrons in such a way that both are able to complete their shells. You can see this illustrated in Fig. 1.1 which shows two hydrogen atoms combining to form a hydrogen **molecule**. The sharing of electrons to produce stable molecules is called **covalent bonding**.

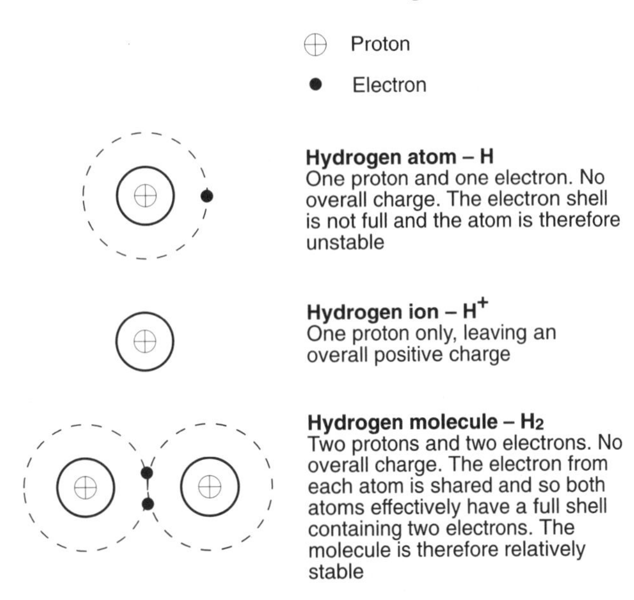

Fig. 1.1 Atomic structure of a hydrogen atom, a hydrogen ion and a hydrogen molecule

The oxygen atom has eight protons and eight neutrons in the nucleus with eight electrons in orbit around it. Two of these electrons fill the inner shell, leaving six in the outer one; this is two less than its maximum of eight. If you study Fig. 1.2 you will see how, by sharing an electron with each of two hydrogen atoms, both the oxygen and hydrogen atoms can complete their electron shells. A water molecule is the result. Fig. 1.2 also shows you how two oxygen atoms and a carbon atom can combine to form a stable carbon dioxide molecule.

An atom such as hydrogen which requires a single electron to complete its outer shell is said to have a **combining power (valency)** of one. Oxygen requires two electrons to fill its outer shell and therefore has a combining power of two. Similarly nitrogen and carbon require three and four electrons, respectively, to complete their outer orbits and therefore have combining powers of three and four. Where a single electron is shared between two

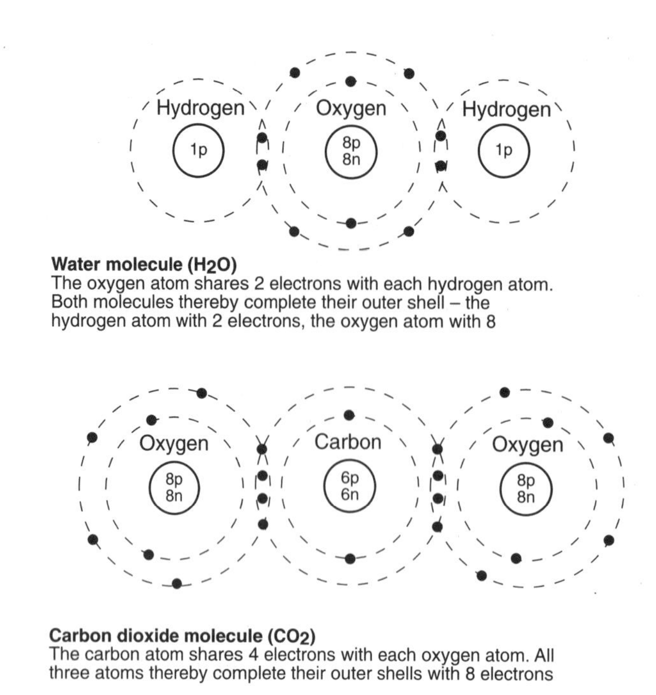

Fig. 1.2 Atomic structure of water and carbon dioxide molecules

atoms the bond is termed a **single bond**. The sharing of two electrons forms a **double bond** and so on. With its combining power of four, carbon readily forms covalent bonds with hydrogen, oxygen and nitrogen to produce a wide variety of molecules. More importantly, it can combine with its own atoms to form long chains linked by single, double and triple bonds. These very long and stable chains of carbon atoms are essential to living organisms.

Ionic bonds

We have seen that atoms can stabilise themselves by sharing electrons. Stability can also be achieved by the loss of an electron (oxidation) or the gain of an electron (reduction). The resulting ions are positively and negatively charged, respectively. Oppositely charged ions attract one another forming a different bond called an **ionic bond**. For example, the positively charged sodium ion (Na^+) forms an ionic bond with the negatively charged chloride ion (Cl^-) to form sodium chloride.

Hydrogen bonds

The electrons orbiting around the nucleus of an atom do not distribute themselves evenly but tend to collect in one position. As a result, this region will be more negative than the rest of the atom. Atoms with this uneven distribution of charge are said to be **polarised**. The negative region of an atom is attracted to the positive region of another atom forming a weak electrostatic bond called a **hydrogen bond**. While individual hydrogen bonds are weak, the vast number of them in large molecules makes them collectively very significant.

Inorganic ions

You may have noticed when reading the 'nutrition information' on packaged food that the list often includes a number of minerals such as sodium, calcium and iron. While these minerals, or inorganic ions, typically make up only 1% of an organism's weight, they are nevertheless essential to its existence. Two groups are recognised – **macronutrients** which are needed in very small quantities and **micronutrients (trace elements)** which are needed in minute amounts. Table 1.1 lists the major macronutrients and their functions.

Table 1.1 *Inorganic ions and their functions in plants and animals*

Macronutrients/ main elements	Functions
Nitrate NO_3^- Ammonium NH_4^+	Nitrogen is a component of amino acids, proteins, vitamins, coenzymes, nucleotides and chlorophyll. Some hormones contain nitrogen, e.g. auxins in plants and insulin in animals.
Phosphate PO_4^{3-} Orthophosphate $H_2PO_4^-$	A component of nucleotides, ATP and some proteins. Used in the phosphorylation of sugars in respiration. A major constituent of bone and teeth. A component of cell membranes in the form of phospholipids.
Sulphate SO_4^{2-}	Sulphur is a component of some proteins and certain coenzymes, e.g. acetyl coenzyme A.
Potassium K^+	Helps to maintain the electrical, osmotic and anion/cation balance across cell membranes. Assists active transport of certain materials across the cell membrane. Necessary for protein synthesis and is a cofactor in photosynthesis and respiration. A constituent of sap vacuoles in plants and so helps to maintain turgidity.
Calcium Ca^{2+}	In plants, calcium pectate is a major component of the middle lamella of cell walls and is therefore necessary for their proper development. It also aids the translocation of carbohydrates and amino acids. In animals, it is the main constituent of bones, teeth and shells. Needed for the clotting of blood and the contraction of muscle.
Sodium Na^+	Helps to maintain the electrical, osmotic and anion/cation balance across cell membranes. Assists active transport of certain materials across the cell membrane. A constituent of the sap vacuole in plants and so helps maintain turgidity.
Chlorine Cl^-	Helps to maintain the electrical, osmotic and anion/cation balance across cell membranes. Needed for the formation of hydrochloric acid in gastric juice. Assists in the transport of carbon dioxide by blood (chloride shift).
Magnesium Mg^{2+}	A constituent of chlorophyll. An activator for some enzymes, e.g. ATPase. A component of bones and teeth.
Iron Fe^{2+} or Fe^{3+}	A constituent of electron carriers, e.g. cytochromes, needed in respiration and photosynthesis. A constituent of certain enzymes, e.g. dehydrogenases, decarboxylases, peroxidases and catalase. Required in the synthesis of chlorophyll. Forms part of the haem group in respiratory pigments such as haemoglobin, haemoerythrin, myoglobin and chlorocruorin.

CARBOHYDRATES

The general formula for carbohydrates is $(CH_2O)_n$:

where $n = 3$ it is called a triose sugar (glyceraldehyde is $C_3H_6O_3$)
where $n = 5$ it is called a pentose sugar (ribose is $C_5H_{10}O_5$)
where $n = 6$ it is called a hexose sugar (glucose is $C_6H_{12}O_6$)

These are all examples of sugars (saccharides). Where there is only one sugar it is called a **monosaccharide**; two monosaccharides can join together to form a **disaccharide**, and many monosaccharide units form **polysaccharides**.

Monosaccharides (single sugars)

e.g. glucose, fructose, galactose
All three have the same basic formula, $C_6H_{12}O_6$, but differ in the arrangement of the atoms

Table 1.2 *Carbohydrates and their functions*

Group of carbohydrates	Name of carbohydrate	Type/composition	Function
Monosaccharides Trioses ($C_3H_6O_3$)	Glyceraldehyde	Aldose sugar	The phosphorylated form is the first formed sugar in photosynthesis, and as such may be used as a respiratory substrate or converted to starch for storage. It is an intermediate in the Krebs cycle.
	Dihydroxyacetone	Ketose sugar	Respiratory substrate. Intermediate in the Krebs cycle.
Pentoses ($C_5H_{10}O_5$)	Ribose/ deoxyribose	Aldose sugars	Makes up part of nucleotides and as such gives structural support to the nucleic acids RNA and DNA. Constituent of hydrogen carriers such as NAD, NADP and FAD. Constituent of ATP.
	Ribulose	Ketose sugar	Carbon dioxide acceptor in photosynthesis.
Hexoses ($C_6H_{12}O_6$)	Glucose	Aldose sugar	Major respiratory substrate in plants and animals. Synthesis of disaccharides and polysaccharides. Constituent of nectar.
	Galactose	Aldose sugar	Respiratory substrate. Synthesis of lactose.
	Fructose	Ketose sugar	Respiratory substrate. Synthesis of inulin. Constituent of nectar. Sweetens fruits to attract animals to aid seed dispersal.
Disaccharides	Sucrose	Glucose + fructose	Respiratory substrate. Form in which most carbohydrate is transported in plants. Storage material in some plants, e.g. *Allium* (onion).
	Lactose	Glucose + galactose	Respiratory substrate. Mammalian milk contains 5% lactose, therefore major carbohydrate source for sucklings.
	Maltose	Glucose + glucose	Respiratory substrate.
Polysaccharides	Amylose Amylopectin } starch	Unbranched chain of α-glucose with 1,4 glycosidic links + branched chain of α-glucose units with 1,4 and 1,6 glycosidic links	Major storage carbohydrate in plants.
	Glycogen	Highly branched short chains of α-glucose units with 1,4 glycosidic links	Major storage carbohydrate in animals and fungi.
	Cellulose	Unbranched chain of β-glucose units with 1,4 glycosidic links + cross bridges	Gives structural support to cell walls.

(i.e. they are isomers); for example, glucose is an aldose sugar (CHO) whereas fructose is a ketose sugar (CO). They all reduce Benedict's reagent, are sweet, soluble and easily transported, and are the main respiratory substrates.

Disaccharides (double sugars)

These are formed by the condensation of any two monosaccharides.
e.g. glucose + glucose = maltose (malt sugar)
glucose + fructose = sucrose (cane sugar)
glucose + galactose = lactose (milk sugar)
These three disaccharides have the basic formula $C_{12}H_{22}O_{11}$.

Some disaccharides, e.g. maltose, will reduce Benedict's reagent but others, e.g. sucrose, are nonreducing sugars. All disaccharides are sweet, soluble in water and are readily converted into monosaccharides by the addition of a water molecule, a process called **hydrolysis**.

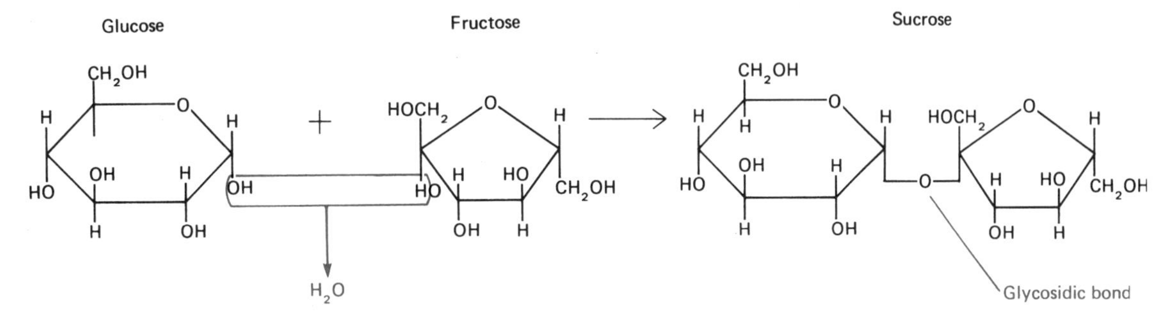

Fig. 1.3 Formation of a disaccharide

LIPIDS

Like carbohydrates, lipids contain carbon, hydrogen and oxygen, although the proportion of oxygen is much reduced in lipids. They are formed as shown in Fig. 1.4.

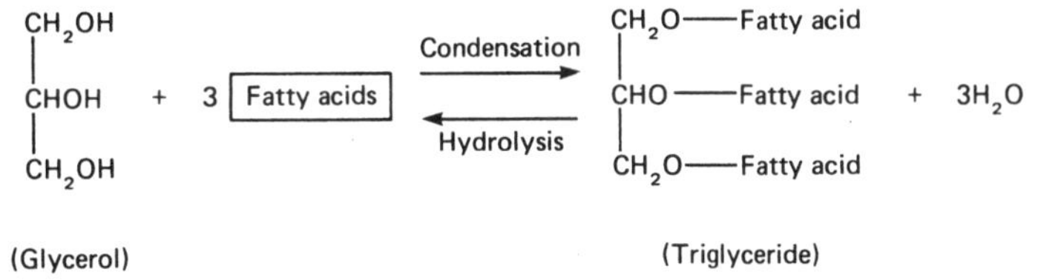

CH_2OH / CHOH / CH_2OH (Glycerol) + 3 Fatty acids ⇌ (Condensation →, ← Hydrolysis) CH_2O—Fatty acid / CHO—Fatty acid / CH_2O—Fatty acid (Triglyceride) + $3H_2O$

Fig. 1.4 Formation of a fat

Some lipids have one of the fatty acids replaced by phosphoric acid. These are called **phospholipids** and are important constituents of cell membranes.

Lipids can be divided into two types: **fats** and **oils**. There is no fundamental difference between the two, it is simply that fats are solid at room temperature whereas oils are liquid. You will probably have heard the terms saturated and unsaturated applied to fats. **Saturated fats** are found largely in animals and contain no double bonds. **Unsaturated fats**, on the other hand, largely come from plants and possess double bonds. There is evidence that saturated fats in the diet are more likely to cause heart disease than unsaturated ones. A fat is often depicted as having a 'head' (the glycerol portion) and a 'tail' (the fatty acid portion, which is often very long). As the head is attracted to water it is described as being **hydrophilic** and the tail, which repels water, as **hydrophobic**.

Lipids serve a variety of functions in living organisms:

1. **energy source** – upon breakdown they yield 38 kJ g^{-1} of energy.

2. **storage** – since lipids store over twice as much energy for the same mass as carbohydrates, they are used when it is important to keep weight to a minimum, e.g. in fruits and seeds.

3. **structural support** – phospholipids are an important component of cell membranes.
4. **protection** – lipids can be stored around delicate organs, e.g. kidneys.
5. **insulation** – especially useful where hair, feathers, etc. are of little use, i.e. underwater (e.g. blubber in whales).
6. **waterproofing** – leaf cuticles are made of waxes; secretion from sebaceous glands in skin.

PROTEINS

These consist of long chains of amino acids, the basic formula of which is shown in Fig. 1.5(a). There are over 20 naturally occurring amino acids, which differ in the composition of the R group. Two amino acids may be linked together by a condensation reaction to form a dipeptide as shown in Fig. 1.5(b).

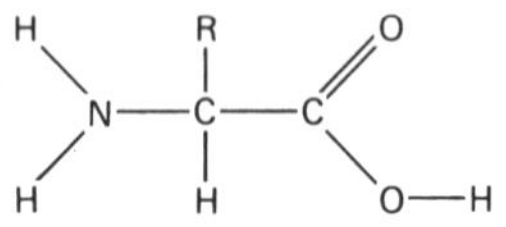

Fig. 1.5(a) Typical amino acid

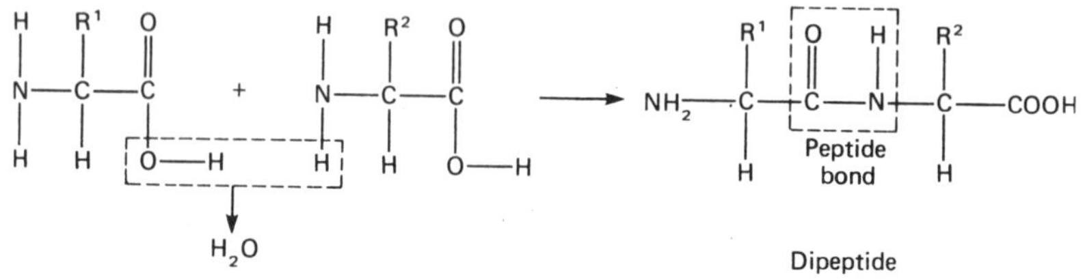

Fig. 1.5(b) Formation of a dipeptide

Many amino acids joined in this way are called **polypeptides** and these in turn are linked to form proteins which may have many thousands of amino acids. Since the amino acids may be joined in any sequence there is an almost infinite variety of possible proteins. The chain of amino acids is referred to as the protein's **primary structure**. This chain is folded (often into a helix) to give the **secondary structure** and this in turn is folded on itself to form the tertiary structure. The combination of a number of polypeptide chains along with associated nonprotein groups results in the **quaternary structure**.

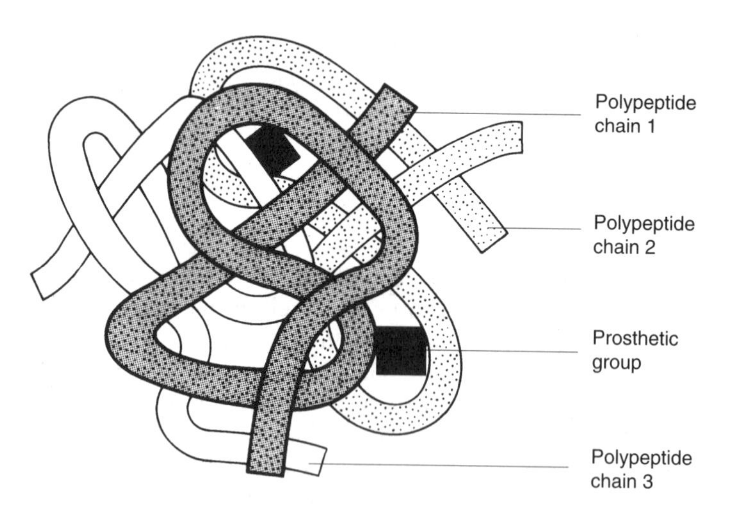

Fig. 1.6 Quaternary structure of a protein

These shapes are due to the fact that proteins are **amphoteric**, i.e. they have both positive and negative charges on them. The attraction of these opposite charges forms weak electrostatic (hydrogen) bonds causing the chain to form a complex three-dimensional structure – **globular proteins**. Ionic bonds, disulphide bridges and hydrophobic interactions all contribute to the final shape of a given protein molecule. All enzymes and some hormones are globular proteins and their functions depend on the precise shape of the protein molecule. Sometimes the protein consists of long parallel chains with cross-links – **fibrous**

proteins. These are insoluble and have structural functions, e.g. collagen in cartilage; keratin in hooves, feathers and hair; actin and myosin in muscle. If a globular protein is heated or treated with a strong acid or alkali the hydrogen bonds are broken and it reverts to a more fibrous nature – a process called **denaturation**. Proteins sometimes occur in combination with a nonprotein substance (**prosthetic group**); these are called **conjugated proteins**, e.g. haemoglobin.

Table 1.3 *Functions of proteins*

Vital activity	Protein example	Function
Nutrition	Digestive enzymes,	
	e.g. trypsin	Catalyses the hydrolysis of proteins to polypeptides
	amylase	Catalyses the hydrolysis of starch to maltose
	lipase	Catalyses the hydrolysis of fats to fatty acids and glycerol
	Fibrous proteins in granal lamellae	Help to arrange chlorophyll molecules in a position to receive maximum amount of light for photosynthesis
	Ovalbumin	Storage protein in egg white
	Casein	Storage protein in milk
Respiration and transport	Haemoglobin	Transport of oxygen
	Myoglobin	Stores oxygen in muscle
	Prothrombin/fibrinogen	Required for the clotting of blood
	Mucin	Keeps respiratory surface moist
	Antibodies	Essential to the defence of the body, e.g. against bacterial invasion
Growth	Hormones, e.g. thyroxine	Controls growth and metabolism
Excretion	Enzymes, e.g. urease, arginase	Catalyse reactions in ornithine cycle and therefore help in protein breakdown and urea formation
Support and movement	Actin/myosin	Needed for muscle contraction
	Ossein	Structural support in bone
	Collagen	Gives strength with flexibility in tendons and cartilage
	Elastin	Gives strength and elasticity to ligaments
	Keratin	Tough for protection, e.g. in scales, claws, nails, hooves, skin
Sensitivity and coordination	Hormones,	
	e.g. insulin/glucagon	Control blood sugar level
	vasopressin	Controls blood pressure
	Rhodopsin/opsin	Visual pigments in the retina, sensitive to light
	Phytochromes	Plant pigments important in control of flowering, germination, etc.
Reproduction	Hormones, e.g. prolactin	Induces milk production in mammals
	Chromatin	Gives structural support to chromosomes
	Keratin	Forms horns and antlers which may be used for sexual display

1.2 CELLULAR ORGANISATION

The cell is the fundamental unit of life and every living organism is composed of them. Modern cell theory states that:

1. all living matter is composed of cells;
2. all new cells arise from other cells;

3. all metabolic reactions of an organism take place in cells;
4. cells contain the hereditary information of the organisms of which they are a part, and this is passed from parent to daughter cell.

All cells are similar in comprising a self-contained and more or less self-sufficient unit, surrounded by a cell membrane and having a nucleus at some stage of their existence. At the same time cells show a remarkable diversity of structure and function. Cells are basically spherical in shape, although modification to suit function leads to a degree of diversity. In size they mostly range from 10–30 μm in diameter. Their size is restricted by:

- the surface area to volume ratio, which must be as large as possible to allow the exchange of metabolic substances;
- the capacity of the nucleus to exercise control over the rest of the cell.

MICROSCOPY

Microscopes magnify objects, allowing structures to be seen which are otherwise well beyond the scope of the human eye. In the light microscope glass lenses focus rays of light, to achieve magnifications of up to 2 000 times. By contrast the electron microscope can achieve magnifications of over 500 000 times using electromagnets to focus beams of electrons.

Table 1.4 *Comparison of light and electron microscopes*

Light microscope	Electron microscope
Advantages	*Disadvantages*
Cheap to purchase (£100–500)	Expensive to purchase (over £1 000 000)
Cheap to operate – uses a little electricity where there is a built-in light source	Expensive to operate – requires up to 100 000 volts to produce the electron beam
Small and portable – can be used almost anywhere	Very large and must be operated in special rooms
Unaffected by magnetic fields	Affected by magnetic fields
Preparation of material is relatively quick and simple, only a little expertise	Preparation of material is lengthy and requires requiring considerable expertise and sometimes complex equipment
Material rarely distorted by preparation	Preparation of material may distort it
A vacuum is not required	A high vacuum is required
Natural colour of the material can be observed	All images in black and white
Disadvantages	*Advantages*
Magnifies objects up to 2 000 times	Magnifies objects over 500 000 times
The depth of field is restricted	It is possible to investigate a greater depth of field

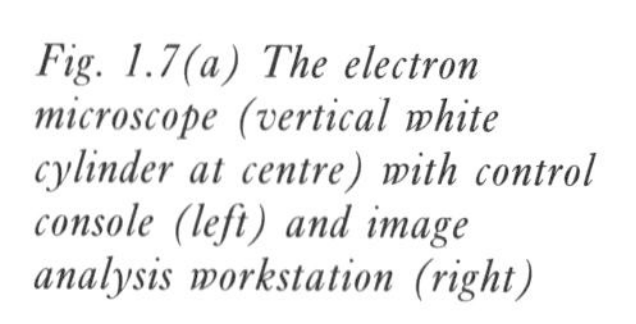

Fig. 1.7(a) The electron microscope (vertical white cylinder at centre) with control console (left) and image analysis workstation (right)

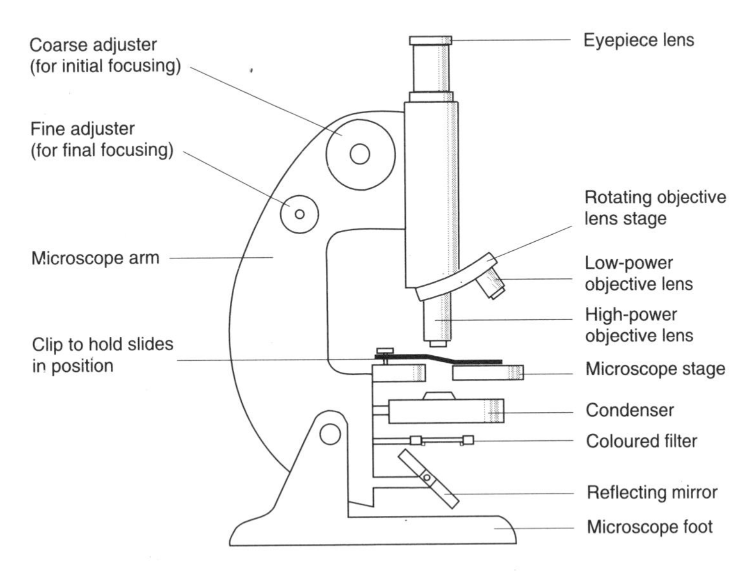

Fig. 1.7(b) The compound light microscope

CYTOLOGY

Cytology is the study of cells. There are two basic types: prokaryotic cells, the structure of which is described in 2.2, and eukaryotic cells.

Table 1.5 *Comparison of prokaryotic and eukaryotic cells*

Prokaryotic cells	Eukaryotic cells
No distinct nucleus; only diffuse area(s) of nucleoplasm with no nuclear membrane	A distinct, membrane-bounded nucleus
No chromosomes – circular strands of DNA	Chromosomes present on which DNA is located
No membrane-bounded organelles such as chloroplasts and mitochondria	Chloroplasts and mitochondria may be present
Ribosomes are smaller	Ribosomes are larger
Flagella (if present) lack internal 9 + 2 fibril arrangement	Flagella have 9 + 2 internal fibril arrangement
No mitosis or meiosis occurs	Mitosis and/or meiosis occurs

Structure of eukaryotic cells

The vast majority of organisms are composed of eukaryotic cells. They are of two basic types: plant cells and animal cells.

Table 1.6 *Plant and animal cell differences*

Plant	Animal
Cellulose cell wall as well as membrane	No cellulose cell wall, only membrane
Pits in cell wall	No pits
Plasmodesmata present	No plasmodesmata
Large vacuole filled with cell sap	Some vacuoles but usually small and numerous
Cytoplasm peripheral	Cytoplasm throughout the cell
Nucleus usually peripheral	Nucleus anywhere in cytoplasm but often central
Two cytoplasmic membranes: outer plasmalemma, inner tonoplast	Only one cytoplasmic membrane
Variety of plastids, e.g. chloroplast, leucoplast	Not normally any plastids
Cilia and flagella absent in higher plants	Cilia common in higher animals
Centrioles absent in higher plants	Centrioles present

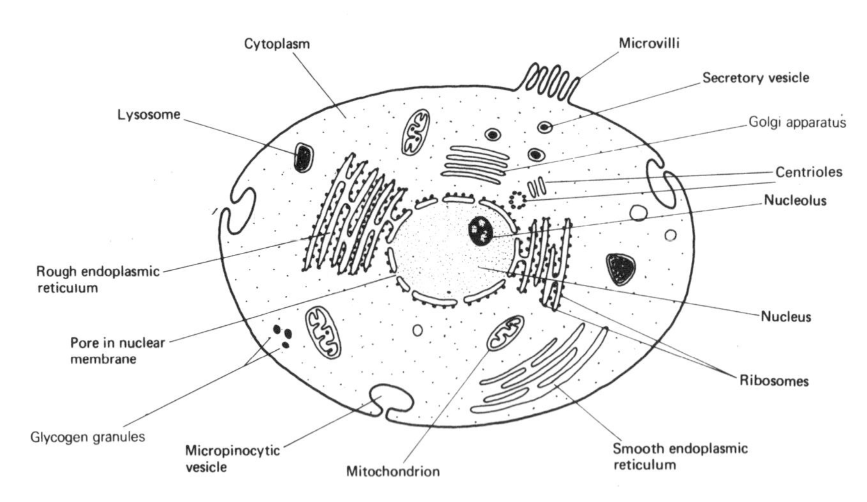

Fig. 1.8 Ultrastructure of a generalised animal cell

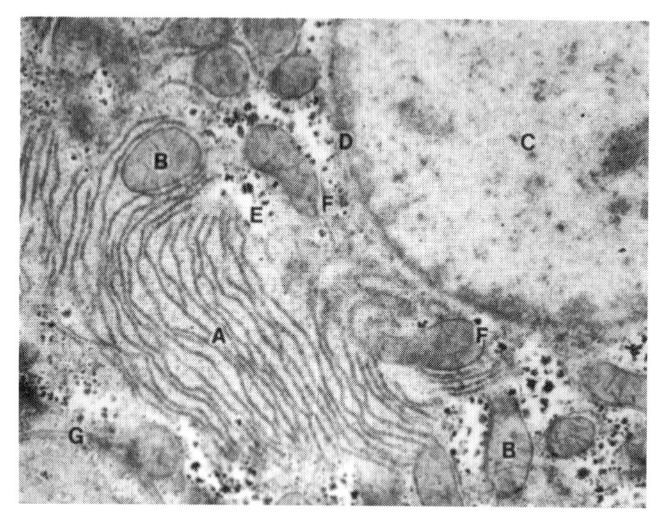

Key: A Rough endoplasmic reticulum
B Mitochondrion
C Nucleoplasm
D Nuclear envelope
E Glycogen granules
F Mitochondrial envelope
G Golgi apparatus

Fig. 1.9 Rat liver cells, magnification 25 000 ×

CELL ORGANELLES

To study cell organelles in detail, the different parts of the cell are often separated – a process called **cell fractionation**. Cells are broken up and placed in a **centrifuge**, an instrument which spins at high speed to exert a force which causes the different parts to settle out at different spin-speeds (**differential centrifugation**).

Cell membrane

The cell membrane acts as a boundary between a cell and its environment. It permits different substances to pass through it at different rates and is therefore described as being **partially permeable**. All cell membranes, whether they surround cells or organelles, are basically the same and are known as **unit membranes**. The generally accepted structure is the **fluid-mosaic model** (see Fig. 1.10) proposed by J J Singer and G L Nicholson in 1972.

Nucleus

If you have ever viewed a eukaryotic cell under a microscope you will probably have noticed the prominent nucleus. It is bounded by a double unit membrane, the nuclear envelope, which contains pores 40–100 nm in diameter. The outer membrane is granular whereas the inner one is smooth. Within the membrane is a mesh of nucleoplasm interspersed with nuclear ribosomes and chromatin (DNA bound to protein). There may also be one or two small, round bodies – the **nucleoli**, which store RNA. The functions of the nucleus are to:

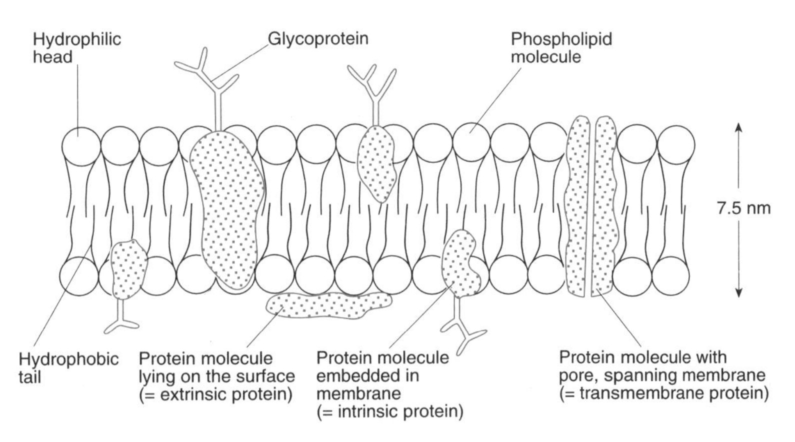

Fig. 1.10 Fluid-mosaic model of membrane structure

1. act as the control centre for the cell;
2. contain the genetic material of the cell;
3. produce ribosomes and RNA;
4. play an essential role in cell division.

Mitochondria

These vary in shape and size but are usually rod shaped, typically being 5 µm in length and 0.2 µm in diameter. The wall consists of two thin membranes, the inner one of which is folded to form **cristae**. Cell respiration takes place in mitochondria, most of the necessary enzymes being attached to the inner membrane and cristae, or occurring in the matrix. Thus cells which expend a lot of energy have numerous mitochondria, each with many cristae, e.g. muscle cells.

Fig. 1.11 Electron micrograph of mitochoɤdria from a hampster's adrenal cortex, clearly showing the cristae

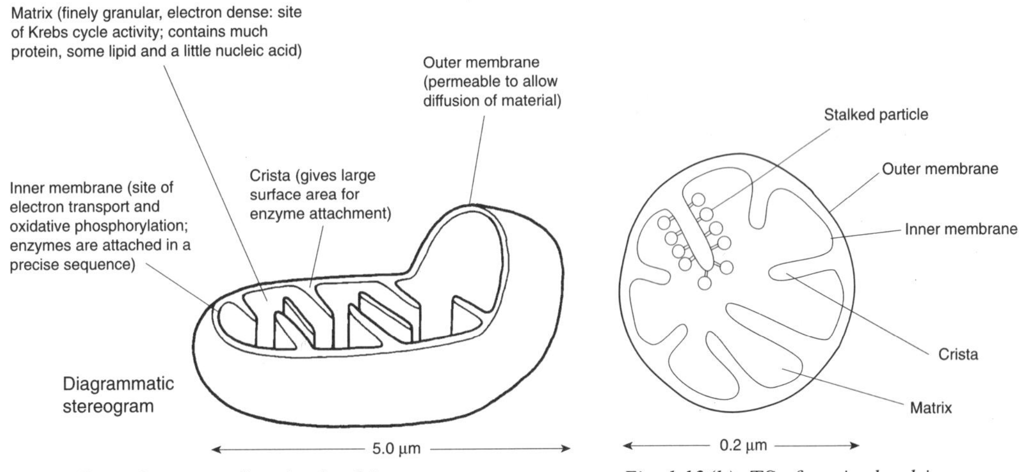

Fig. 1.12(a) Structure of a mitochondrion

Fig. 1.12(b) TS of a mitochondrion

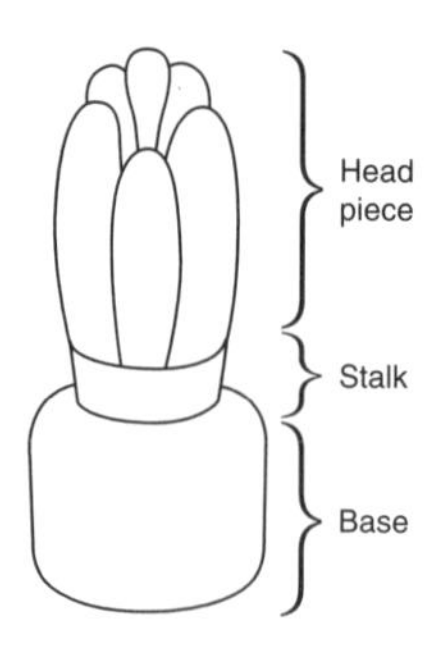

Fig. 1.12(c) Stalked particle

Endoplasmic reticulum

If you study an electron micrograph of a cell you will readily observe in the cytoplasm a network of membranes. This is the **endoplasmic reticulum (ER),** which forms a cytoplasmic skeleton. An extension of the outer nuclear membrane, the ER forms a series of sheets which enclose flattened sacs called **cisternae** (see Fig. 1.13). Where the ER is lined with ribosomes it is known as **rough endoplasmic reticulum**; where ribosomes are absent the term **smooth endoplasmic reticulum** is used. The functions of ER are to:

1. form a transport network throughout the cell;
2. provide a large surface area for chemical reactions;
3. play a role in protein synthesis (rough ER);
4. collect and store manufactured material;
5. form a structural skeleton to help maintain the shape of the cell;
6. produce lipids and steroids (smooth ER).

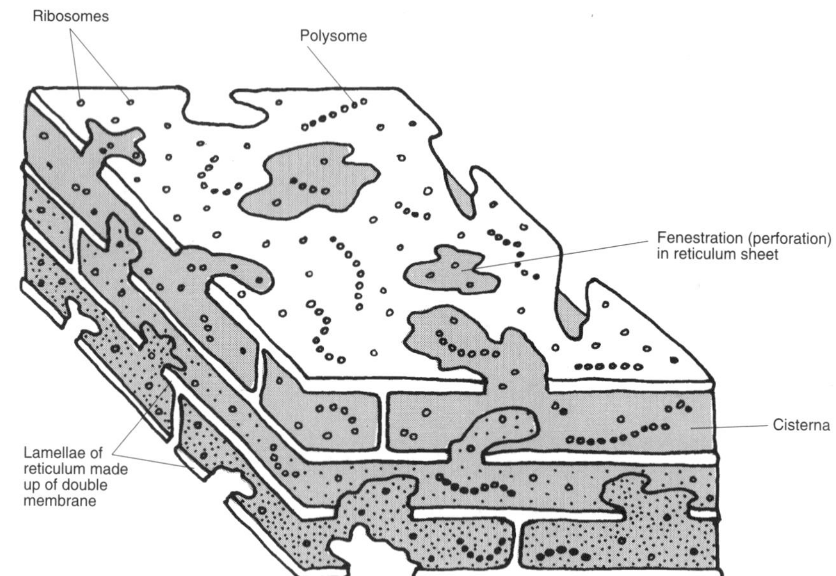

Fig. 1.13(a) Structure of rough endoplasmic reticulum

Golgi apparatus (dictyosome)

Similar in structure to smooth ER, the Golgi apparatus is composed of membrane-bound flattened sacs piled up in stacks. The Golgi apparatus is especially well developed in secretory cells. Its functions include:

1. production of glycoproteins;
2. secretion of carbohydrates involved in the production of new cell walls;
3. production of secretory enzymes;
4. transport and storage of lipids;
5. formation of lysosomes.

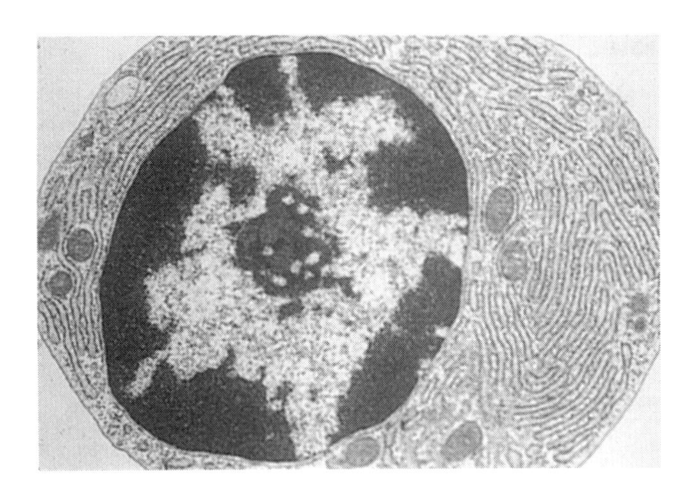

Fig. 1.13(b) Electron micrograph of plasma cell showing network of rough endoplasmic reticulum around the nucleus

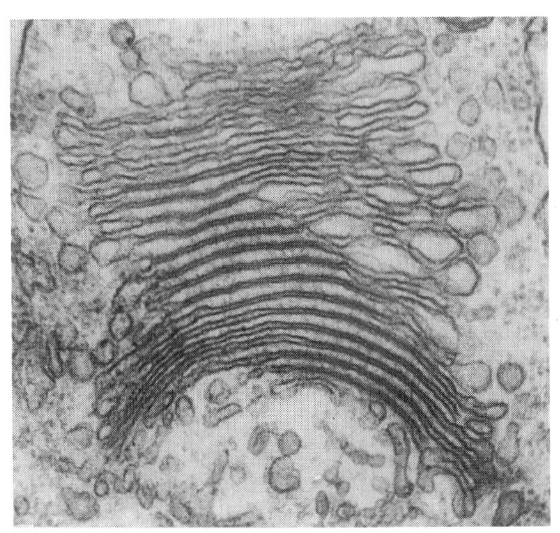

Fig. 1.14 Electron micrograph of Golgi apparatus, magnification 50 000 ×

Lysosomes

These are spherical organelles 0.1–1.0 µm in diameter, which contain around 50 enzymes, mostly hydrolases, in acid solution. Their functions include:

1. the destruction of unwanted or worn-out cell organelles;
2. the digestion of material engulfed by a cell, e.g. bacteria engulfed by white blood cells;
3. the release of enzymes outside the cell (**exocytosis**) to digest external material;
4. the complete digestion of a cell after its death (**autolysis**).

Ribosomes

These are small cytoplasmic granules about 20 nm in diameter in eukaryotic cells (80S type) but slightly smaller in prokaryotic cells (70S type). In groups they are known as **polysomes**. Ribosomes are made up of small RNA molecules and protein and are important in protein synthesis (see 3.1).

Peroxisomes (microbodies)

These are small, spherical, membrane-bound bodies 0.5–1.5 µm in diameter. They contain enzymes, in particular catalase which breaks down the highly toxic cellular by-product hydrogen peroxide into water and oxygen, thereby preventing the cell being poisoned. Peroxisomes are found in large numbers in actively metabolising cells, such as liver cells.

Vacuoles

This is a general term for any fluid-filled sac surrounded by a membrane. In plant cells there is normally a single, large central vacuole surrounded by a membrane called a **tonoplast**. In plant cells this vacuole stores amino acids and sugars as well as certain waste, it contains coloured pigments and, most importantly, supports the herbaceous parts of a plant by providing an osmotic system which creates a pressure potential.

Storage granules

These include starch grains within plant cells, glycogen granules in animal cells and oil or lipid droplets in both. All act as energy stores.

Centrioles

In animal cells two centrioles are found at right angles to each other. They are hollow cylinders about 0.2 μm in diameter and have the same basic structure as the basal bodies of cilia. They function in the formation of the spindle at cell division (see 3.2) and also give rise to cilia and flagella.

Cilia and flagella

Both may be concerned with locomotion in acellular organisms but cilia are also widespread in higher animal groups where they move material within an organism, e.g. cilia move mucus in the respiratory tract of mammals. Cilia are up to 25 μm long and flagella 1 000 μm long; both have an average diameter of less than 0.3 μm. They have the same basic structure. At the base of each cilium is a basal body composed of nine peripheral fibres but not the central two. Movement of cilia and flagella requires ATP, and the bending process is initiated at the base and transmitted towards the tip.

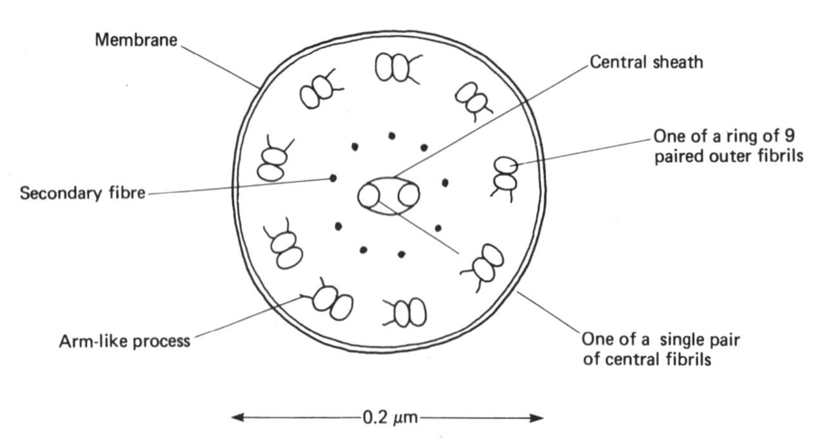

Fig. 1.15 TS of cilium

Microtubules

These slender, unbranched tubes around 24 nm in diameter are widespread in eukaryotic cells. Their functions include:

1. providing an internal cellular skeleton (**cytoskeleton**);
2. providing routes in cells along which materials move;
3. forming a framework on which the cellulose cell wall of plant cells is laid down;
4. aiding cell division by forming the spindle;
5. a major component of cilia and flagella.

Microfilaments

These very thin strands about 6nm in diameter are made up of the protein actin and sometimes myosin. They are probably involved in movement within cells.

Microvilli

Around 0.6 μm in length, the microvilli are tiny, finger-like projections of the cell membrane of certain cells such as those in the kidney tubule and the epithelium of the intestines. Collectively they form the **brush border** of these cells and help to increase their surface area, aiding absorption.

Cellulose cell wall

You will probably have observed that plant cells, when seen through a microscope, have a distinct boundary layer. This is the cell wall – a characteristic feature of plant cells. It is made up of fibrils of the polysaccharide cellulose, which is often impregnated by other polysaccharides such as pectin and lignin. The function of the cell wall is to:

1. provide support in herbaceous plants;
2. allow movement of water through and along it;
3. act as a waterproofing layer when impregnated with other substances such as lignin.

Chloroplasts

These are found only in plant cells. Their shape and size vary with species but there is always a double unit membrane surrounding a matrix (**stroma**) in which are stacks of lamellae (**grana**). The lamellae hold the chlorophyll in the most suitable position for photosynthesis. The stroma contains numerous starch granules and enzymes for the reduction of carbon dioxide.

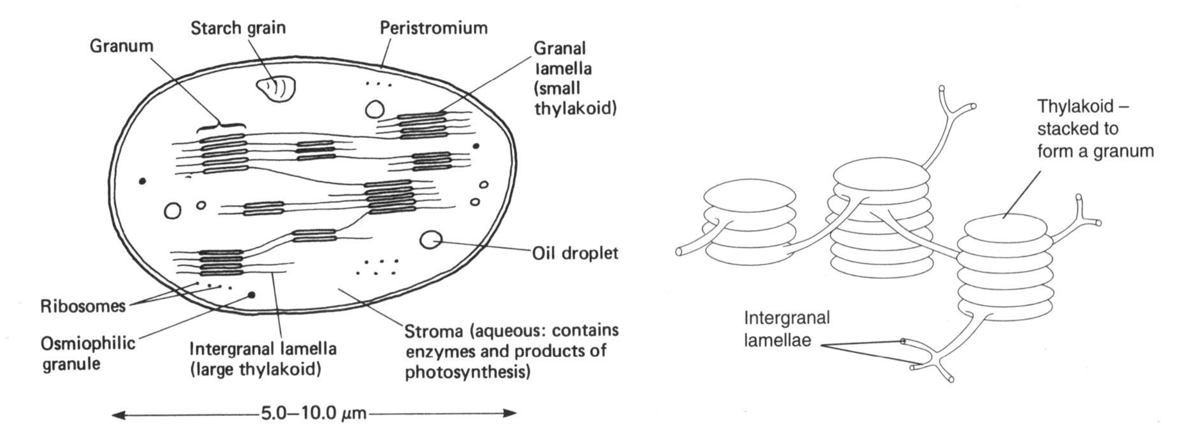

Fig. 1.16(a) Structure of a chloroplast

Fig. 1.16(b) Structure of a group of grana

TRANSPORT OF MATERIALS INTO AND OUT OF CELLS

The movement of materials in and out of cells is achieved by **diffusion, osmosis, active transport, phagocytosis** and **pinocytosis**.

Diffusion

This is the process by which the particles of a substance move from a region where it is highly concentrated to a region where its concentration is lower. Generally a slow process, the rate of diffusion is faster if:

1. the difference in concentration between the two regions (**concentration gradient**) is made greater;
2. the distance between the two regions is decreased;
3. the area over which diffusion occurs is increased;
4. the molecules diffusing are small and fat-soluble;

5. the pores in the cell membrane are numerous and large.

Facilitated diffusion is a more rapid form of diffusion, in which the molecules move through channels in the membrane.

Neither type of diffusion involves any expenditure of energy.

Osmosis

Osmosis is the net movement of water molecules from a region of their higher concentration to a region of their lower concentration, through a **partially permeable** membrane. Imagine the following situation:

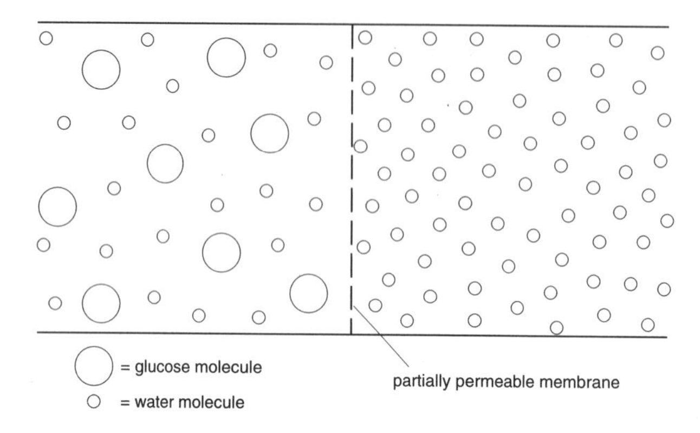

Fig. 1.17 (a)

Both the water (solvent) and glucose (solute) molecules move randomly as a result of their kinetic energy, but only the water (solvent) molecules are small enough to cross the partially permeable membrane. They will do this until equal numbers of water molecules are present on either side of the membrane.

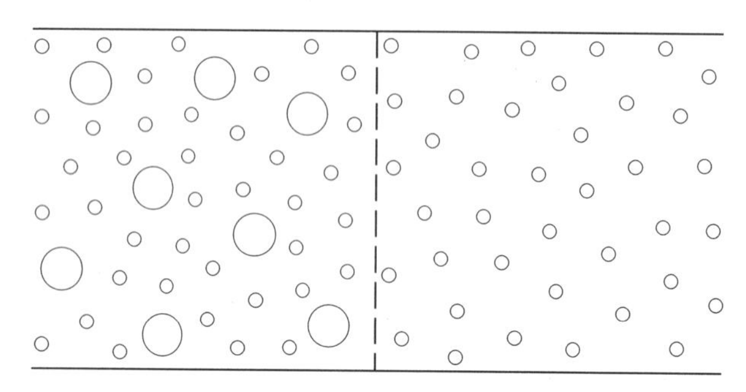

Fig. 1.17 (b)

In theory a dynamic equilibrium now exists. In practice however, the glucose molecules on the left of the membrane impede the water molecules on that side from moving to the right side of the membrane. As those on the right are not impeded they continue to move to the left faster than those moving in the opposite direction.

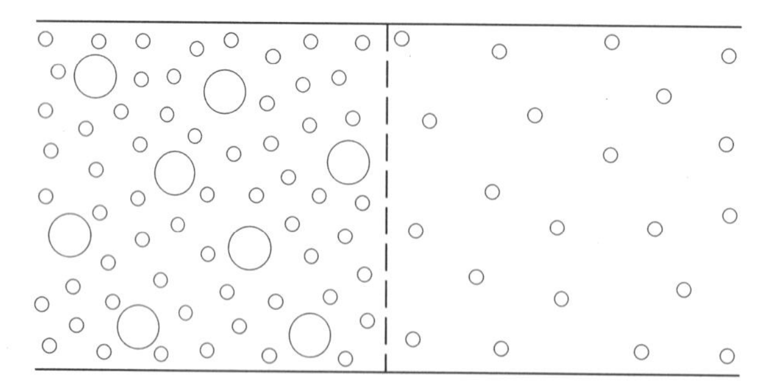

Fig. 1.17 (c)

Water molecules accumulate on the left of the membrane until their concentration is high enough to offset the blocking effect of the glucose. The probability of a water molecule moving from right to left is now the same as one moving from left to right and a dynamic equilibrium is established. More details of osmosis are given in 7.3.

Active transport

This process requires energy and enables materials to be moved *against* a concentration gradient, ie. from a region of low concentration to one which is higher. The proteins spanning the membrane are thought to convey materials from one side to the other. Because energy is used, cells carrying out active transport usually have a high respiratory rate, many mitochondria and a high concentration of ATP. Any factor which reduces or stops respiration, eg. cyanide, will reduce or stop active transport.

Phagocytosis

This is the process by which a cell can obtain particles which are too large to be taken in by other means. The cell invaginates to form a depression in which the particles are contained. This then pinches off to form a vacuole. An illustrated account is given in 7.2.

Pinocytosis

This is similar to phagocytosis except that the vacuoles are smaller and liquids rather than solids are taken in.

Both phagocytosis and pinocytosis involve the taking of materials into the cell in bulk. They are therefore examples of **endocytosis**. The removal of materials from the cell in bulk is called **exocytosis**.

1.3 TISSUE ORGANISATION

What is a tissue?

A tissue is a group of usually similar cells which, together with their intercellular substance, perform a particular function. The study of tissues is called **histology**.

ANIMAL TISSUES

Epithelial tissues

Associated with its role as a covering tissue subject to mechanical damage, epithelial tissue has the following characteristics:

1. cells frequently attached to basement membrane;
2. adjacent cells joined together by intercellular cement;
3. there may be connecting bridges of cytoplasm between the cells;
4. cells may be built up into a large number of layers to resist abrasion, e.g. epidermis of skin;
5. rapid division of epithelial layer to resist abrasion and maintain only a thin layer of cells, e.g. intestines;
6. resistance to abrasion may be increased by presence of keratin, e.g. epidermis of skin.

The following are types of epithelial tissue.

Columnar epithelium: tall, narrow cells often with cilia which beat with a definite rhythm, transporting mucus and other particles. Found in nasal cavities, trachea, oviducts, ventricles of brain.

Squamous epithelium: sometimes called 'pavement' epithelium. Flattened, single layer of thin nucleolated cells. Found where rapid diffusion is essential, e.g. Bowman's capsule, alveoli, endothelium.

Compound (stratified) epithelium: many layers of cells; the ones nearest the basement membrane are usually columnar and those farthest away from it are often flattened and dead, possibly impregnated with keratin. Found where there is considerable mechanical stress, e.g. epidermis of skin, oesophagus, anal canal, vagina.

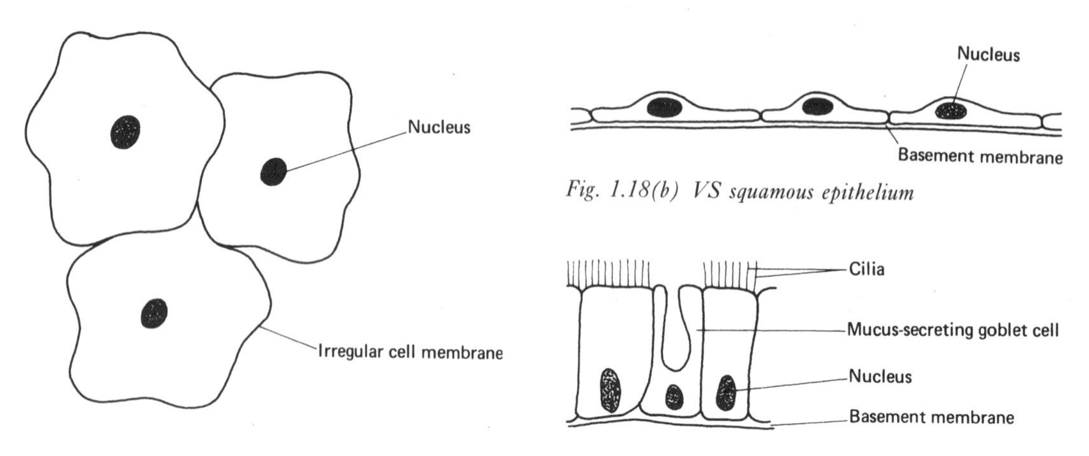

Fig. 1.18(a) Squamous epithelium, surface view

Fig. 1.18(b) VS squamous epithelium

Fig. 1.18(c) VS ciliated columnar epithelium

Connective tissues

The distinguishing characteristic of connective tissue is the presence of a variety of cells which are embedded in a large quantity of intercellular substance called the **matrix**. This type of tissue includes supporting tissues such as cartilage and bone as well as blood, which serves a largely transport role.

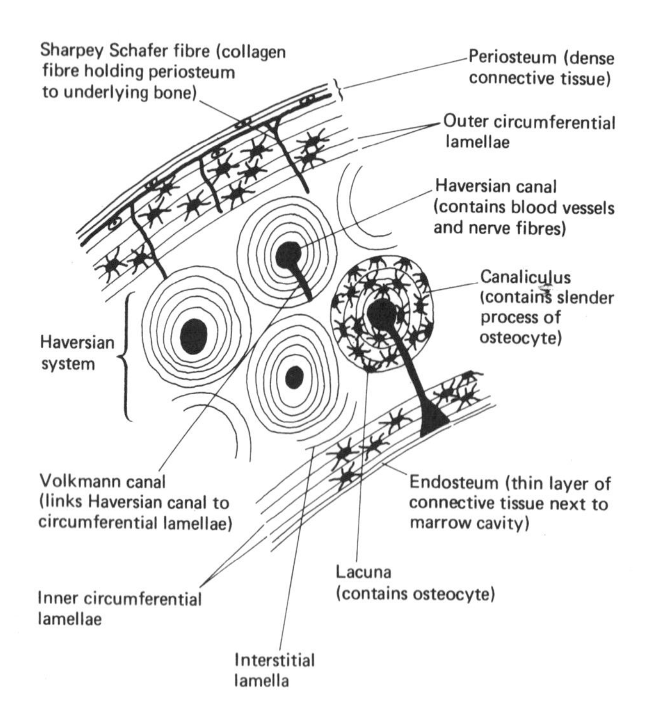

Fig. 1.19 TS compact bone

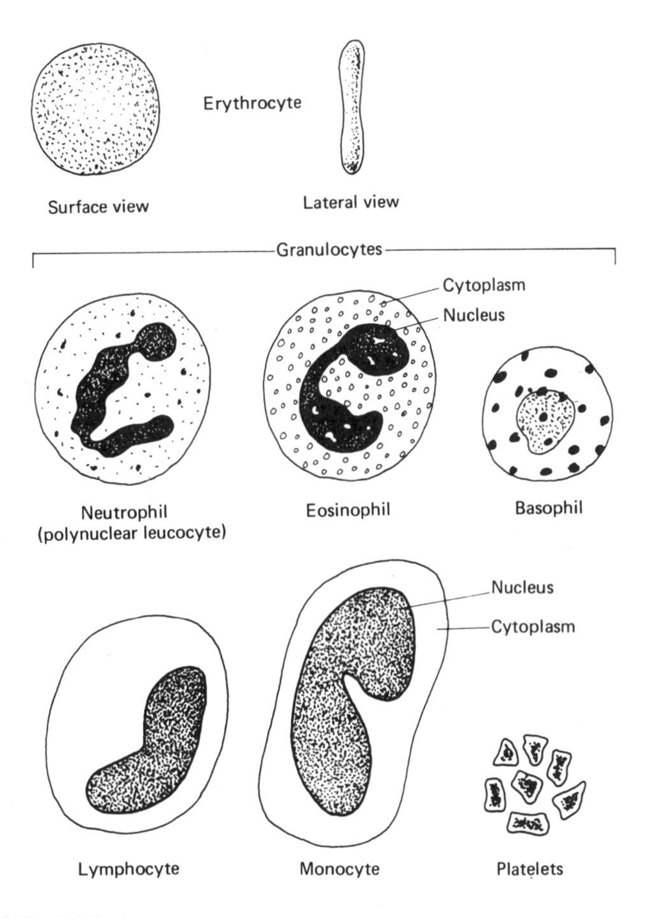

Fig. 1.20 Types of blood cell

Muscular tissue

All muscle cells can contract. The main type in mammals is **voluntary (skeletal) muscle.** This forms the bulk of body muscle and is composed of long fibres (**myofibrils**) made up of fine threads called **myofilaments**. These comprise the proteins **actin** and **myosin** whose arrangement is responsible for the myofilaments' striped appearance. The tissue is under voluntary control – hence its name.

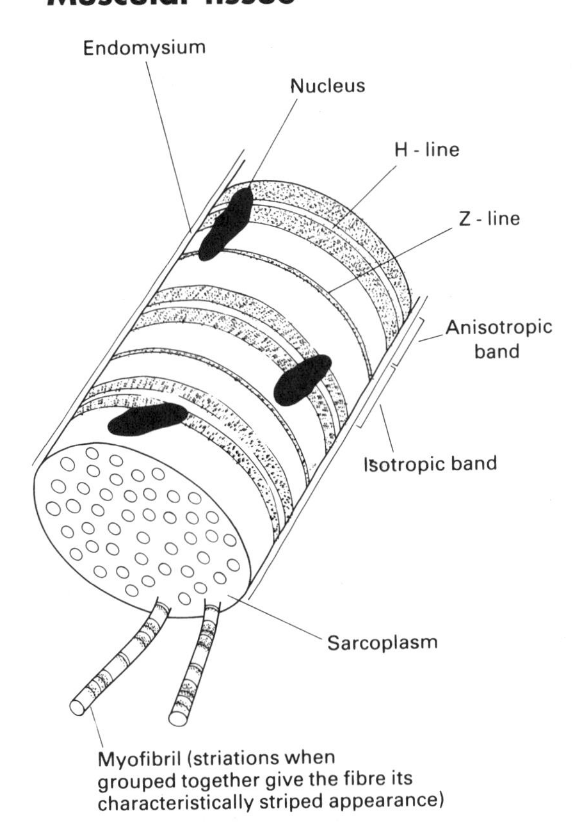

Fig. 1.21(a) Voluntary muscle fibre

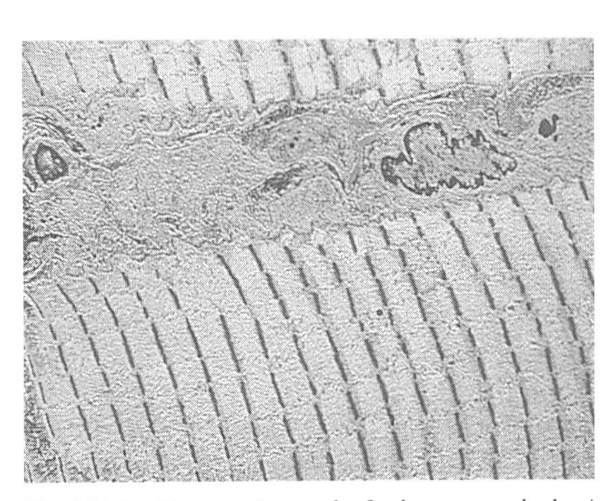

Fig. 1.21(b) Electron micrograph of voluntary muscle showing the banding pattern of the myofibrils, magnification 1800 ×

Nervous tissue

This has highly developed properties of irritability and conductivity. It comprises closely packed nerve cells known as **neurones**.

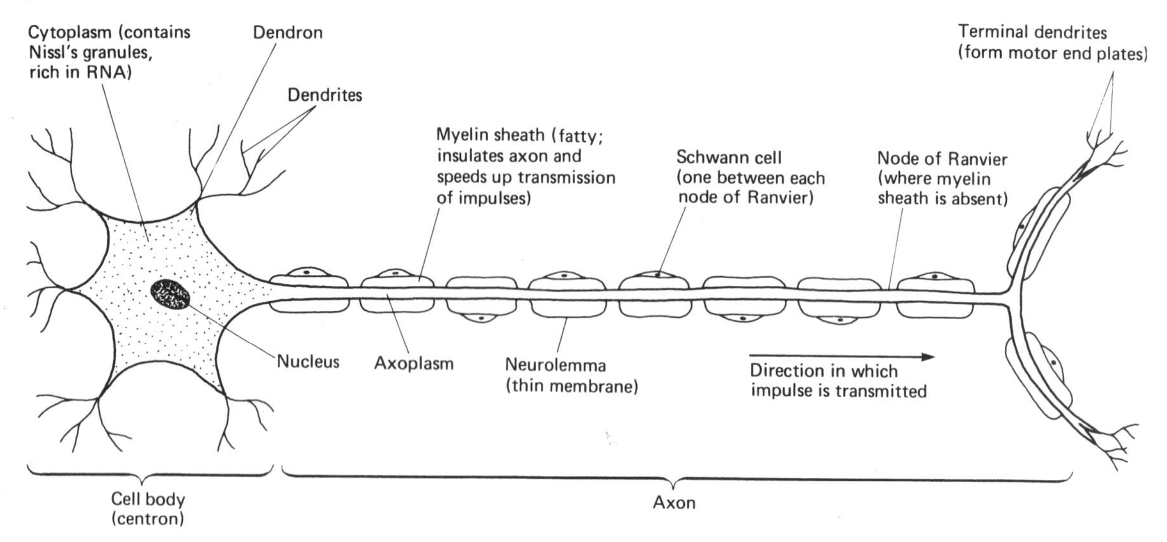

Fig. 1.22 Effector (motor) neurone

PLANT TISSUES

Parenchyma

These unspecialised cells are the major component of the ground tissue and are present in vascular tissues. They are potentially meristematic.

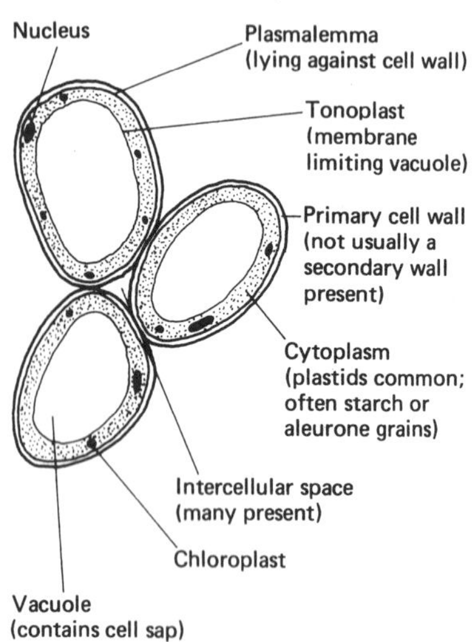

Fig. 1.23(a) TS parenchyma

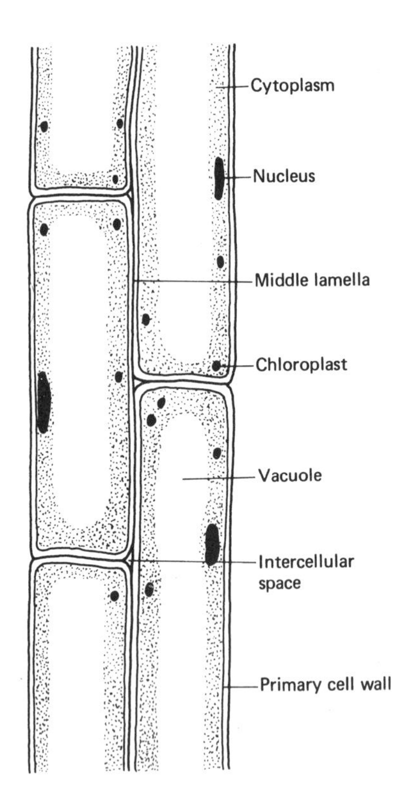

Fig. 1.23(b) LS parenchyma

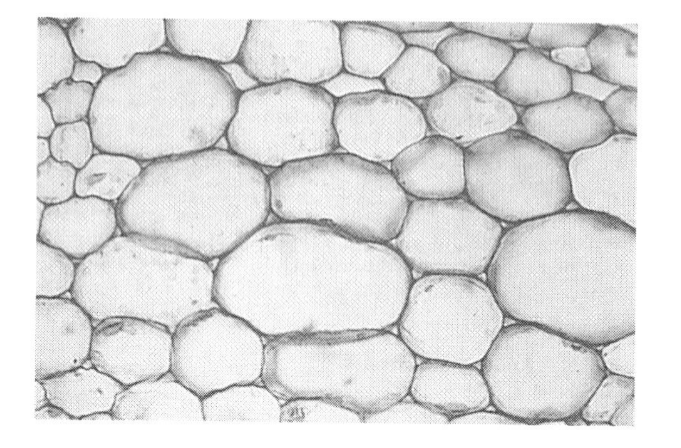

Fig. 1.23(c) TS parenchyma, magnification 250 ×

Collenchyma

The primary cell wall is unevenly thickened with cellulose and pectic substances but not lignin. Pits are present in the walls. Their function is support, especially in younger stems where plasticity is necessary to allow for changing growth requirements.

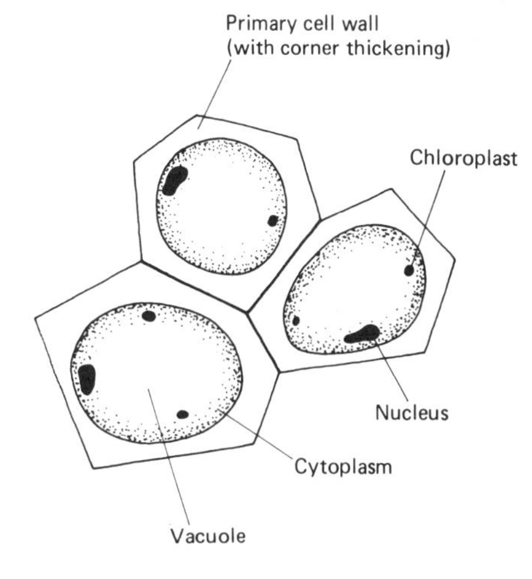

Fig. 1.24(a) TS collenchyma

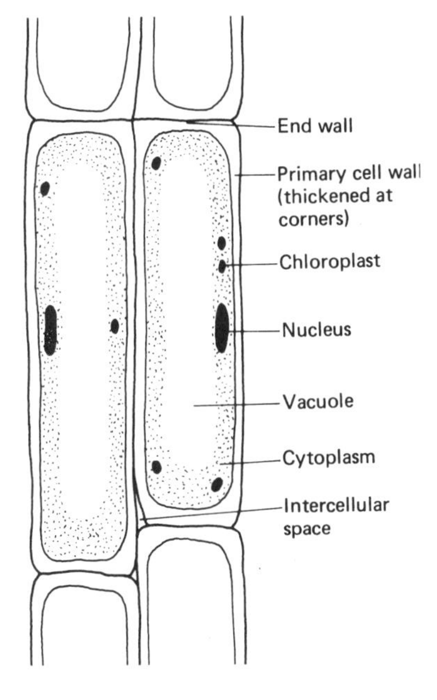

Fig. 1.24(b) LS collenchyma

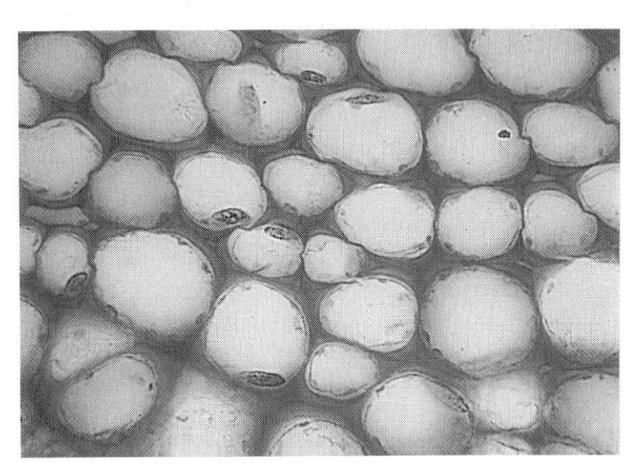

Fig. 1.24(c) TS collenchyma, magnification 350 ×

Sclerenchyma

This is a supporting tissue with a secondary cell wall of lignin deposited on the primary cell wall of cellulose. The two basic types are sclereids and fibres but the differences between them are not always clear-cut. The pits in the cell walls may be simple or bordered.

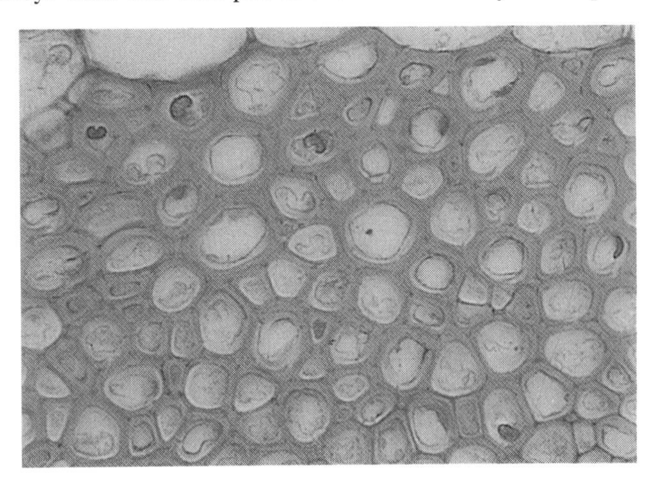

Fig. 1.25 TS sclerenchyma, magnification 450 ×

Fibres are often found in the vascular bundles of dicotyledons or around the vascular bundles of monocotyledons. They are often grouped in strands. Jute, hemp and flax are economically important fibres.

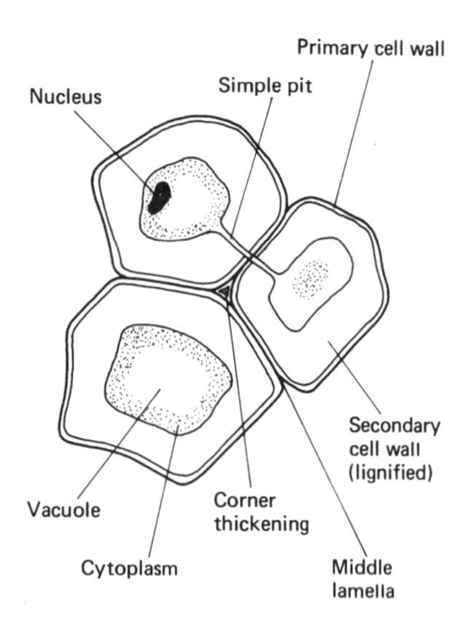

Fig. 1.26(a) TS fibre

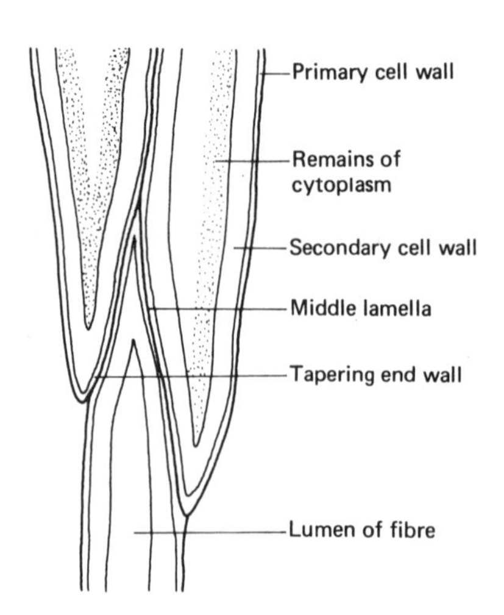

Fig. 1.26(b) LS fibre

Xylem

Primary xylem is formed from the embryo and the resultant meristems. Secondary xylem develops later, during secondary thickening. Xylem is made up of four main elements: vessels, tracheids, fibres and paranchyma.

Vessels These cells have one or more perforations at each end so that water can move easily from cell to cell. Cell walls may be simple or perforated by bordered pits.

Tracheids These cells have no open ends and the pits occur in pairs so that the water can pass easily through the thin pit membrane.

Fibres These are long cells whose secondary walls are commonly lignified. Pits are frequent. Fibres are primarily used for support.

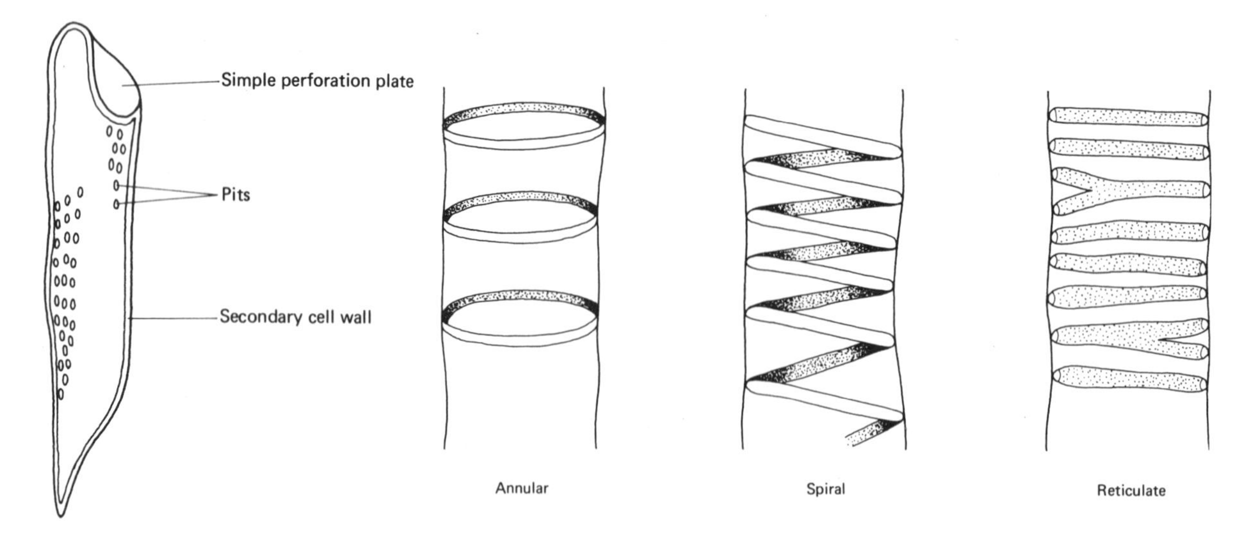

Fig. 1.27(a) Vessel

Fig. 1.27(b) Thickenings of secondary cell walls in primary xylem

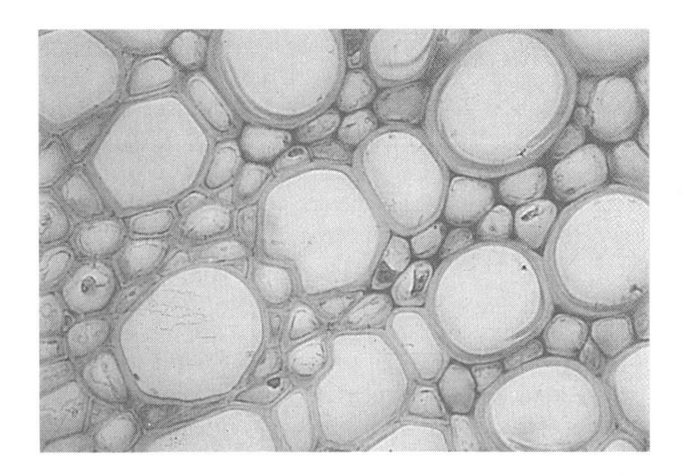

Fig. 1.27(c) TS xylem, magnification 600 ×

Phloem

Primary phloem develops from the procambium and secondary phloem from the vascular cambium. Phloem is made up of four main components: sieve tubes, companion cells, sclerenchyma and parenchyma.

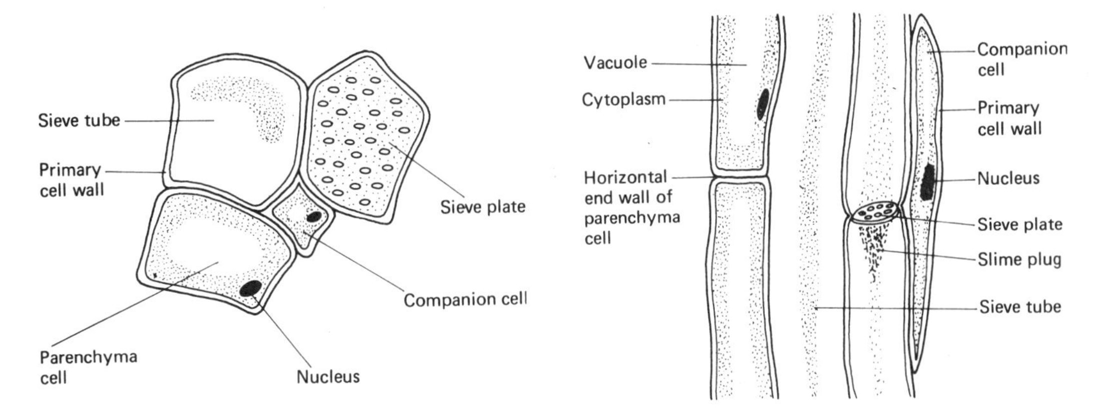

Fig. 1.28(a) TS phloem

Fig. 1.28(b) LS phloem

Fig. 1.28(c) TS phloem, magnification 300 ×

Sieve tubes These are the most highly specialised phloem cells. Modified pits called sieve plates form on adjacent cells; most are in a vertical series but some lateral sieve plates do occur. As the phloem ages callose blocks the sieve plate and the sieve tube dies.

Companion cells These form from sieve tubes early in their development and contain a nucleus whereas the sieve tube does not; when the sieve tube dies so does the associated companion cell. The walls between the two are thin and densely pitted.

Chapter roundup

We have seen in this chapter that, far from being a structureless mass of cytoplasm, an individual cell is in fact a very complex collection of interrelated organelles which work together to permit the cell to carry out its particular functions. Cells are extremely diverse in both their structure and in the proportion of organelles they contain – but, whether in terms of its external appearance or internal structure, the composition of any cell is determined by the particular function it performs. This specialisation of cells to perform one particular function leads to a division of labour in which one cell devotes all its energies to carrying out just a few tasks in order to become proficient at them, rather than being a 'jack of all trades, but master of none'. This, however, leads to dependence on other cells to perform those tasks it has ceased to carry out. Human society structures itself in much the same way: a doctor may specialise in medical matters, but depend on the motor mechanic to repair his car; the motor mechanic can service his own vehicle but relies on the doctor to diagnose and relieve any ailments he has.

Similar cells often group together as tissues in order to pool their energies and perform a task more efficiently, often with the assistance of related tissues, e.g. muscle cells group together, along with connective tissues such as blood, in order to be able to contract effectively. Again human society mirrors this trait: doctors group together in hospitals with nurses, radiographers and ancilliary workers to provide an efficient service. In the next chapter we shall see how this diverse variety of cells is combined into an even more diverse variety of organisms.

Illustrative questions and worked answers

1 (a) In what ways do lipids differ from carbohydrates? (3)
(b) Using examples to illustrate your answer, describe the functions of lipids in organisms. (10)
(c) Why do many organisms store lipids rather than carbohydrates? (5)
Time allowed 35 minutes

Tutorial note

This style of question is common to all examination boards. It requires wide-ranging knowledge of both carbohydrates and lipids, but the form of the question is designed to test the ability to select relevant points, make comparisons and understand at least one important underlying principle. You must pay particular attention to the mark distribution and deploy your time accordingly. The 18 marks available over the 35 minutes allowed make the calculation especially easy – almost exactly 2 minutes for each mark.

Take care in part (a) to emphasise *differences;* it is all too easy to simply describe carbohydrates and lipids, leaving the examiners to draw out the differences for themselves. It is a fundamental rule of examiners that they must not do the work for the candidate. They cannot draw conclusions or make comparisons for you – it is your job, not simply to state the factors, but to present them in the required form. As part (a) carries only 3 marks, a few clear, fundamental differences should be stated rather than a long series of minor ones.

The key words in part (b) are ' examples to illustrate' and you should ensure that each function is supported by at least one relevant example. Be as precise as you can in naming organisms – 'ladybird' is preferable to 'beetles', which is preferable to 'insects', which is most certainly better than 'invertebrates' or 'animals'. However, it is often quite acceptable to use a group name, e.g. 'mammals', provided the point applies to all members of the group. Indeed, to specify an individual species such as 'an aardvark' implies that this is the only mammal displaying the characteristic in question.

With 5 marks available you should include some detail in your answer to part (c) rather than make one or two general points.

Suggested answer

(a)

Lipids	Carbohydrates
Smaller proportion of oxygen in a molecule	Large proportion of oxygen
Ester	Aldehyde or ketone
Do not mix with or dissolve in water	Many are soluble in water
Yield much energy on oxidation	Yield half as much energy on oxidation
Do not form long chain polymers	Polysaccharides are long chain polymers

(b) Each gram of lipid, when oxidised, yields 38 kJ of energy. This makes them a very rich source of energy for use in the respiration of almost every organism.

Being such a concentrated energy source, lipids are particularly valuable as an energy store, especially where mass needs to be kept to a minimum. Examples include the seeds of plants which need to be as light as possible if they are to be effectively dispersed by the wind or animals. Many plant seeds are therefore rich in lipids, e.g. sunflower, linseed. Animals also prefer fat as a store since its low mass for a given energy yield is of advantage as the animal must carry the store with it as it moves around. Fat stored by an animal often performs a secondary function. It can be stored around the kidneys in mammals where it acts as a packing material protecting the organ from physical damage.

Lipids also have good insulatory properties and their storage just beneath the skin of mammals and birds helps to maintain body temperature. The insulatory role of lipids is secondary to that of fur and feathers in most mammals and birds, but in aquatic species such as whales or penguins fur and feathers are inefficient. In these species, lipid, e.g. in the form of blubber, is the major insulatory material.

The fact that lipids do not mix with water gives them excellent waterproofing properties. The sebaceous glands in mammals secrete a fatty substance which waterproofs the skin. In many plants, especially those living in dry conditions, a waxy cuticle reduces transpiratory losses. This layer is especially thick in species such as the holly and certain cacti.

All cell membranes contain phospholipids and these perform an important structural role as well as making the cell permeable to some substances but not others.

(c) Lipids have less than half the mass of carbohydrates for a given quantity of energy stored. This makes them a preferable store where the organism, or part of it, has to move from place to place. As they do not mix with water, lipids can be stored without risk of them being dissolved out of their place of storage which is a problem with some carbohydrates. Unlike most carbohydrates, lipids can also serve a secondary function such as insulation, protection or waterproofing.

2 In the space below draw a diagram to show the structure of a mitochondrion as revealed by the electron microscope.

(a) Label four of its component parts. (2)

(b) Add a scale to your diagram which shows the approximate size of the organelle. (1)

(c) How would you proceed to isolate mitochondria from the cells of living tissue such as the brain or the liver? (2)

(d) Having obtained a suitable sample of mitochondria in (c), briefly describe how you would demonstrate any one of their functions experimentally. (4)

Time allowed 10 minutes

Tutorial note

As this is a short structured question your answer needs to be detailed but concise. As always, mark allocation should determine the time spent on each section – approximately 1 minute per mark in this case. Diagrams should be large, clear and properly labelled using single lines without arrowheads. In this style of question the need for connected prose is not as vital as in an essay question. Clear, concise notes are normally an acceptable means of conveying the facts. Nevertheless, all writing must be legible and the meaning must be clear and unambiguous.

Suggested answer

(a) and (b) A simple labelled TS or LS as illustrated in either Fig. 1.12(a) or Fig. 1.12(b) complete with scale is needed (remember you have only around 3 minutes to complete this part).

(c) Cells from the brain or liver are broken up physically and placed in a centrifuge, which is spun at a speed which causes all organelles and debris larger than mitochondria to settle out. The deposit is discarded and the supernatant is spun again at a slightly faster speed – sufficient to cause the mitochondria to settle out. This time the deposit is retained and the supernatant liquid discarded.

(d) The major function of mitochondria is the conversion of pyruvate into carbon dioxide and water with the liberation of energy, in a process called the Krebs cycle. One method of demonstrating this function is to add pyruvate to the isolated mitochondria. The liberation of carbon dioxide would indicate that the mitochondria were capable of carrying out the Krebs cycle. A further confirmation would be the addition of cyanide to the mitochondria and pyruvate. As an inhibitor of cytochrome oxidation, the cyanide should prevent any carbon dioxide being produced. This indicates that the mitochondria must possess these enzymes and therefore be the site of the electron transport pathway as well as the Krebs cycle.

Question bank

1 Structure and function are closely related. By reference to (a) voluntary muscle tissue and (b) parenchyma tissue in plants, discuss how far this statement is true.
Time allowed 40 minutes

Pitfalls

You should ensure that structure and function are related throughout – take care not just to describe the structure and the function of each tissue without showing how each influences the other.

Do not limit yourself to structures visible under the light microscope – subcellular organelle structure and function should also be included.

Remember that 'tissues' can include associated parts, e.g. blood system, intercellular spaces.

Points

In each instance it is essential to show clearly how the structure permits a specific function to be carried out efficiently.

(a) Striated (skeletal) muscle

Structure	Function
Elongated fibres	Allow considerable contraction
Parallel fibres	Give maximum contractile effect
Fibre ends tapered and interwoven	Provide strength
Large number of mitochondria	Provide large amount of ATP
Actin and myosin arrangement	Allows contraction by filaments sliding over each other
Rich supply of blood vessels	Provide adequate supply of oxygen and glucose
Myoglobin present	To store oxygen for release when blood oxygen levels are low
Motor end plates	Allow stimulation of muscle
Fibres arranged in motor units	To permit variable degree of contraction

(b) Parenchyma cells

Structure	Function
Unspecialised tissue	Variety of functions
Many intercellular spaces	Diffusion of gases
Isodiametric cells	Packing material
Thin cellulose cell walls	Permit passage of materials
Transparent cell wall	Permits entry of light for photosynthesis
Permeable walls	Allow water entry for turgidity
Large cells/large vacuoles	Provide storage space
Chloroplasts present	Allow photosynthesis
Chromoplasts present	In petals provide colour to attract insects for pollination
Leucoplasts present	To store starch

2 Discuss the location of membranes and their functions within cells.
Time allowed 35 minutes

Pitfalls

You must be especially careful to discuss all membranes in cells and not just the cell membrane, i.e. include the membranes which surround or make up organelles. At the same time the plasma membrane and its various adaptations (microvilli, plant root hairs) must not be ignored.

Remember to state where each membrane is found and its precise function. A diagram showing the structure of the cell membrane is *not* needed and would simply reduce the time available for answering the question properly.

Do not restrict yourself to one or two forms of membrane but try to include a wide range of examples.

Points

Distribution	Function
Plasma membrane	Partial permeability; allows osmosis, diffusion, active transport; limits cell
Nuclear membrane	Limits DNA; allows mRNA out
Mitochondria: outer	Allows glycolytic products in
Mitochondria: inner	Attachment of respiratory enzymes
Endoplasmic reticulum	Cellular transport; attachment of ribosomes
Chloroplast: outer	Allows photosynthetic products out and substrates in
Chloroplast: lamellae	Reservoir of chlorophyll, carotenes, etc.
Golgi apparatus	Sorting of ER-synthesised material; storage of glycoprotein; synthesis of polysaccharides (e.g. cellulose in plants)
Lysosomes	Limit autolytic enzymes
Tonoplast	Limits cell sap
Pinocytic/phagocytic vesicles	Uptake of materials
Other specialised membranes:	
Root hairs	Increase surface area
Microvilli (e.g. kidney tubule)	Increase surface area
Myelin sheath membrane	Insulation of nerve fibre
Neurilemma	Diffusion of Na^+ and K^+ allowing depolarisation

3 The general formula for a disaccharide is

A. $C_{12}H_{22}O_{11}$
B. $C_5H_{10}O_5$
C. $C_3H_6O_3$
D. $C_{12}H_{24}O_{12}$
E. $C_6H_{12}O_6$
Time allowed 1 minute

Pitfalls

Take care not to confuse a monosaccharide and a disaccharide. Monosaccharides are largely hexoses (C_6) and pentoses (C_5); disaccharides consist of two monosaccharides joined together with the loss of water.

Do not be tempted by answer D; it is two monosaccharides ($C_6H_{12}O_6$) combined *but* without the loss of a water molecule (H_2O).

Points

The correct answer is A. $C_6H_{12}O_6 + C_6H_{12}O_6 - H_2O = C_{12}H_{22}O_{11}$

CHAPTER 2

VARIETY OF ORGANISMS

Units in this chapter

Chapter objectives

There are over two million different living organisms which must be sorted into groups to aid our study of them. In this chapter you will learn about the way plants and animals are categorised and how these schemes of classification have to be updated as we learn more and more about evolution. You will find out some of the main features which help us to put organisms into groups and how we name them so that scientists can communicate information about them clearly and unambiguously. Since living organisms evolved from nonliving material, it is not surprising that the dividing line between these groups can be thin. For example, viruses possess features of both living and nonliving things.

You will also see that acellular organisms are far from being the simple structures they are sometimes thought to be and that they show considerable specialisation within the confines of a single cell. However, they can never become very large since there is a limit to the volume of cytoplasm that may be controlled by a single nucleus. The evolution of multicellular organisms began in water, where the unicells lived, and over the course of millions of years they gradually became adapted to a more difficult habitat – the land. As we go through the classification of the various plant and animal groups you should try to look out for these adaptations and consider the whole way of life of the organisms concerned. In addition, you will learn something about the economic importance of the various organisms and how our lives are influenced by many of them.

2.1 PRINCIPLES OF CLASSIFICATION

Before we can study living organisms it is necessary to sort them into groups of a manageable size. This sorting is called **classification**. The classification system used should be universal so that scientists from different parts of the world can find out about organisms easily and can convey information about them to each other. We now know that each organism has evolved from another in a 'continuum'. This means that organisms do not really fall into distinct categories and the division of them into groups has been devised by man simply for his own convenience. There is not, therefore, a 'correct' scheme of classification although there is one which is more widely accepted than most – **natural classification**.

Natural classification

It is important that we try to find a system of classification that most closely reflects the true evolution of organisms. In order to do this we need to consider a number of different factors when trying to decide which category an animal or plant fits into. If organisms are to be placed in the same group they should be as similar as possible to each other. These similarities should now be based on biochemical and chromosome studies as well as morphology and anatomy which have been referred to for many years. It is important that any anatomical features used to classify an organism must be homologous rather than analogous.

Homologous features are ones which have a similar origin, structure and position, regardless of their function in the adult.

Analogous features are ones which have a similar function in the adult, but which are not homologous, e.g. wings of birds and wings of butterflies.

Taxonomic ranks

The study of classification is called **taxonomy**. It is convenient to distinguish large groups of organisms from smaller subgroups and a series of rank names has been devised to identify the different levels within this hierarchy. The rank names used today are largely derived from those used by Linnaeus over 200 years ago. The largest groups are known as **phyla** and the organisms in each phylum have a body plan radically different from organisms in any other phylum. Diversity within each phylum allows it to be divided into **classes**. Each class is divided into **orders** of organisms which have additional features in common. Each order is divided into **families** and at this level differences are less obvious. Each family is divided into **genera** and each genus into **species**.

Binomial nomenclature

Every organism is given a scientific name according to an internationally accepted system of nomenclature, first devised by Linnaeus. The name is always in Latin and is in two parts. The first name indicates the genus and is written with an initial capital letter; the second name indicates the species and is written with a small initial letter. These names are always distinguished in text by the use of italics or underlining. For example, the scientific name for man is *Homo sapiens*, which fits into the series of ranks as follows:

phylum	Chordata
class	Mammalia
order	Primates
family	Hominidae
genus	*Homo*
species	*sapiens*

These are the ranks of classification to which every specimen must be assigned but a taxonomist may use any number of additional categories within this scheme, e.g. suborder, superfamily, subgenus.

2.2 LOWER ORGANISMS

We will now look at four groups of organisms which are neither plants nor animals. These are the viruses, which cannot be fitted into the classification of living organisms, the Prokaryotae, bacteria which have unusual cells that lack nuclei and the usual membrane-bounded structures, and the Protoctista and the Fungi, both of which are made of eukaryotic cells but are not true plants or animals.

VIRUSES

Viruses are very different from any of the groups which follow. Outside living cells they are inert particles, known as **virions**, but when they enter living cells they show the characteristics of living organisms. Viruses are tiny, ranging in size from 20–300 nm, so they cannot be seen using a light microscope. For this reason their existence was not known until the end of the nineteenth century and it was not until 1935 that they were isolated and crystallised.

Viruses enter living cells and, once inside, multiply with the assistance of the host cells. Viruses contain very few enzymes and so, as intracellular parasites, they use the host's enzymes for their own metabolism. They are highly specific to both the host and a particular tissue within that host.

There are many different shapes of virus, although many are rod shaped or polyhedral. Each virus is composed of a core of nucleic acid surrounded by a protein coat, or **capsid**. Some types of viruses also develop a membrane layer outside the capsid, called an **envelope**. Viruses which invade animal cells may contain either RNA or DNA: plant viruses have RNA and those which invade bacteria (**bacteriophages**) have DNA.

Viruses are unable to move and so rely on passive dispersal, or a vector, to move them between host cells. Some viruses, such as the T_2 phage which infects *Escherichia coli*, immediately kill the host cell which they enter and are known as **virulent phages**. In others, known as **temperate phages**, the host and virus may exist together for many generations.

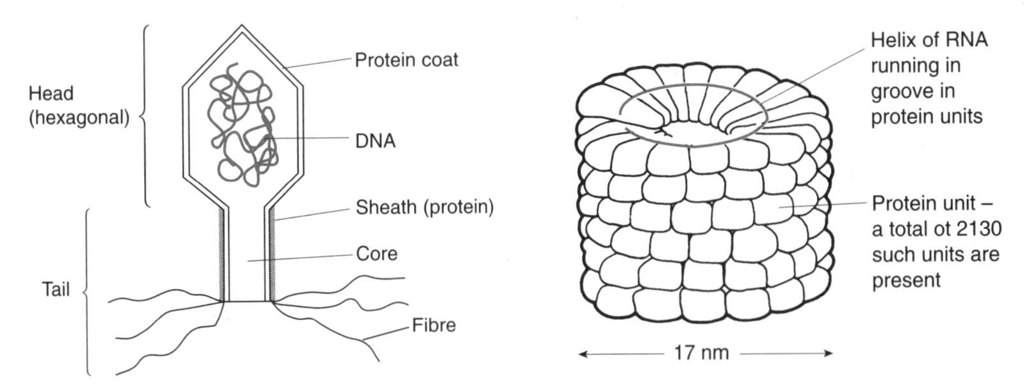

Fig. 2.1 (a) Structure of a bacteriophage

Fig 2.1 (b) Tobacco mosaic virus (TMV)

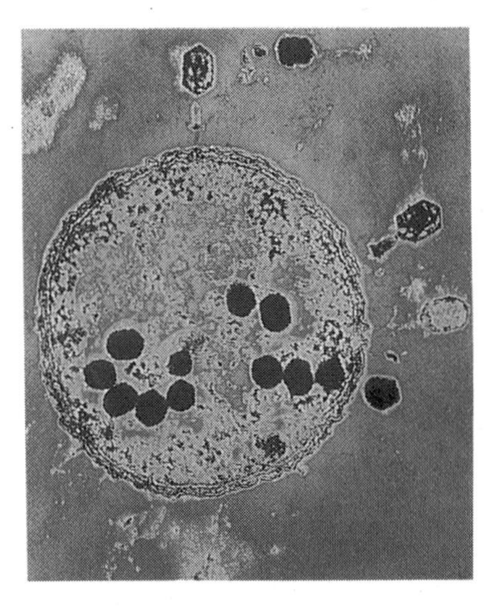

Fig. 2.1(c) Electron micrograph of T_2 bacteriophages (dark ovals) within the bacterium Escherichia coli

Life cycle of a virulent phage

The phage becomes attached to the surface of the bacterial cell and its DNA is injected into the host cell. Immediately replication of the bacterial DNA, and hence enzyme production, ceases and all of the host systems are taken over by the viral DNA to produce more strands of viral DNA. The host enzyme and synthetic systems are used to produce protein coats for this DNA and eventually the wall of the host cell ruptures (lysis) releasing about 200 new phages. The whole process may be completed in one hour. It is summarised in Fig. 2.2.

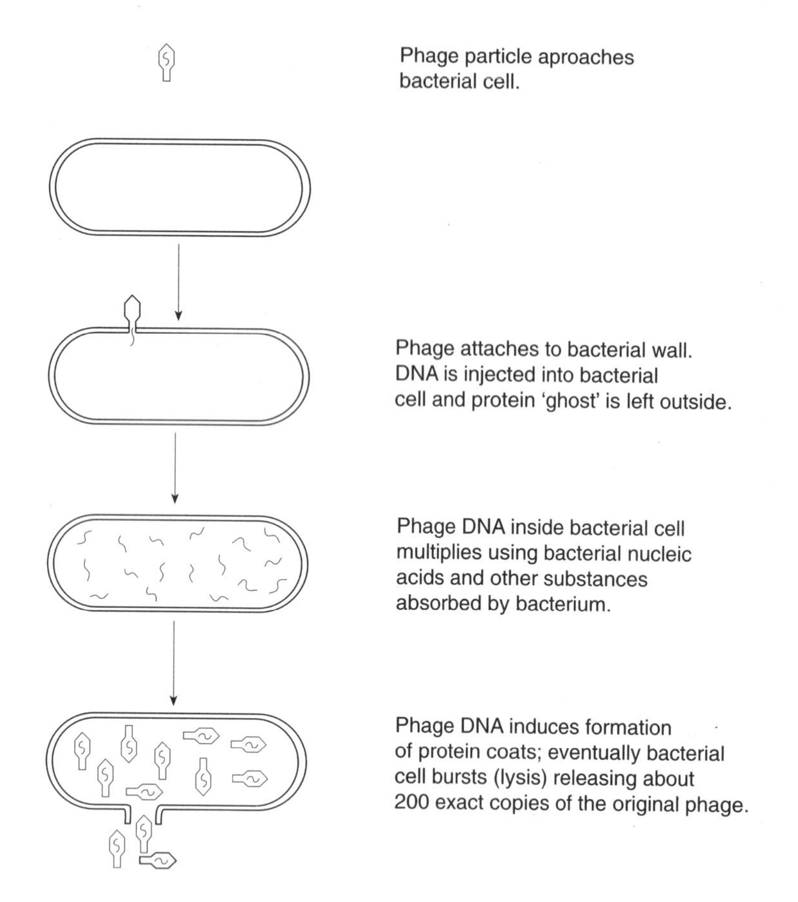

Fig. 2.2 Life cycle of a virulent phage

Life cycle of a temperate phage

In temperate phages the takeover of the host cell is much less rapid and the host and phage may exist together for many generations. Host DNA may become incorporated in the viral DNA and this DNA is carried to the next host, thereby resulting in new characteristics. This process of **transduction** is an important method by which antibiotic resistance spreads throughout a population of bacteria.

Retroviruses

In order to understand what retroviruses are you may need to refer to 3.1, which explains protein synthesis. Retroviruses are RNA viruses which possess an enzyme called **reverse transcriptase**, which can catalyse the formation of a strand of DNA from a strand of RNA. The new DNA then synthesises a complementary strand of DNA and a complete chromosome forms which can be incorporated into the DNA of the host cell, thus transforming it. The details of this process are given in Fig. 2.3.

The best-known retrovirus is the HIV virus which causes AIDS. HIV infects T-helper cells, a type of lymphocyte which helps both B-lymphocytes and T-lymphocytes to carry out their functions. Without these the body's immune system is rendered ineffective, not only against HIV, but also against other infections. Hence AIDS victims are often killed by opportunist organisms which take advantage of impaired resistance.

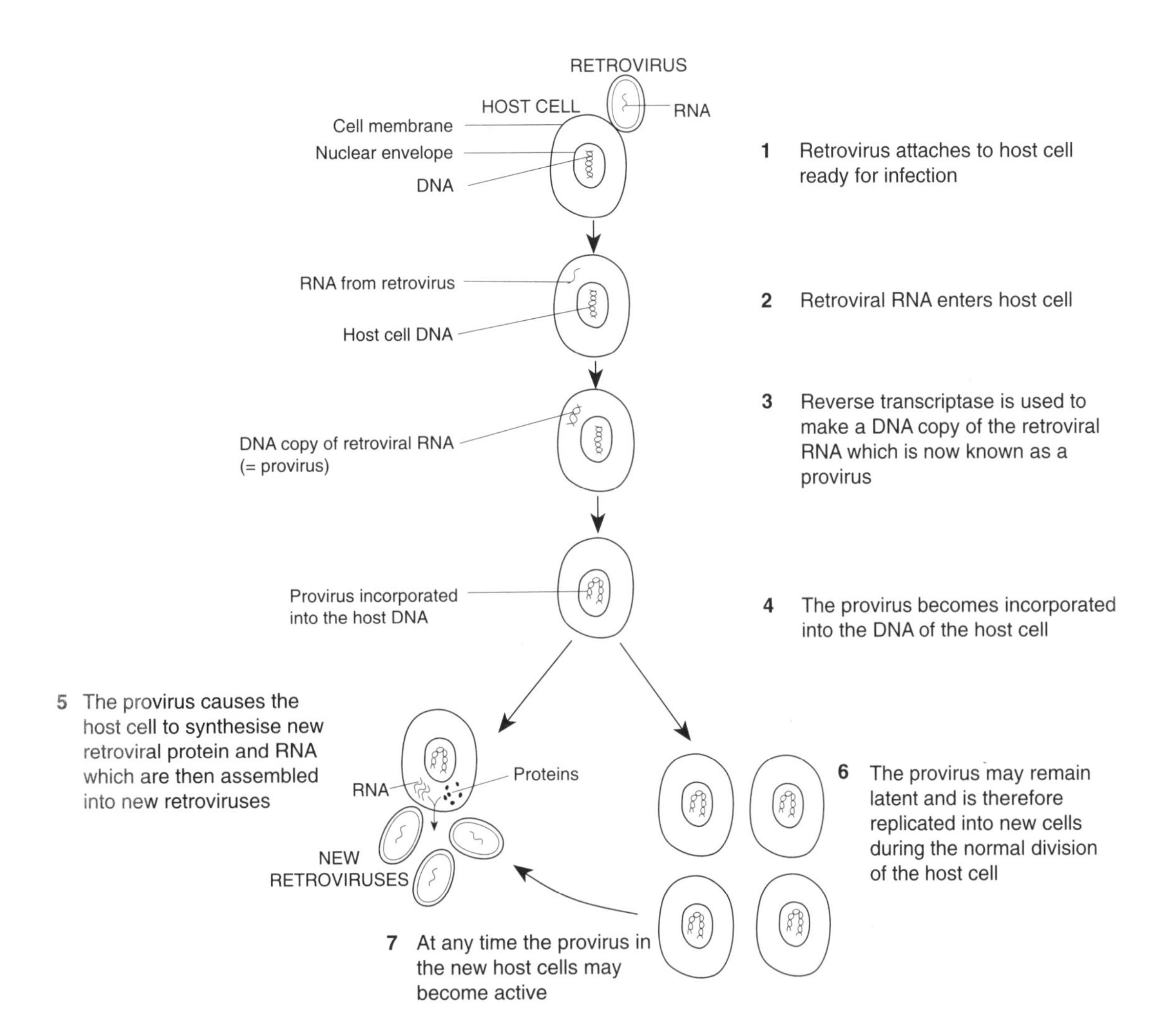

Fig. 2.3 Life cycle of a retrovirus

PROKARYOTAE

Bacteria are the smallest cellular organisms and the most abundant. They may respire aerobically or anaerobically and exhibit a variety of feeding mechanisms. They range in size from 0.5–4.0 μm. They are generally distinguished from each other by their shape. Spherical bacteria are known as **cocci** (singular = coccus), rod shaped as **bacilli** (singular = bacillus) and spiral shaped as **spirilla** (singular = spirillum). Cocci stick together in chains (streptococcus) or in clusters (staphylococcus). Fig. 2.4(a) shows the structure of a typical bacterial cell.

The cells of bacteria differ in a number of respects from those of other organisms. They are known as prokaryotic cells and have the following characteristics:

1. no distinct nucleus;
2. DNA not incorporated in chromosomes but comprising a single, circular strand;
3. no spindle forms at cell division;
4. membrane-bounded organelles (e.g. Golgi, ER, mitochondria) are absent;
5. cell wall of protein and polysaccharide.

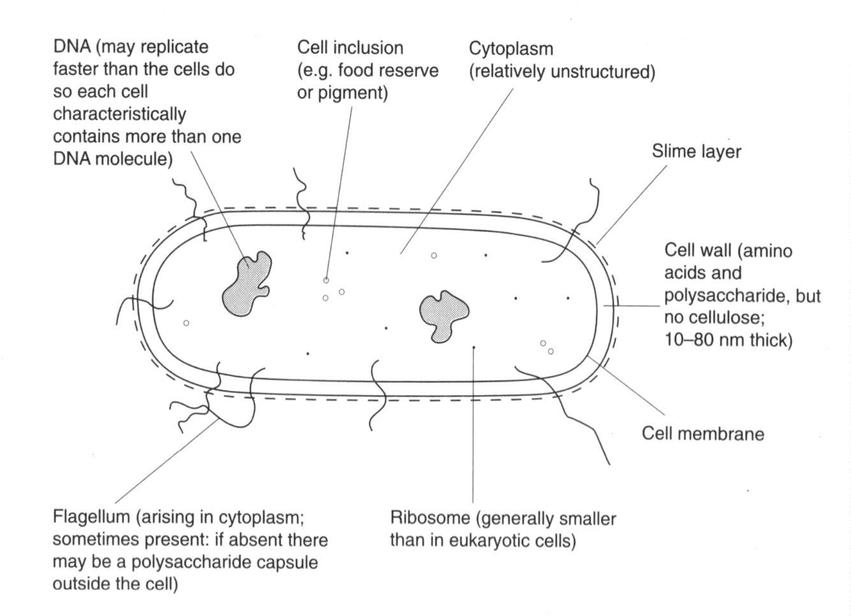

Fig. 2.4(a) Generalised bacterial cell

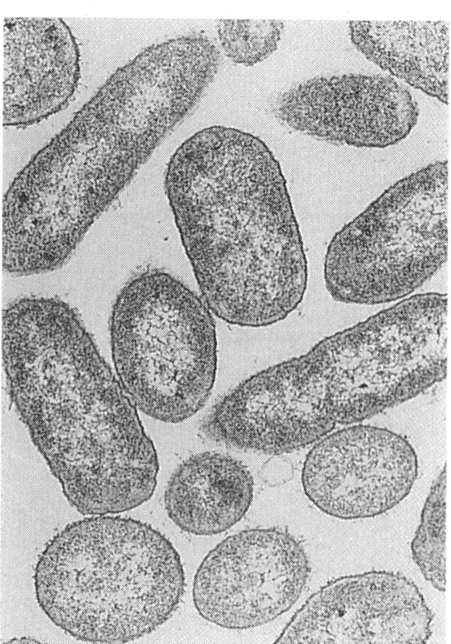

Fig. 2.4(b) Electron micrograph of the rod-shaped bacterium Escherichia coli

Bacterial cells differ from each other in the nature of their cell wall. In some forms the glycoprotein is supplemented by large molecules of lipopolysaccharide. Such cells are not stained by gentian violet and are said to be **Gram negative**. Those without the lipopolysaccharide combine with dyes like gentian violet and are said to be **Gram positive**. Gram-positive bacteria are more susceptible to antibiotics and lysozyme than are Gram-negative ones.

Bacteria are often classified on the basis of their method of obtaining energy:

- Most obtain energy from the oxidation or breakdown of living or nonliving organic matter, i.e. they are **heterotrophic**. They are parasitic, saprobiontic or mutualistic.
- A few obtain energy from the oxidation of inorganic materials and use this energy to synthesise their own foods, i.e. they are **chemoautotrophic**. For example, iron bacteria oxidise ferrous compounds to ferric hydroxide and release energy; colourless sulphur bacteria oxidise hydrogen sulphide to sulphur and release energy.
- Some bacteria can use the energy of sunlight for the manufacture of food, i.e. they are **photoautotrophic**. For example, green and purple sulphur bacteria contain bacteriochlorophyll and photosynthesise using hydrogen sulphide (not water) as a source of hydrogen; sulphur, not oxygen, is a by-product.

Bacteria reproduce by binary fission (see 4.1), one cell being capable of giving rise to over 4×10^{21} cells in 24 hours. Bacteria may also produce thick-walled spores, which are highly resistant, often surviving drought and extremes of temperature.

The importance of bacteria

1. Bacteria cause the breakdown and recycling of plant and animal remains, in particular the recycling of essential elements such as carbon, nitrogen and phosphorus. The same processes account for the bacterial decomposition of sewage.
2. They form symbiotic relationships with other organisms; for example, bacteria in the human gut synthesise some of the vitamin B complex and others break down cellulose in the guts of herbivores.
3. They are involved in food production, e.g. yoghurt, some cheeses, vinegar, coffee and tea.

4. Manufacturing processes such as tanning leather, retting flax to make linen and making soap powders involve bacteria.
5. They are a source of antibiotics, e.g. streptomycin.
6. They are easily cultured and therefore used for research.
7. Pathogenic bacteria are intercellular parasites and the symptoms of a disease are often caused by the toxins they produce. Some bacteria infect plants, e.g. *Xanthomonas phaseolus* causes common blight of beans. Human diseases caused by bacteria include whooping cough (*Bordetella pertussis*), some forms of pneumonia, leprosy, syphilis, tuberculosis, diphtheria, typhoid (*Salmonella typhi*), cholera and scarlet fever.

PROTOCTISTA

These are eukaryotic organisms which are neither plants, animals nor fungi. They are often unicells or assemblages of similar cells. This kingdom includes all nucleated algae and all protozoa. We shall look at algae as examples of this group.

Algae

The algae are an artificial, but convenient, group whose subdivisions are based mainly on structural and biochemical differences associated with photosynthesis. Although thought of primarily as an aquatic group, the algae are found almost everywhere, except in sandy deserts and permanent ice and snow. As primary producers they are a major component of phytoplankton, and derivatives of alginic acid, derived from brown seaweeds, find application in many commercial processes. For example, they are used as thickeners in food; in cosmetics; in latex formation; as gelling agents in confectionery, meat jellies and dental impression powders; as emulsifiers in ice cream, polishes, processed cheese and synthetic cream; as surface films on glazed tiles; as medical gauzes; and in some types of sausage casings. In coastal areas seaweed may be used as animal fodder and as fertiliser. Algae aid the oxidation of sewage and thus promote the growth of aerobic bacteria. They are used in percolation beds in water purification plants, but in reservoirs they cause problems by blocking filters and giving the water an unpleasant taste.

Table 2.1 *Classification of algae*

Phylum	Characteristics	Examples
Chlorophyta (green algae)	Chlorophyll is main photosynthetic pigment; chiefly found in fresh water	*Spirogyra, Ulva*
Phaeophyta (brown algae)	Multicellular; contain chlorophyll and fucoxanthin (brown pigment)	*Ectocarpus, Fucus*

Fig. 2.5 Fucus spiralis *(spiral wrack seaweed)*

FUNGI

This kingdom comprises eukaryotic organisms with a protective wall which is not made of cellulose but is often of hemicellulose, chitin and protein. They are all heterotrophic and are usually organised into fine tubular strands known as hyphae which may be grouped as a mycelium. There are about 80 000 species in the Kingdom Fungi. They help to maintain soil fertility by recycling many important minerals and decomposing the organic matter of both plants and animals. Industrial uses of fungi include the extraction of enzymes such as invertases and drugs like steroids and ergotamine; cheese making; and baking and brewing (yeast is a fungus). Although few animal diseases are caused by fungi, there are many examples of plant infections, such as *Puccinia graminis*, which causes rust of wheat and *Phytophthora infestans*, which causes late blight of potatoes. Stored food can be damaged by moulds; dry rots attack buildings and mildews affect cotton, wool and manufactured goods.

Table 2.2 *Classification of Fungi*

Phylum	Characteristics	Examples
Oomycota	Aseptate mycelium; asexual spore is a zoospore; sexual stage is an oospore	*Phytophthora* – causes late blight of potatoes *Peronospora* – downy mildew
Zygomycota	Aseptate, branching mycelium; asexual reproducton by sporangia or conidia; sexual spore is a zygospore	*Mucor*
Ascomycota	Septate hyphae; asexual reproduction by conidia or budding; sexual reproduction by ascospores	*Saccharomyces* (yeast), *Penicillium*
Basidiomycota	Septate hyphae; hyphae often massed into extensive three-dimensional structures (puffballs, toadstools, bracket fungi); sexual reproduction by basidiospores	*Agaricus* – mushroom

Fig. 2.6 Mould fungus on surface of peaches

2.3 BIOTECHNOLOGY

Biotechnology involves the production of materials by biological agents (commonly microorganisms) through the application of scientific and engineering principles. It is not something new – ancient civilisations used microorganisms in the manufacture of wines, beers, vinegar and cheeses.

Biotechnology is now a multimillion pound industry, whose expansion occurred during the Second World War, when explosives and antibiotics were mass-produced with the aid of microorganisms. Before we look at the methods used to produce a diversity of materials by biotechnology, we will first look at how microorganisms grow.

Growing microorganisms

If a culture of a microorganism, such as a bacterium, is left to grow, a growth curve as illustrated in Fig. 2.7 is typically observed.

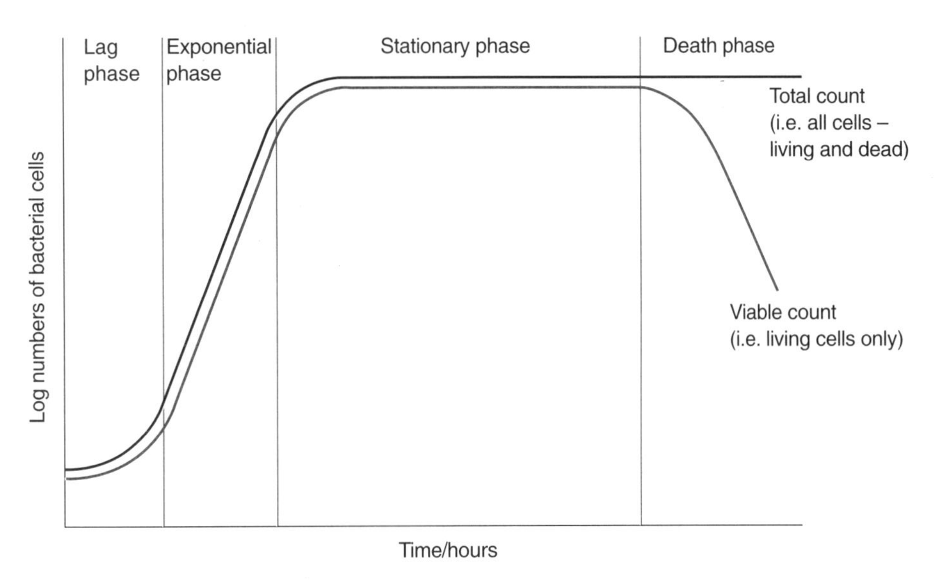

Fig. 2.7 Bacterial growth curve

- **Lag phase** – initially growth is slow while the bacteria mobilise their stored food and their enzymes, but the rate of growth is increasing towards the end of this phase.
- **Exponential phase** – provided that nutrients are in good supply growth is very rapid, with cell numbers doubling as often as every ten minutes.
- **Stationary phase** – as the nutrient supply is used up and as waste products accumulate, the rate of growth slows until the rate at which new cells are produced is equal to the rate at which dead ones are broken down.
- **Death phase** – as nutrients run out altogether and the waste products poison the cells, the number of living ones rapidly declines, although the overall number of cells remains constant.

Factors affecting growth

❶ **Nutrients** – as with all cells, microorganisms consist mostly of carbon, hydrogen, oxygen and nitrogen with smaller quantities of sulphur and phosphorus. These six basic nutrients are essential to growth. In addition, substances needed in much smaller, but still significant, amounts are potassium, iron, calcium and magnesium (= **macronutrients**). Needed in smaller quantities still are copper, zinc, cobalt, manganese and molybdenum (= **micronutrients**). Also required are a group of chemicals known as **growth factors**, which include vitamins, amino acids, purines and pyrimidines.

❷ **Temperature** – most microorganisms grow most rapidly in the 20 – 45°C range.

❸ **pH** – microorganisms can survive a wider range of pH than plant or animal cells, being able to tolerate acidity as low as pH 2.5 and alkalinity as high as pH 9.

4. **Oxygen** – most microorganisms are aerobic and therefore need to be supplied with oxygen to grow (**obligate aerobes**). Some find oxygen toxic and grow badly in its presence (**obligate anaerobes**), while others grow better when oxygen is present, but continue to do so in its absence (**facultative anaerobes**).
5. **Water** – all microorganisms require water to grow, and in order to absorb it the water potential of the medium on which they exist must be higher (less negative) than their cell contents. Much use is made of a concentrated medium (one with a low or more negative water potential) to preserve food, e.g. bottling of jam or fruit in sugar, or salting of meat. In these circumstances the medium dehydrates and thereby kills the microorganisms, which might otherwise spoil the food.
6. **Light** – some microorganisms are photosynthetic and therefore require light.

Fermentation

The mass production of materials from the growth of specific microorganisms is termed fermentation and the vessel in which they are grown is called a **fermenter**. There are a number of different designs, one of which, the stirred-tank fermenter, is shown in Fig. 2.8.

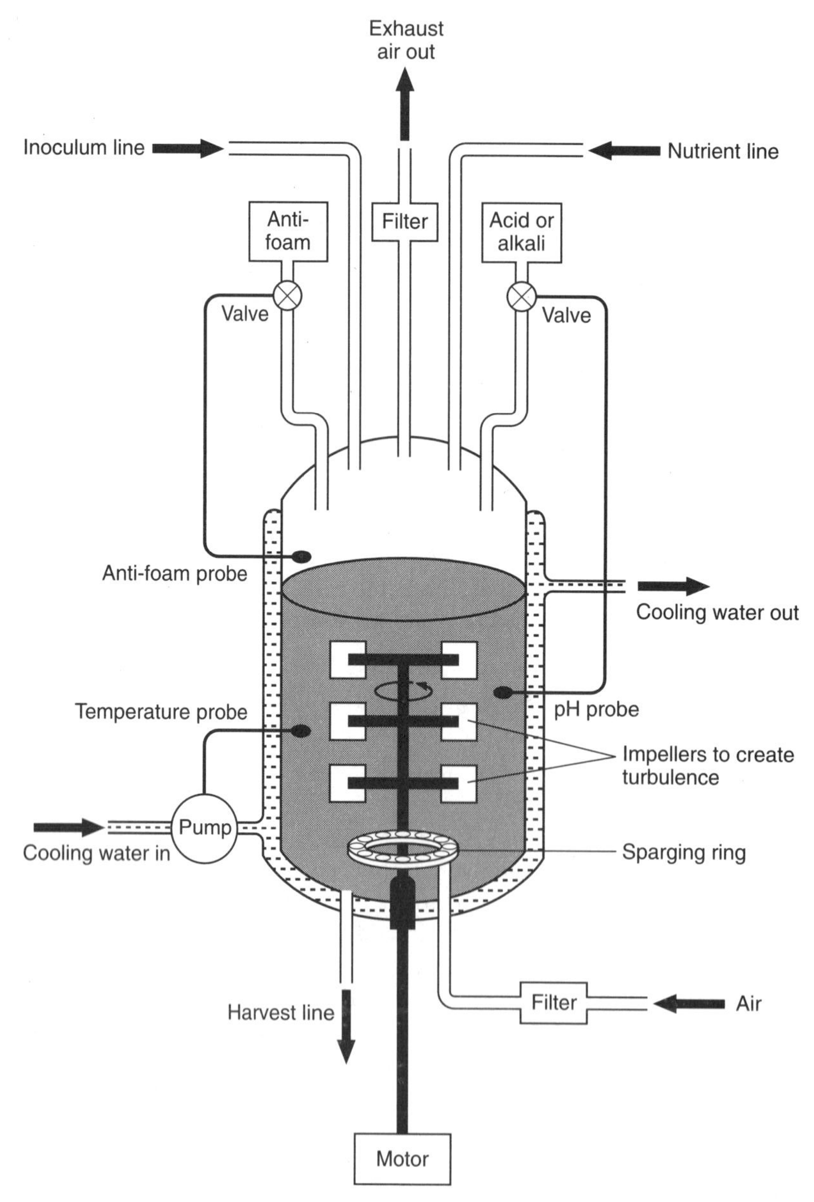

Fig. 2.8 A stirred-tank fermenter

Fermenters can be operated in two main ways.

1. **Batch cultivation** – the microorganisms and the nutrient medium are added to the fermenter and growth continues until enough product has been formed, at which point the fermenter is emptied and the product is extracted. The fermenter is then cleaned, sterilized and filled again with a new batch.
2. **Continuous cultivation** – the fermenter is set up as for batch cultivation, but is allowed to run continuously, with a constant supply of raw materials being added and the product continually being removed.

Sterilization

The conditions in a typical fermenter are inevitably suitable for the growth of most microorganisms. In order to obtain a relatively pure product efficiently, it is essential that only the required organism is grown and contamination by others is avoided. Sterilization of the vessel and its associated pipework is carried out by passing steam through them for a period of time before filling. Both the vessel and pipework are highly polished to reduce the number of pits and crevices which might harbour microorganisms. The nutrient medium is usually heat sterilized and the air supply to the fermenter is filtered and/or heat sterilized.

USES OF BIOTECHNOLOGY

The range of materials produced through biotechnology is vast, due almost entirely to the major expansion which took place as a result of the development in the 1980s of recombinant DNA technology (see 3.1).

Drug production

One of the first substances manufactured in quantity using biotechnology was penicillin, during the Second World War. Since that time the range of pharmaceuticals produced by similar methods has expanded considerably and includes other antibiotics, such as streptomycin and chloramphenicol.

Recombinant DNA technology has made it possible to introduce human genes into microorganisms and so use them to mass-produce human hormones, such as insulin (details of which are given in 3.1). Other hormones manufactured in this way include cortisone, the sex hormones testosterone and oestradiol, and erythropoietin used in the treatment of anaemia. Bovine somatotrophin (BST) is an example of a non-human hormone which is produced for injection into cows in order to increase milk yield.

Food production

Biotechnology can play a role, either directly or indirectly, in the production of a variety of foods.

1. **Wine and beer** – yeasts are used to ferment hexose sugars, such as glucose, into alcohol, according to the equation:

   $$\underset{\text{hexose sugar}}{C_6H_{12}O_6} \longrightarrow \underset{\text{ethanol}}{C_2H_5OH} + \underset{\text{carbon dioxide}}{2CO_2}$$

   The range of beverages produced in this way is very diverse and depends on the source of the hexose sugar and the type of yeast used to ferment it. Wine, for example, is fermented from grapes, normally by using the yeast *Saccharomyces ellipsoideus,* while a British bitter beer is fermented from barley by using the yeast *Saccharomyces cerevisiae.*
2. **Baking** – the baking of bread also uses the yeast *Saccharomyces cerevisiae.* Cereal grain is crushed to form flour, exposing the starch. Water is added to produce a dough and in so doing the natural amylases are activated and the starch is hydrolysed to glucose. The yeast then uses this glucose as a respiratory substrate, producing carbon dioxide as a by-product. The gas forms small bubbles, which expand when the dough is heated in an oven, thus causing the bread to rise.

❸ **Cheese and yoghurt** – cheese is formed when lactic acid bacteria ferment lactose in milk as follows:

$$\underset{\text{lactose}}{C_{12}H_{22}O_{11}} + \underset{\text{water}}{H_2O} \longrightarrow \underset{\text{lactic acid}}{4\ CH_3CHOHCOOH}$$

The acid so produced curdles the milk, which separates into a solid curd and a liquid whey. The curd is heated in the range 32 – 42 °C and some salt is added, before it is pressed into moulds. The cheese is left to ripen for various periods depending on the type of cheese. Caerphilly, for example, is left for just a fortnight whereas a mature Cheddar may wait a year.

Yoghurt is made from pasteurized milk with much of the fat removed. To this, lactic acid bacteria are added and the mix is then incubated at around 45 °C for five hours.

❹ **Single Cell Protein (SCP)** – this comprises the cells of microorganisms (or their products) grown for human or animal consumption. The cells can be grown on waste material, making them an economic source of food. A wide variety of microorganisms can be used in the manufacture of SCP. One product, **Pruteen**, is made by the bacterium *Methylophilus methylotrophus* acting on methanol, a waste material of certain chemical processes. Another product is **mycoprotein**, produced when the fungus *Fusarium graminearum* is grown on flour waste.

❺ **Enzymes** – fermentation techniques are used to produce a wide range of enzymes. While not foods themselves, they are essential to the manufacture of a variety of foods, e.g. pectases are used to clear fruit juices and proteases are used in tenderising meat.

❻ **Fuel production** – with a finite supply of natural oil there is pressure on the producing companies to find other sources of fuel. Two schemes already make alcohol and methane as an alternative to fossil fuels:

- **Gasohol** – in Brazil, where cane sugar is in plentiful supply, it can be fermented into alcohol by yeast in much the same way as in the production of wines and beers. The alcohol is distilled to concentrate it and it is then mixed with petrol and burned as a fuel in motor vehicles. All cars in Brazil have been converted to burn the new fuel. The alcohol could be used alone, but the petrol is added to discourage drinking of the fuel. The waste material produced when the sugar has been extracted from the cane is called **bagasse**, and it is burned as a fuel in the distilleries, making the process even more efficient and less wasteful. As yet, other schemes have not proved as economic as that in Brazil, but as oil supplies diminish and the price rises, the fermentation of waste sawdust, straw, paper and vegetable matter may well become economically attractive.

- **Biogas** – human sewage and domestic rubbish form a plentiful supply of materials which can be fermented by microorganisms. These convert the waste to methane gas (**biogas**) which can be burned as a fuel in the same way as natural, domestic gas. Some sewage farms in Great Britain are already self-sufficient in power through using methane produced in this way. One advantage of biogas production is that it does not require complex equipment; small domestic biogas digesters are common in India and China. All family sewage and organic wastes are collected in the digester and the resultant methane gas is used for cooking, lighting and heating the home.

OTHER BIOTECHNOLOGY PRODUCTS

Microorganisms are used to produce many other substances which have commercial, industrial and medical uses. Perhaps best known of these are the enzymes in biological washing powders. As many clothing stains are proteins, e.g. blood and some food stains, protease enzymes are added to washing powders and liquids. These are typically produced by microorganisms such as *Aspergillus oryzae*. Lipases too are used to remove greasy marks. Other products of biotechnology and their uses are given in Table 2.3.

Table 2.3

Product	Uses	Produced by
Cellulases	Brighteners used in washing powders	*Trichoderma spp*
Acetone and butanol	Solvents	*Clostridium sp*
Gellan	Food thickener	*Pseudomonas spp*
Glutamic acid	Flavour enhancer in food	*Corynebacterium sp*
Vitamin B_{12}	Diet supplement	*Propioni bacterium sp*
Streptokinase	Treatment of thrombosis	*Streptomyces spp*
Cyclosporin	Immunosuppressant drug	*Cotypocladium sp*

CELL AND TISSUE CULTURE

As the use of microorganisms in producing valuable materials has expanded, attention has turned to using plant and animal cells and tissues in the same way. Although they are more difficult to grow, some progress has been made. One area of cell culture already making a valuable contribution is the manufacture of antibodies, in particular, single cell antibodies (**monoclonal antibodies**).

As the β– lymphocytes which produce antibodies are hard to grow outside the body, the production of monoclonal antibodies in any quantity was impossible until recently, when the desired β–lymphocytes were fused with cancer cells. Cancer cells divide rapidly and these new **hybridoma cells** were found to do so as well, giving a continual source of cells producing a monoclonal antibody.

Monoclonal antibodies have many uses, including estimating the quantity of a chemical in a mixture, a technique called **immunoassay**. They are therefore used to detect the hormones present in urine (pregnancy testing kits), in detecting drugs in urine (e.g. for athletes) and in detecting the human immunodeficiency virus (AIDS test for HIV).

2.4 PLANTS

The Kingdom Plantae includes all those organisms we think of as typical plants, like mosses, ferns, flowering plants and trees. They are all multicellular organisms made up of eukaryotic cells. Their cells are bounded by cellulose walls and they are able to photosynthesise.

We will now look at the major features by which the three main phyla are recognised. These phyla are the Bryophyta, the Filicinophyta and the Angiospermophyta.

BRYOPHYTA

There are over 23 000 species of bryophytes and although they are mostly terrestrial they have to live in damp places because water is essential for fertilisation. None of the bryophytes have true roots. Instead they have rhizoids for anchorage and absorption of water occurs over the whole surface of these small plants. There are two main stages in their life cycle: they alternate between a spore-producing sporophyte stage and a gamete-producing gametophyte stage. There are only two classes: the mosses (Musci) and the liverworts (Hepaticae). Their main classification features are given in Table 2.4.

Table 2.4 *Classification of the Bryophyta*

Phylum	Characteristics
Bryophyta	No true roots, body anchored by filamentous rhizoids Clear alternation of generations, with gametophyte independent, sporophyte dependent Gamete-producing cells surrounded by jacket of sterile vegetative cells No asexual spores produced

Class	Characteristics	Examples
Hepaticae (liverworts)	Thalloid gametophyte is flat ribbon or leafy shoot; found in moist, shady places	*Riccia, Marchantia, Pellia*
Musci (mosses)	More conspicuous; better able to withstand drought; gametophyte has two growth stages: filamentous protonema and upright plant with spirally arranged leaves; more complex capsule	*Polytrichum, Funaria, Mnium*

Bryophytes have little importance except as early colonisers of bare land, helping to prevent soil erosion and enrich the ground for the growth of larger plants. Peat, used for fuel where alternatives are scarce, is formed from dead, compressed bog mosses.

Fig. 2.9(a) Moss

Fig. 2.9(b) Liverwort

FILICINOPHYTA

This phylum includes all the true ferns and there are about 11 000 different species. Ferns have large leaves called fronds, which are coiled when in bud. Their sporangia (spore-producing structures) are grouped in clusters called sori. Most living ferns are small and have no direct economic importance to man, although they are significant ground cover plants in moist areas. The larger ferns which formed the dominant terrestrial vegetation for about 70 million years from the Devonian to the Permian periods contributed greatly to the coal deposits now so useful to man. Like the bryophytes this group shows alternation of generations between sporophyte and gametophyte stages. In the ferns the sporophyte is clearly dominant and, since it is better adapted to life on land, this has enabled ferns to have a lower water requirement than the mosses and liverworts.

Fig. 2.10 Fern

Table 2.5 *Classification of the Filicinophyta*

Phylum	Characteristics	Examples
Filicinophyta	Prominent frond-like leaves; sporangia in clusters (sori); underground rhizomes	*Dryopteris* – fern *Pteridium* – bracken

ANGIOSPERMOPHYTA

These are the dominant plants of the world today with over 250 000 species, of which about three-quarters are dicotyledons and one-quarter are monocotyledons. They vary in form from simple duckweed through herbaceous and shrubby plants to trees such as the chestnut and oak. Their evolution has closely paralleled that of the insects on which many species depend for pollination. They are extremely well suited to life on land both in their morphology (e.g. efficient water-carrying xylem vessels) and in their reproduction (e.g. seeds enclosed in an ovary). You should be aware that they still show alternation of sporophyte and gametophyte generations, although the latter are severely reduced to no more than a few cells retained within the sporophyte. Angiosperms are man's most important food plants; they include the cereals, vegetables, succulent fruits and sugar cane. As well as providing food for man's domesticated animals in grasses and clover, they are a source of oils, insecticides and drugs. In addition, since they cover such a large proportion of the land, they are significant primary producers and replenish the oxygen in the atmosphere. Two classes of angiosperms are recognised, the Dicotyledonae and the Monocotyledonae, and their characteristic features are given, together with those of the phylum as a whole, in Table 2.6.

Table 2.6 *Classification of the Angiospermophyta*

Phylum	Characteristics	
Angiospermophyta	Flowers produced Ovules develop inside ovary Ovary wall develops into fruit Vessels in xylem	
Class	**Characteristics**	**Examples**
Monocotyledonae	Embryo has a single cotyledon; leaves usually show parallel venation; flower parts typically in multiples of 3; stem contains scattered vascular bundles; no secondary growth	*Endymion non-scriptus* (bluebell) – family Liliaceae *Triticum* (wheat) – family Graminae
Dicotyledonae	Embryo has two cotyledons: leaves have a network of veins; flower parts often in multiples of 4 or 5; stem contains ring of vascular bundles; secondary growth occurs	*Ulex* (gorse) – family Leguminosae *Helianthus* (sunflower) – family Compositae *Prunus domestica* (plum) – family Rosaceae

2.5 ANIMALS

The Kingdom Animalia includes all the multicellular organisms which do not photosynthesise and which show some degree of nervous coordination. In this summary of animal classification we consider only the major phyla and some of their subdivisions. You should consult your own syllabus carefully to see which ones you need to study since the examination boards differ widely in their requirements.

PHYLUM CNIDARIA

Animals in this phylum are all aquatic and predominantly marine. They are composed of two cellular layers separated by a rather jelly-like mesogloea. The whole body is radially symmetrical and occurs in two forms: a jellyfish-like **medusoid phase** and a hydroid or **polyp phase**.

All cnidarians are carnivorous, feeding only on living prey by means of unique stinging cells called **nematoblasts**. The most economically important cnidaria are the corals, whose skeletons of calcium carbonate provide habitats for many other animals and, in the Indian and Pacific Oceans, may form coral islands, atolls and fringing reefs. The main features of this phylum and its three classes are given in Table 2.7.

Table 2.7 *Classification of the Cnidaria*

Phylum	Characteristics	
Cnidaria	Two cell layers, separated by mesogloea Radially symmetrical with tentacles Single body cavity (enteron; gastrovascular cavity) with only one opening to exterior Nematoblasts (stinging cells) Polymorphism common (hydroid or medusoid forms)	
Class	**Characteristics**	**Examples**
Hydroza	Typically hydroid and medusoid forms	*Hydra, Obelia, Physalia* – Portuguese man-of-war
Anthozoa	No medusoid stage	*Actinia* – sea anemone, corals
Scyphozoa	Reduced hydroid stage	*Aurelia* – jellyfish

Fig. 2.11(a) Jellyfish

Fig. 2.11(b) Sea anemone

PHYLUM PLATYHELMINTHES

The platyhelminthes are a group of flatworms with a definite head and bilateral symmetry. Many are free living but some of the most important are parasites of man or his domesticated animals. Examples include *Schistosoma*, the blood fluke that causes bilharzia, *Taenia*, a tapeworm, and *Fasciola*, a fluke causing liver rot in sheep. Details of the classification of this phylum are given in Table 2.8.

Table 2.8 *Classification of the Platyhelminthes*

Phylum	Characteristics
Platyhelminthes (flatworms)	Unsegmented Bilaterally symmetrical (prerequisite for efficient locomotion) Dorso-ventrally flattened Acoelomate Gut, where present, branched with one opening

Class	Characteristics	Examples
Turbellaria	Free living; ciliated epidermis	*Planaria*
Trematoda	Endoparasitic with one vertebrate and one invertebrate host	*Fasciola* – liver fluke *Schistosoma* – blood fluke
Cestoda	Parasitic with two vertebrate hosts; body divided into proglottids	*Taenia* – tapeworm

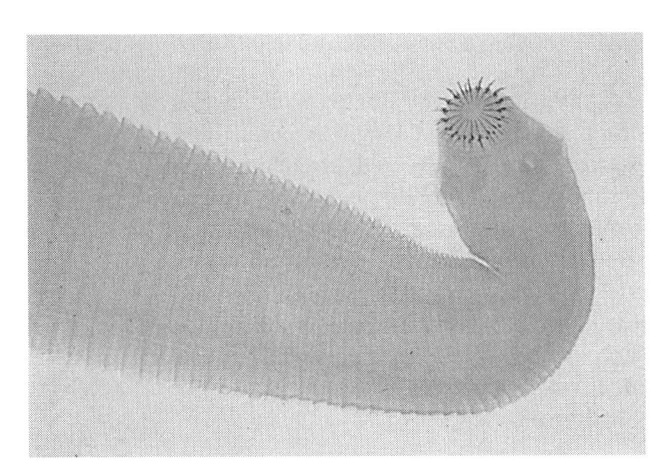

Fig. 2.12(a) Tapeworm

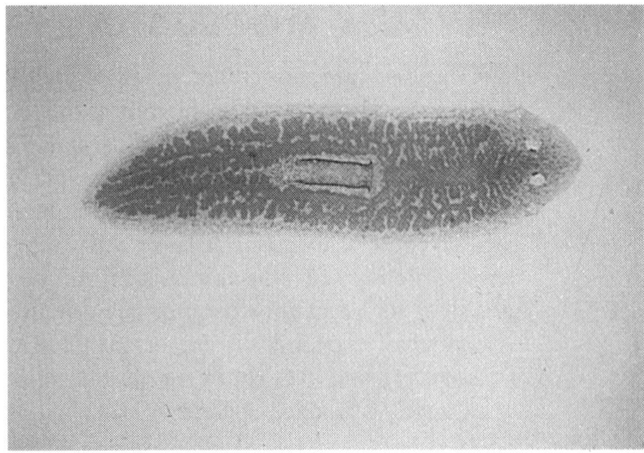

Fig. 2.12(b) Planarian

PHYLUM ANNELIDA

The annelids are the 'true worms'. There are about 9000 species of them living in the sea, in fresh water and in the soil. In this group we see for the first time the development of two features which led to an increase in the size and complexity of multicellular animals – the coelom and metameric segmentation. The **coelom** is a fluid-filled space surrounded by mesoderm which provides space in which organs can develop, allowing more cell surfaces to be exposed for diffusion. **Metameric segmentation** is seen in the annelids, arthropods and chordates. In its most primitive form the body is seen to comprise a linear series of similar segments all of the same age. The segmental arrangement includes both internal and external structures with repetition of muscles, blood vessels and nerves in each segment. This repetition of structures gives potential for specialisation.

Although all of the annelids contribute to food chains, the most important is the earthworm, which contributes greatly to soil formation and improvement in the following ways:

1. tunnels improve aeration and drainage
2. dead vegetation is pulled into the soil where decay by saprobionts takes place
3. mixing of soil layers
4. addition of organic matter by excretion and death
5. secretions of gut neutralise acid soils
6. improving tilth by passing soil through gut

The characteristic features of the annelids are given in Table 2.9.

Table 2.9 *Classification of the Annelida*

Phylum	Characteristics
Annelida (segmented worms)	Coelom well developed Metamerically segmented Chaetae usually present Thin cuticle of collagen

Class	Characteristics	Examples
Oligochaeta	Obvious clitellum; few chaetae	*Lumbricus* – earthworm *Allolobophora* – earthworm
Polychaeta	Many chaetae on parapodia	*Nereis* – ragworm *Arenicola* – lugworm
Hirudinea	Ectoparasitic; suckers	*Hirudo* – leech

PHYLUM MOLLUSCA

The molluscs are unsegmented animals with a head, foot and visceral hump; many species have a calcareous shell. This phylum is the second largest in the animal kingdom, comprising about 100 000 living species. Molluscs also have a very long fossil record, stretching back to the Precambrian period. Most of the species alive today are marine although there are also terrestrial and freshwater examples. The main features of two classes, the Gastropoda and the Pelycopoda, are given in Table 2.10, but you should remember that octopuses and squids are also molluscs. Molluscs are significant sources of food for man (mussels, whelks, oysters) as well as for other animals, but they can also cause harm. Snails are intermediate hosts for a number of economically important parasites such as *Schistosoma*, causing bilharzia, and *Fasciola*, causing liver rot in sheep. The shipworm *Teredo* is a pelycopod which bores holes in submerged wooden structures, such as jetties and boats.

Table 2.10 *Classification of the Mollusca*

Phylum	Characteristics	
Mollusca	Unsegmented Head, muscular foot and visceral mass Mantle secretes shell	
Class	**Characteristics**	**Examples**
Gastropoda	Asymmetrical due to torsion; single shell, often coiled; feed using radula	*Littorina* – periwinkle *Helix* – snail
Pelycopoda	Shell in two halves; filter feeders	*Mytilus* – muscle *Cardium* – cockle

PHYLUM ARTHROPODA

This phylum is the largest and most successful in the animal kingdom with over 800 000 species described. All arthropods share the basic characteristics of an exoskeleton and jointed appendages, but within the phylum there are five main subdivisions: the Crustacea, Chilpoda, Diplopoda, Insecta and Arachnida. The Crustacea is now recognised to include such a diverse range of animals that it is designated a superclass containing a number of classes. Table 2.11 attempts to summarise the main classification features of the Arthropoda and its subdivisions; you should consult your syllabus to see which you are expected to know.

Table 2.11 *Classification of the Arthropoda*

Phylum	Characteristics	
Arthropoda	Metamerically segmented Coelomate Chitinous exoskeleton Jointed appendages Compound eyes (often)	
Superclass	**Characteristics**	**Examples**
Crustacea	Calcite hardens exoskeleton; cephalothorax; gills; two pairs of antennae	*Daphnia* – water flea *Oniscus* – woodlouse *Carcinus* – crab

Class	Characteristics	Examples
Chilopoda	Distinct head with one pair of poison jaws; similar legs along length of body – one pair per segment	*Lithobius* – centipede
Diplopoda	Two pairs of legs per apparent body segment	*Iulus* – millipede
Insecta	Usually winged as adults; compound eyes; three pairs of legs; body divided into head, thorax and abdomen	*Locusta* – locust *Musca* – fly *Apis* – bee *Libellula* – dragonfly
Arachnida	Four pairs of legs; no true jaws; no compound eyes	*Araneus* – spider *Scorpio* – scorpion

By far the most significant group of arthropods is the Insecta with over 750 000 species described. Insects are mainly terrestrial and, although there are some freshwater forms, none are truly marine. The insect body plan can become specialised for many different modes of life and therefore the group shows great adaptive radiation; insects are widely distributed because of their ability to fly. They can be beneficial or harmful to man and some examples of each category are given in Table 2.12.

Fig. 2.13(a) Woodlouse (Crustacea)

Fig. 2.13(b) Centipede (Chilopoda)

Fig. 2.13(c) Millipede (Diplopoda)

Fig. 2.13(d) Fruit fly (Insecta)

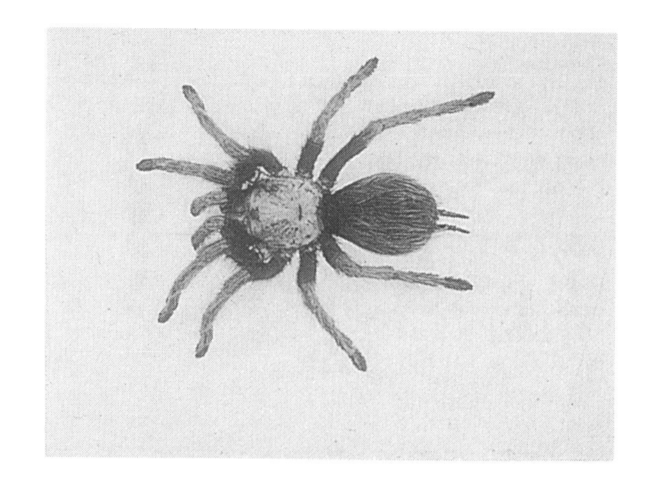

Fig. 2.13(e) Tarantula (Arachnida)

Table 2.12 *Beneficial and harmful effects of insects*

Beneficial	Harmful
Use of insect products: silk, honey, bees' wax; cochineal, shellac	Destroy leaves and fruit of plants, e.g. locusts, boll weevils, fruit flies
Commercial production of fruits dependent on insect pollination	Transmit disease, e.g. mosquitoes (malaria and yellow fever); houseflies (typhoid and dysentry); fleas (bubonic plague and typhoid); tsetse flies (sleeping sickness)
Biological control of harmful organisms, e.g. ladybirds eat aphids	Annoy or harm domestic animals, e.g. lice
Recycling of dead and decaying plant and animal remains, e.g. by beetles and flies	Cause destruction of materials in and around houses, e.g. cockroaches (spoil food); moth larvae (feed on carpets and clothing); termites (bite into wooden furnishings or gnaw through the wooden supports of houses)
Food source of many economically important animals, e.g. trout	

PHYLUM CHORDATA

This phylum includes all the well-known vertebrates. Superficially it seems to include a vast range of different body forms but they are all united by internal and developmental characteristics. The notochord and visceral (pharyngeal) clefts may not be retained in the adult form but all adult chordates have a hollow, dorsal nerve cord and most have a post-anal tail. All the classes we are going to look at belong to the subphylum **Vertebrata** which has evolved during the last 500 million years to become the dominant group of animals of land, sea and air. In terms of number of species they do not rival the arthropods but their biomass is much greater and they are ecologically dominant. The vertebrates will be the subject of much of the rest of this book.

The characteristics of the chordates and the five classes of vertebrates are given in Table 2.13 as well as some examples of the main orders of mammals.

Table 2.13 *Classification of the Chordata*

Phylum	Characteristics
Chordata	Notochord – at least in embryo CNS tubular and dorsal Ventral heart Paired visceral clefts at some stage in life history Post-anal tail (usually)

Class	Characteristics	Examples
Chondrichthyes (cartilagenous fish)	Cartilagenous skeleton; no operculum; no swimbladder	*Scyliorhinus* – dogfish
Osteichthyes (bony fish)	Bony skeleton and scales; operculum; swimbladder	*Clupea* – herring
Amphibia	Dependent on fresh water for development; smooth, glandular skin	*Rana* – frog *Triturus* – newt
Reptilia	Dry skin with epidermal scales; eggs with thick leathery shells	*Lacerta* – lizard *Natrix natrix* – grass snake
Aves (birds)	Endothermic; epidermal feathers; beak; pectoral limbs modified to form wings; eggs with shells	*Troglodytes troglodytes* – wren
Mammalia	Endothermic; hairy, glandular skin; diaphragm; secondary palate; mammary glands (except in monotremes); heterodont dentition	*Talpa* – mole (order Insectivora) *Plecotus* – long-eared bat (order Chiroptera) *Rattus* – rat (order Rodentia) *Canis* – dog (order Carnivora) *Delphinus* – dolphin (order Cetacea)

The importance to man of other vertebrates is enormous. As well as being parts of food webs in almost every habitat, many are a direct food source for man and may be farmed. The most important are fish, some birds and mammals such as cattle, sheep and pigs. Birds and small mammals consume large quantities of insect pests and weed seeds. However, rodents, especially rats, may damage crops and stored food. Mammals also carry disease vectors, for example bubonic plague transmitted by fleas of rats. They may also be secondary hosts for parasites such as tapeworms which infect man and his domesticated animals. Some snakes are poisonous to man. Useful products of vertebrates include leather, fur and wool for clothing, as well as items such as glue and fertiliser.

Chapter roundup

In this chapter we have looked at the need for a classification system and some of the factors upon which a successful classification is based. You should have realised that in sorting living organisms into groups we are trying to reflect, as far as possible, the true evolutionary relationships between those organisms. You have seen that viruses are on the border between living and nonliving things, that acellular organisms carry out all the characteristics of living organisms within the confines of a single cell and that, in general, plants and animals have become more complex as they have adapted to life on land. We have also considered a few of the interrelationships beween man and other organisms and these will be expanded upon in later chapters. We have started to look at the dominant plants and animals – the angiosperms and the mammals – and these two groups will form the basis of the detailed studies of physiology which follow.

Illustrative questions and worked answers

1 (a) Give three reasons why a frog and a mouse are classified in the same phylum.
(b) Give three structural differences between a frog and a lizard.
(c) State three features of birds that have contributed to the success of this group.
Time allowed 7 minutes

Tutorial note

This structured question tests straightforward recall of learned facts and should therefore present you with few problems if you are well prepared. The only difficulty you may face is in selecting the best facts and giving adequate detail. The answers must be short so you should think carefully before writing anything and choose your words with utmost care. In part (a), for example, a frog and a mouse both have a notochord at some stage of their life history and this answer would doubtlessly be accepted. To use the term 'vertebral column' instead of 'notochord' would also be acceptable, but it is unlikely that 'backbone' would be an adequate alternative at this level. You should always use A-level terminology and detail and not expect GCSE-standard answers to bring the same credit. Keep to obvious similarities that are of sufficient biological importance (see suggested answer). You must avoid trite or superficial similarities such as 'they both have limbs, brains and internal skeletons'.

In part (b) the two key words are 'structural differences'. Any features that are even remotely common to both should be ignored as should any functional differences. Some typical differences are listed in the suggested answer. When you are asked to give differences between two groups it is always useful to present your answer in the form of a table. This will ensure that you always mention both the groups under consideration and that it is always clear to which you are referring.

Answers to part (c) must only include features that have clearly made birds *as a group* successful. Migration is commonly associated with birds but in fact only a minority of bird species exhibit this behaviour and so it cannot be responsible for the success of the birds as a whole. Again you should remember to use A-level terminology, preferring 'endothermic' to 'warm blooded'.

Suggested answer

(a) You should choose any three from the following list:

1. dorsal hollow nerve cord
2. cranium
3. notochord or vertebral column
4. ventral heart
5. post-anal tail
6. pharyngeal gill slits
7. jaws present

(b) You should select any three pairs from this table of differences:

Frog	Lizard
Scales absent	Scales present
Four digits on forelimb	Five digits on forelimb
Webbed feet	No webbed feet
Tongue attached anteriorly	Tongue attached posteriorly
Ventricle of heart not divided	Ventricle partly divided
Jacobson's organ absent	Jacobson's organ present

(c) There are a number of suggestions in the following list from which you should select three. The earlier features are of greater importance.

1. feathers for flight and insulation
2. endothermic allowing the maintenance of a high metabolic rate essential for flight
3. chalky, waterproof egg
4. parental care and nest building well developed
5. forelimbs modified to form wings
6. excretion of uric acid to conserve water

7. internal fertilisation
8. hollow bones, reducing weight for flight
9. air sacs

2 Write an essay on the arthropods.
Time allowed 35 minutes

Tutorial note

This style of essay question requires a very thorough knowledge of the group under consideration since little guidance is given by the title. Regardless of the invitation to 'write an essay on', 'write a general account of', 'make notes on' or 'discuss', the method of approaching the question is similar. Sometimes alternatives are given in this type of question and, if so, you should be careful not to exceed the number of groups required.

Although this question requires information to be restricted to one major group, the scope of the essay beyond that point is quite open. Examiners expect a broad essay incorporating all aspects of the group and it is useful to organise your answer under the following headings:

1. classification/taxonomy
2. morphology/anatomy
3. physiology
4. ecology/distribution
5. economic importance/relevance to man

It is essential to spend some time writing a plan for your answer because the examiners will certainly give marks for the way in which it is organised as well as for the facts themselves. You will also lose credit if you fail to give sufficient attention to spelling and grammar.

Information on the arthropods is so vast that it is impossible to say exactly what should be included. It is best to consider that there may be a mark for each clear, relevant, detailed biological fact, supported by an example. In other words, there would be a large number of possible responses all warranting maximum marks, unlike many other questions where you are working towards one specific answer. It is essential to maintain the balance of the essay because, whatever system of marking the examiners use, credit will certainly be lost if this is not done. A useful way to ensure that you do this is to consider that on a 25 mark essay on the arthropods there would be a maximum of 5 marks available on any one of the five listed topics. If your essay was about the angiosperms or the fungi it might be realistic to put more emphasis on economic importance and perhaps less on classification. As plant and animal groups are frequently taught under a general heading of classification, it is a common failing of candidates to restrict the answer to taxonomy and classification which, however well answered, would bring a maximum of about 5 marks.

At the same time as ensuring that the answer incorporates all aspects of the group, it is essential that clear, detailed biological facts are included, supported where possible with specific examples and the use of appropriate biological terms. Diagrams should only be used in an essay when they are relevant and where they are quicker and clearer than a written account, or add something new. Never repeat in prose the information shown on a diagram; this unnecessary duplication is time consuming and marks for a particular point can only be given once. All diagrams should be fully labelled and supplemented by annotations.

Suggested answer

1. Classification/taxonomy: characteristics of the phylum and the main subdivisions. Include some specific examples and, if possible, some reference to fossil members of the group.
2. Morphology/anatomy: some details of the body plan of typical representatives of the group – possibly an insect and a crustacean; reference to diversity of form.

3. Physiology: try to give some idea of the range within each of the following headings:
   (a) reproduction – metamorphosis, parthenogenesis
   (b) respiration – tracheae, gills, lungbooks (spiders)
   (c) nutrition – carnivore/herbivore/omnivore, parasitic
   (d) excretion – urea or uric acid
   (e) sensitivity – range of sense organs, reference to compound eyes
   (f) locomotion – walking, swimming, flying
4. Ecology:
   (a) habitats – terrestrial, freshwater, marine
   (b) distribution
   (c) behaviour – e.g. social or solitary
   (d) relationships to other organisms – e.g. symbiosis
   (e) position in food chain – trophic level
5. Economic importance:
   (a) beneficial types – both directly, e.g. as a food source, and indirectly, e.g. pollination
   (b) harmful types – stings and bites and also as vectors of disease

While insects are the most numerous group in terms of number of species, deserving slightly more attention than any of the other classes, the essay should not be confined to them.

Question bank

1 The wing of a housefly and of a bat are
A both homologous and analogous
B analogous not homologous
C homologous not analogous
D neither homologous nor analogous
Time allowed 1 minute

Pitfalls

You will doubtlessly realise that both the housefly and the bat use their wings for flight. However, in order to answer this question you will also need to know something about the structure and origin of the wing in each of these animals. The wing of the bat is based upon the same pattern of bones as the forelimb of most vertebrates – the pentadactyl plan. The wing of the housefly does not contain bone and has a totally different origin. Having established this point in your mind, you now have to remember the meaning of the terms homologous and analogous and this is where you may have problems. Organs are said to be homologous when they are fundamentally similar in origin, structure and position, regardless of their function in the adult. Analogous organs have a similar function in the adult but may have totally different origins.

2 Below is a diagram of a virus attached to a bacterial cell.
(a) Label parts A–D
(b) Name the type of virus shown.
(c) Briefly describe the remaining stages in the sequence.
(d) Give the names of two human diseases caused by viruses.
(e) Name one plant disease caused by a virus and state why viral diseases of plants are usually spread by insect vectors such as aphids.
Time allowed 10 minutes

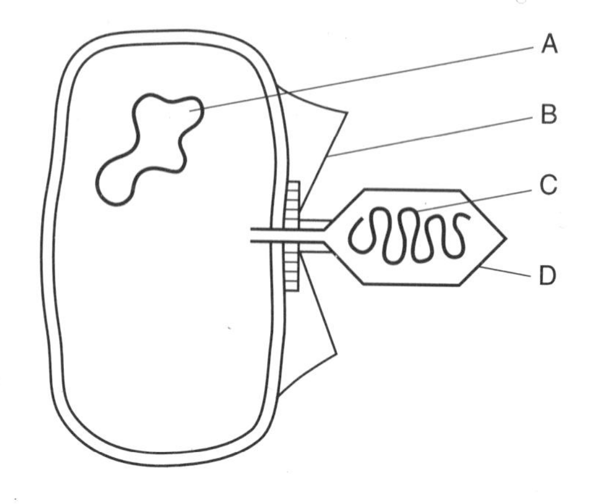

Pitfalls

As with many classification questions, this one is largely straightforward recall of information, and the biggest problem you might face is simply that you might not have learnt your work thoroughly enough to answer adequately. In part (a) both label A and label C point to DNA so be sure to distinguish them clearly, i.e. A is the circular DNA of the bacterial cell whereas C is the DNA of the virus. Part (c) requires you to follow the sequence of events through to the end, i.e. until new viral cells are released from the bacterium. In part (e) the role of the vector in entering plant cells should not be forgotten, as viruses cannot easily penetrate the cell walls of plant cells unaided.

Points

(a) A = Circular DNA of the bacterial cell
B = Fibre tail of the virus (bacteriophage)
C = DNA of the virus (bacteriophage)
D = Protein coat of the virus (bacteriophage)

(b) Bacteriophage

(c) The DNA from the bacteriophage is injected into the bacterial cell. Once inside, the phage DNA multiplies using the resources of the bacterial cell. Having caused the bacterial cell to form around 200 copies of itself, the bacteriophages finally cause the cell to burst (lysis), thus releasing the new phages.

(d) Choose from poliomyelitis, mumps, measles, influenza, Acquired Immune Deficiency Syndrome (AIDS).

(e) Choose the example from tobacco mosaic, potato virus X, barley yellow dwarf virus, turnip yellow mosaic virus.

Viruses cannot easily penetrate the cellulose cell wall of plants and therefore they often infect via an aphid's pointed mouthpart, which penetrates plant cells when the aphid feeds. As the aphid's mouthpart normally enters the phloem this has the added advantage for the virus of being deposited in the moving fluid of the phloem, thus enabling the virus to spread fairly rapidly throughout the plant host.

3 Below is a diagram of a simple laboratory fermenter.

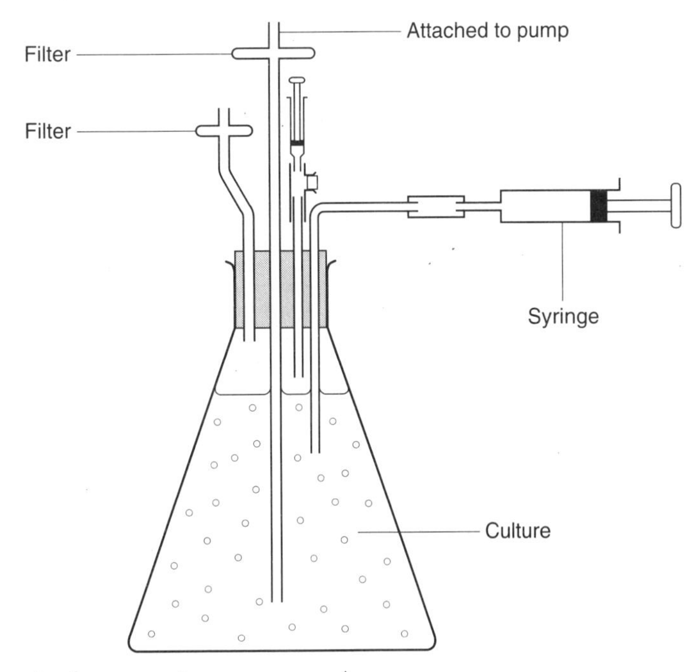

(a) What is the function of
(i) the filter
(ii) the syringe
(iii) the pump

(b) Describe how you would set up the fermenter before adding any microorganisms. Include any precautions that you would take.

(c) How might you modify the fermenter in order to increase the rate of growth of microorganisms?

(d) How do industrial fermenters differ from the one shown?

Pitfalls

In (a) there is more than one function for each part listed. For example, in (i) there are two filters; one to prevent the fermenter contents being contaminated by microorganisms from the atmosphere, and the other to prevent the atmosphere being contaminated by microorganisms from the fermenter.

In part (b) the precautions are primarily to serve two purposes; to maintain sterility and to keep the apparatus airtight enough to prevent escape of gases other than through the filter, but not so airtight that it explodes. Remember, sterility begins before you set up the apparatus.

Points

(a) (i) The filter helps to maintain sterility in the fermenter by preventing entry and escape of microorganisms.

(ii) The syringe permits samples from the fermenter to be removed for analysis and/or for specific substances e.g. more nutrients, to be added to the fermenter.

(iii) The pump adds air to the fermenter; this is important in order to provide oxygen for aerobic microorganisms. The flow of air also acts like a stirrer in mixing up the contents of the fermenter.

(b) The work surface must be disinfected and both the apparatus and nutrient solution must be sterilized (e.g. in an autoclave or pressure cooker at 121 °C under a pressure of 103 kPa) *before* the equipment is assembled. The sterile vessel should be filled with the sterile nutrient solution and then the syringe, filters, inlet and outlet tubes should be added. Silicone grease can be used on joints to ensure airtightness, but while the bung should be airtight, it should be loose enough to allow it to be pushed out in the event of a pressure build-up caused by an excess of carbon dioxide gas being produced.

In part (c) growth can be increased by:

- thermostatic control of the fermenter's temperature at the optimum for the microorganisms being grown;
- use of a buffer solution to provide the optimum pH;
- use of a stirrer to mix nutrients and keep the microorganisms from settling out;
- use of a diffuser to give more aeration and increased dissolving of oxygen.

In part (d) comparison of the fermenter shown here with that in Fig. 2.8 indicates that the industrial fermenter:

- has a cooling jacket to maintain an even temperature;
- has mechanical impellers to mix the contents;
- is made of stainless steel;
- is bigger;
- has an acid/alkali reservoir to control pH;
- has the facility to add an anti-foaming agent;
- can control automatically parameters such as temperature, pH, pressure and foaming through constant monitoring by appropriate probes;

Furthermore, in the industrial fermenter:

- the addition of nutrient is carefully controlled;
- a sparging ring diffuses the incoming air.

CHAPTER 3

CELL DIVISION AND GENETICS

Units in this chapter

Chapter objectives

In this chapter we will unravel some of the mysteries of why offspring are recognisably similar to their parents but rarely identical to them; why characteristics seem to disappear in one generation only to reappear in the next; and how hereditary information is passed from generation to generation. In addition, we will explore the methods by which new species arise through the gradual process of evolutionary change. You will acquire the ability to solve genetic problems and make predictions about the likely offspring of given parents.

3.1 DNA AND THE GENETIC CODE

NUCLEIC ACIDS

Nucleic acids are of large molecular mass and are composed of units called **nucleotides**. Each nucleotide comprises three parts:

1. phosphoric acid (phosphate) – H_3PO_4
2. pentose sugar – either ribose ($C_5H_{10}O_5$) or deoxyribose ($C_5H_{10}O_4$)
3. organic base – pyrimidine (single ring structure), e.g. cytosine, thymine and uracil
   – purines (double ring structure), e.g. adenine and guanine

Apart from combining in chains to make nucleic acids, nucleotides make up other biologically important molecules as illustrated in Table 3.1.

There are two main nucleic acids: ribonucleic acid (RNA) and deoxyribonucleic acid (DNA).

Table 3.1

Molecule	Abbreviation	Function
Deoxyribonucleic acid	DNA	Contains the genetic information of cells
Ribonucleic acid	RNA	All three types play a vital role in protein synthesis
Adenosine monophosphate Adenosine diphosphate Adenosine triphosphate	AMP ADP ATP	Coenzymes important in making energy available to cells for metabolic activities, osmotic work, muscular contractions, etc.
Nicotinamide adenine dinucleotide Flavine adenine dinucleotide	NAD FAD	Electron (hydrogen) carrier important in respiration, in transferring hydrogen atoms from the Krebs cycle along the respiratory chain
Nicotinamide adenine dinucleotide phosphate	NADP	Electron (hydrogen) carrier important in photosynthesis, for accepting electrons from the chlorophyll molecule and making them available for the photolysis of water
Coenzyme A	CoA	Coenzyme important in respiration, in combining with pyruvate to form acetyl coenzyme A and transferring the acetyl group into the Krebs cycle

RIBONUCLEIC ACID (RNA)

RNA is a single-stranded chain of nucleotides which contain the organic bases adenine, guanine, cytosine and uracil, but never thymine. Its basic structure is given in Fig. 3.1(c). There are three types of RNA: messenger, transfer and ribosomal.

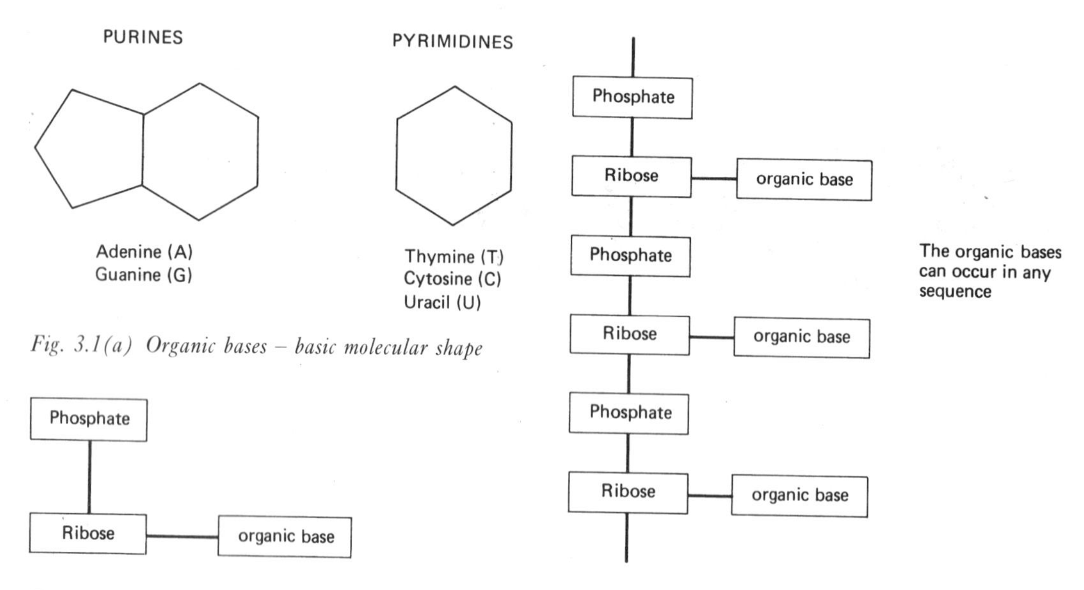

Fig. 3.1(a) Organic bases – basic molecular shape

Fig. 3.1(b) Basic structure of nucleotide

Fig. 3.1(c) Portion of an RNA molecule

- **Messenger RNA** (mRNA) is composed of up to thousands of nucleotides formed into a helix. It is manufactured in the nucleus as the mirror image of one strand of DNA. There is an immense variety of types and they act as templates for protein synthesis.
- **Transfer RNA** (tRNA) is composed of around 80 nucleotides formed into a cloverleaf shape. Again it is manufactured in the nucleus by DNA, but there are only

just over 20 types, each one capable of carrying a specific amino acid. Its role is therefore the transport of amino acids during protein synthesis.

- **Ribosomal RNA** (rRNA) is a large molecule which can form both single and double helices. Although manufactured in the nucleus by DNA, it is found in the cytoplasm where it plays a number of roles in protein synthesis.

DEOXYRIBONUCLEIC ACID (DNA)

DNA is a double-stranded chain of nucleotides which contain the organic bases adenine, guanine, cytosine and thymine, but never uracil. Each chain is immensely long, being composed of many millions of nucleotide units. Its complex structure was elucidated by James Watson and Francis Crick in 1953 and is best described as a twisted ladder in which the deoxyribose and phosphate molecules form the uprights and the organic bases the rungs. The 'rungs' of this ladder must be of the same length so it follows that, as both purines and pyrimidines are found in DNA, each 'rung' must be made up of a purine linked with a pyrimidine. Two purines would make the 'rung' too long; two pyrimidines would make it too short (see basic molecular shape). Analysis of DNA shows the quantity of adenine and thymine to be the same and the quantity of guanine and cytosine to be the same, indicating that they form the organic base pairs. The structure of DNA is illustrated in Fig. 3.2.

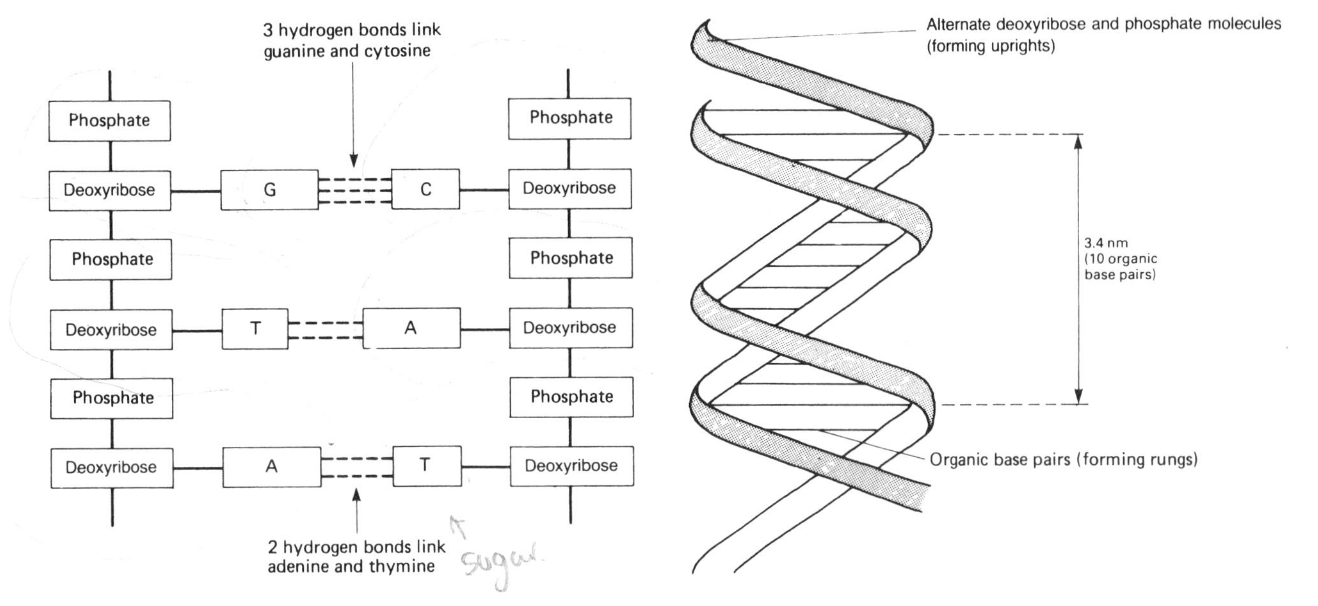

Fig. 3.2(a) Basic DNA structure – helix unwound

Fig. 3.2(b) Double helical structure of DNA

Evidence that DNA is the hereditary material

DNA is now widely recognised as the hereditary material of cells. Evidence for this comes from a number of sources:

1. Chromosome analysis – chromosomes are made of DNA and protein only, and as chromosomes can be seen to play a role in cell division, it seems likely that one or the other of these substances is the hereditary material.

2. Constancy of DNA in the cell – the amount of DNA remains constant for all cells except gametes, which have half the usual amount. These facts are consistent with the expected changes in the quantity of hereditary material during cell division.

3. Metabolic stability of DNA – DNA is extremely stable, a characteristic that is essential to any material which is passed from generation to generation over millions of years.

4. Mutagenic effects and DNA – agents such as X-rays and certain chemicals which are known to cause inherited mutations can also be shown to alter the structure of DNA.

5. Experimental evidence – this provides clear proof that DNA is the hereditary material. Three major examples are as follows: (a) experiments on bacterial transformation (Griffith 1928), (b) identification of the transforming principle (Avery, McCarty and MacLeod 1944), (c) transduction experiments (Hershey and Chase 1952). You should consult a suitable textbook for details of the work.

Table 3.2 *Main differences between RNA and DNA*

DNA	RNA
Organic bases are adenine, thymine, guanine, cytosine	Organic bases are adenine, uracil, guanine, cytosine
Double helix	Single strand
Deoxyribose (H)	Ribose (OH)
Mostly nuclear	Throughout the cell
More stable	Less stable
Permanent	Temporary
Insoluble	Soluble
One basic type	Three types: messenger, transfer, ribosomal
Concentration constant	Concentration varies according to cell type
Adenine, thymine/cytosine, guanine ratio about equal	Adenine, uracil/cytosine, guanine ratio more variable
Very large molecular mass (100 000 – 120 000 000)	Much smaller molecular mass (20 000 – 2 000 000)

Replication of DNA

As the genetic material of living organisms, it follows that DNA must be able to replicate itself exactly if information is to be passed from cell to cell and from generation to generation. The accepted method for this process is called **semi-conservative replication** (see Fig. 3.3).

Evidence for this method of replication was provided by Meselsohn and Stahl (see Fig. 3.4). Many generations of *Escherichia coli* cells were grown in cultures containing ^{15}N (heavy nitrogen) to ensure that their DNA contained ^{15}N. They were then transferred to a medium containing only ^{14}N (light nitrogen). The next two generations of *Escherichia coli* each had their DNA separated by centrifugation and the presence of light and heavy nitrogen in the DNA was detected by its absorption of ultraviolet light. The results showed the first generation of bacteria to have DNA containing equal proportions of ^{14}N and ^{15}N. In the second generation half the bacteria had DNA containing only ^{14}N, the remainder having DNA with equal proportions of ^{14}N and ^{15}N.

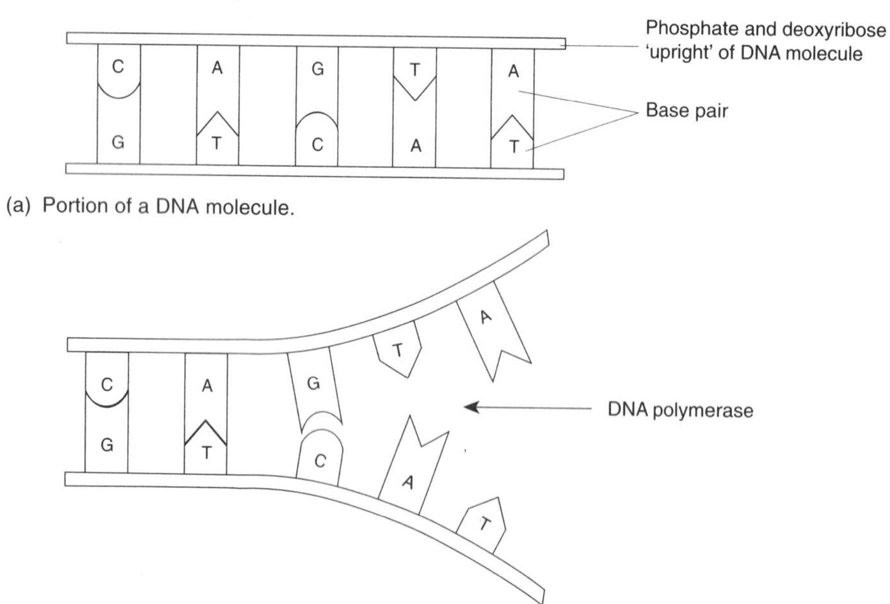

(a) Portion of a DNA molecule.

(b) The enzyme DNA polymerase breaks down the bonds between the base pairs, causing the two strands of the DNA molecule to separate.

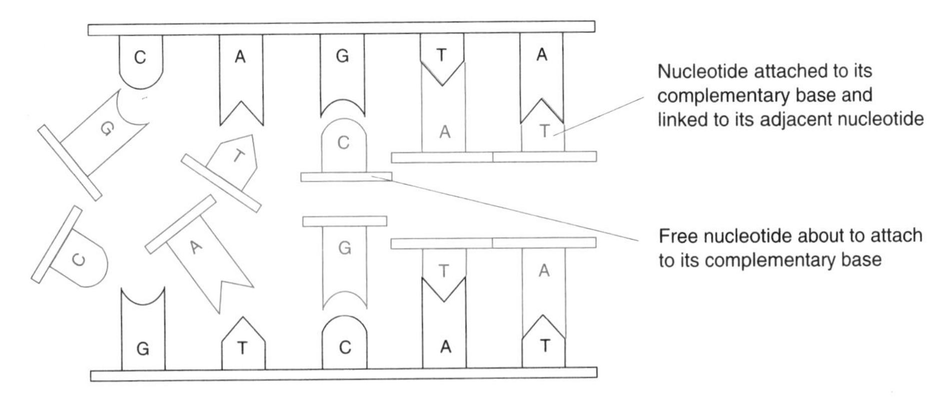

(c) Free nucleotides bond with their complementary bases on each strand of the DNA molecule.

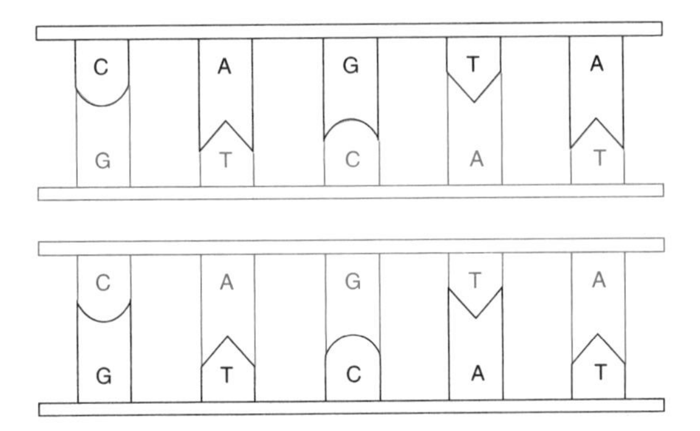

(d) When the nucleotides are lined up they join together to form a polynucleotide chain. Two identical strands of DNA are thus formed. Each new DNA molecule retains half of the original DNA material and this form of replication is therefore described as semi-conservative.

Fig. 3.3 Replication of DNA

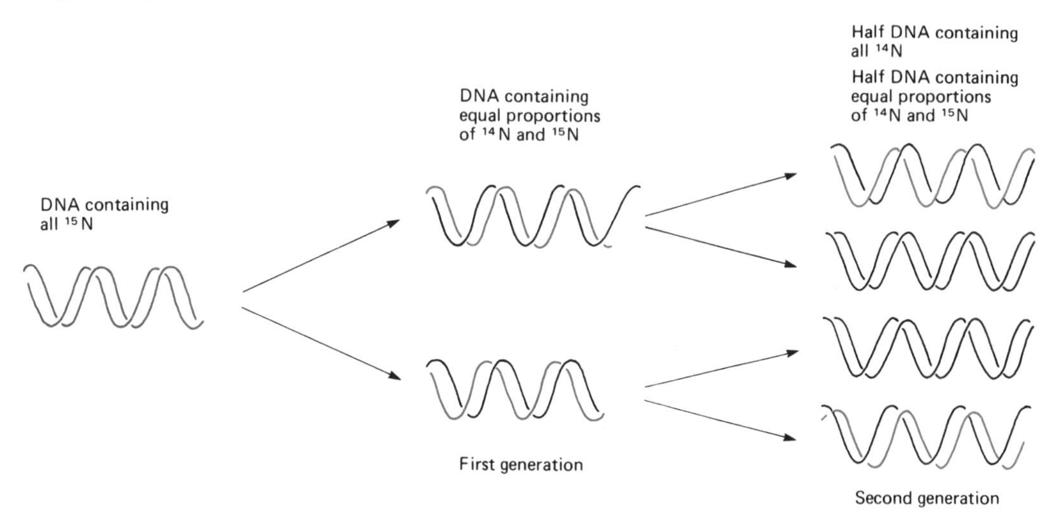

Fig. 3.4 Evidence for semi-conservative replication of DNA

THE GENETIC CODE

Having now seen the structure of the DNA molecule and how it replicates, you may wonder how exactly it can store information and determine the nature of cells and hence the structure of all living organisms. The answer lies in the sequence of bases on the DNA molecule. Although there are only four bases – adenine, guanine, cytosine and thymine – the immense length of the DNA molecule means that there is an almost infinite variety of combinations of these bases. DNA is thus a set of instructions written in a language comprising just four letters (A, G, C and T).

How then are these instructions translated into cell structure? Cells are made up of a variety of chemicals, but most of these are the same regardless of the organism or the cell in question. Carbohydrates, lipids and other organic molecules are much the same in all organisms. Inorganic ions and water are identical in all living things although the proportions of each may vary. It is only in a cell's proteins that real differences arise. Each cell has its own, often unique, blend of proteins which determines its individual structure and function. Most chemicals in cells, including proteins, are produced by the action of enzymes and these too are proteins. If it is DNA that is determining which proteins, especially enzymes, are produced, then this would explain how DNA determines the structure of cells and hence organisms.

Proteins are extremely large molecules of almost infinite variety and yet they are made up of just 20 different amino acids. If the bases on the DNA were codes for amino acids, then the sequence of these bases would determine their precise order in a protein and hence its properties. But, with only four different bases, how can 20 amino acids be coded for? Clearly the code for each must comprise more than one base – in fact three is the minimum required. It is therefore a **triplet code** of three bases which determines each amino acid. By an ingenious series of experiments Nirenberg determined the exact code for each one. You might like to look up details of Nirenberg's experiments in a suitable textbook.

A triplet code produces 64 different possible combinations and, with just 20 amino acids commonly found in proteins, it follows that some amino acids have more than one code – up to six in fact. Nirenberg found that some triplets of bases did not code for any amino acids. These turned out to be **stop** or **nonsense** codes which, in effect, separate the instructions for one protein from those for the next. The codes are the same in all living organisms.

THE SYNTHESIS OF PROTEINS

There are four main stages to the synthesis of a protein.

1. **Formation of amino acids** Plants absorb nitrates from the soil and reduce them to amino (NH_2) groups. These may then be combined with appropriate carbohydrates to give an amino acid. Animals obtain nine of their amino acids in their diet (**essential amino acids**) but can synthesise the remainder (**non-essential amino acids**) for themselves.

2. **Transcription** This is the process by which the base codes on a DNA molecule are transcribed as a sequence of bases on a messenger RNA (mRNA) molecule. The DNA in the nucleus unwinds at a specific point along its length which codes for the required protein. The region which unwinds will be used to make a polypeptide, and is termed a **cistron**. Free nucleotides, which are abundant in the cell, arrange themselves alongside their opposite base pair (except that uracil replaces thymine on the mRNA). The enzyme **RNA polymerase** zips together the nucleotides to give rise to the mRNA. Only one of the DNA strands (the **sense strand**) is copied. The mRNA then peels off the DNA and diffuses out of the nucleus via a nuclear pore. Each triplet of bases on the mRNA is referred to as a **codon**.

3. **Amino acid activation** Amino acids which are to become part of the new protein are firstly attached to transfer RNA (tRNA) with the aid of energy derived from ATP. Each amino acid has a specific type of tRNA which differs in the sequence of bases at one end of the molecule. This is known as the **anticodon**. At the other end of the molecule is a sequence of bases (CCA) which is the same in all types of tRNA. The amino acid is attached to this triplet of bases.

4. **Translation** Having left the nucleus the mRNA becomes bound to a string of ribosomes, collectively called a **polysome**. The two codons at one end of the mRNA attract the complementary anticodons of tRNA with their attached amino acids. A ribosome acts as a framework holding the two tRNA molecules together and bringing their two attached amino acids into close proximity. In this position a peptide bond forms between them producing a dipeptide (Fig. 3.6(a)). The ribosome then moves

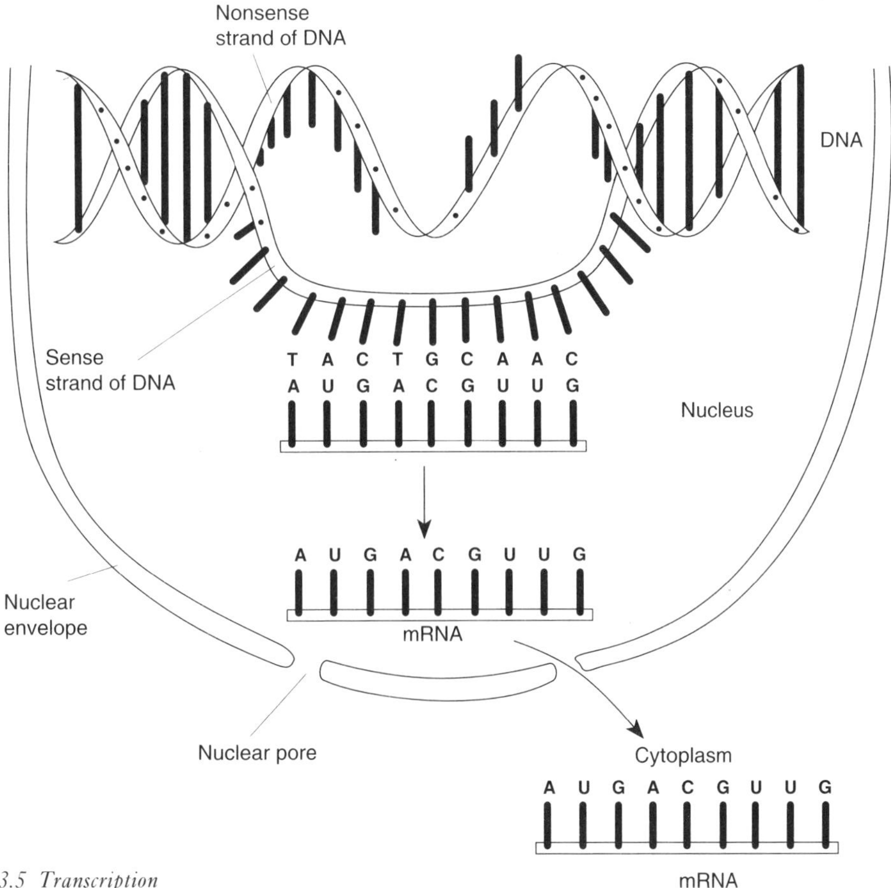

Fig. 3.5 Transcription

along the mRNA to the point where the third codon has attracted its complementary anticodon on tRNA. A third amino acid is thus attached to the dipeptide (Fig. 3.6(b)).

The ribosome continues to pass along the mRNA attaching further amino acids together until a stop (nonsense) code is reached, at which point the now completed polypeptide is 'cast off'. Several ribosomes pass along the mRNA simultaneously so that many polypeptides are formed at the same time (Fig. 3.6(c)). The polypeptides are formed into helical structures and combined with different polypeptides and prosthetic groups to give rise to the complete protein.

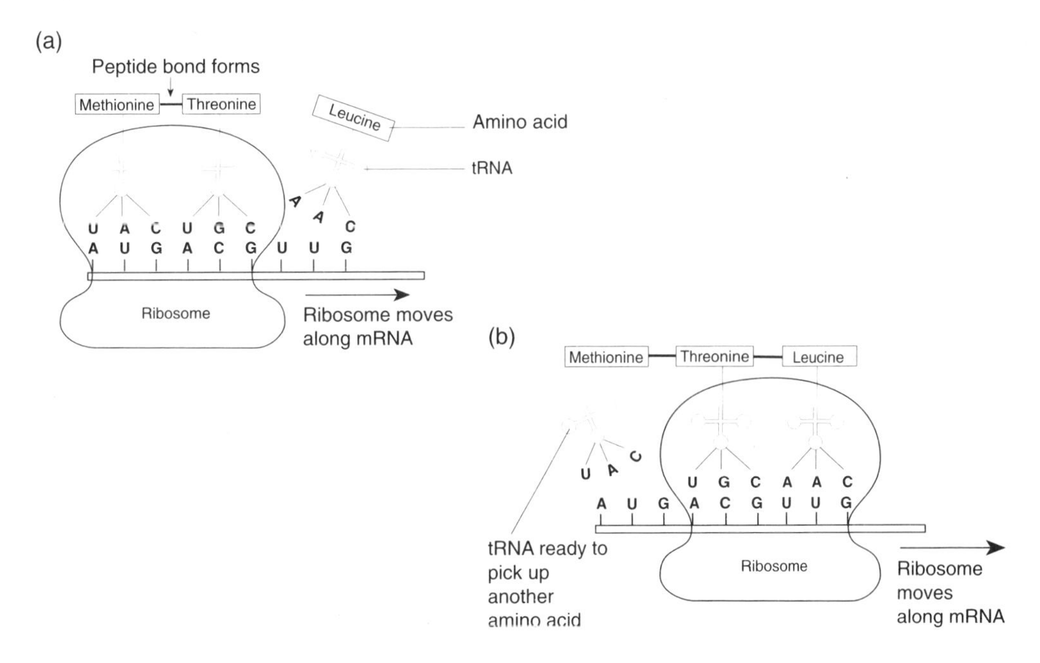

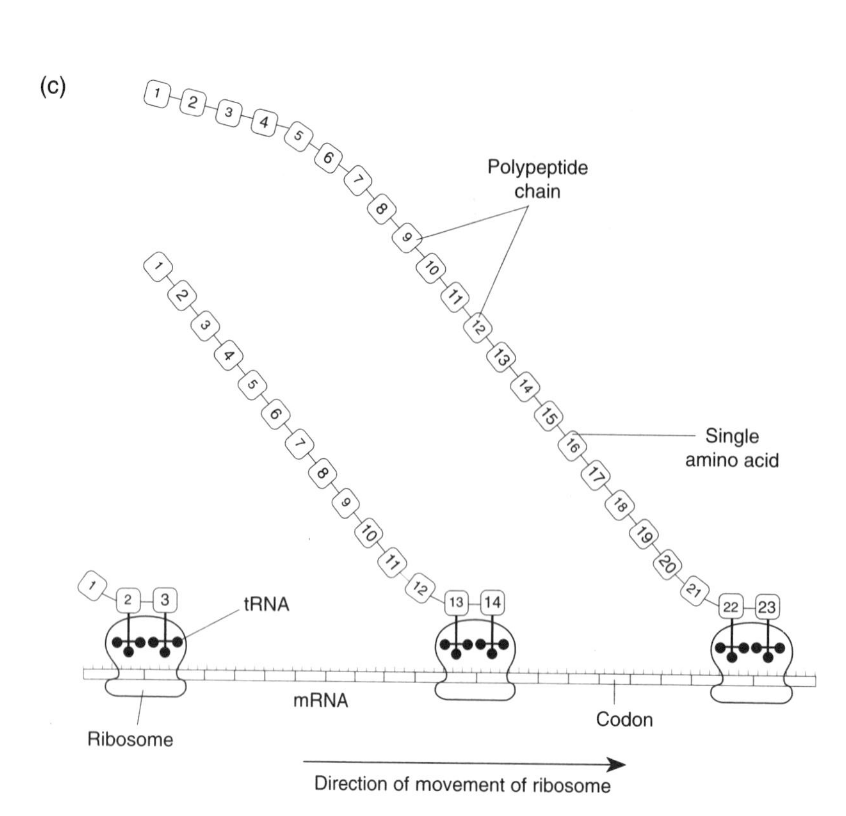

Fig. 3.6 Translation

HOW GENES ARE EXPRESSED AND CONTROLLED

Experiments have confirmed the theory that **one gene specifies one polypeptide**, but not all polypeptides need to be produced continuously. How then is a gene switched on and off so that polypeptides can be produced as and when they are needed? Research by two scientists, Jacob and Monod, suggests there are two mechanisms: **enzyme induction** and **enzyme repression**. Both mechanisms involve the concept of a group of genes acting together as a unit, called an **operon**. The genes comprising the operon are:

- **Structural genes** – these code for the polypeptides which make up an enzyme;
- **An operator gene** – this turns on and off the structural genes.

Another gene, the **regulator gene**, while not part of the operon, affects production of the enzyme by forming a protein called a **repressor**. The repressor may combine with the operator gene and so cause it to turn off the structural genes. Enzyme induction and repression can be explained in terms of these elements.

Enzyme induction

The bacterium *Escherichia coli* normally respires the sugar glucose. If, however, it is grown on a medium containing only lactose it produces two new enzymes, one which allows lactose to be absorbed and the other which permits it to be respired. How then are these enzymes produced in response to lactose being present?

The enzymes in question are produced by a set of structural genes which are controlled by their operator gene. In the absence of lactose the operator gene is repressed by a repressor molecule produced by the regulator gene. The structural genes are therefore switched off and the enzymes they code for are not produced. When lactose is present, however, it combines with the repressor molecule, making it unable to act on the operator gene. The operator gene therefore switches on the structural genes and these produce the mRNA which, in turn, helps produce the enzymes which absorb and respire lactose.

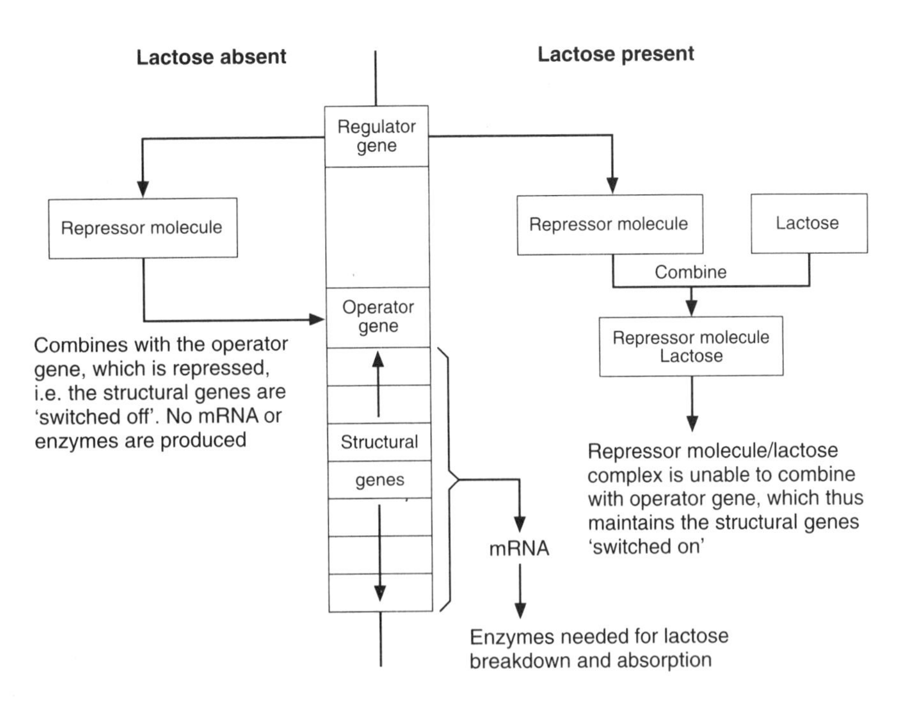

Fig. 3.7 Enzyme induction

Enzyme repression

This is the opposite to enzyme induction in so far as the structural genes are normally switched on rather than off. In the case of the bacterium *Escherichia coli*, it requires a regular supply of the amino acid tryptophan, which it normally makes from raw materials using the enzyme tryptophan synthetase. If, however, tryptophan is provided in the medium, there is no necessity to use energy in producing it and therefore the bacterium ceases to make tryptophan synthetase. In this case the repressor molecule from the regulator gene is in an inactive form and therefore does **not** repress the operator gene. This means that the structural genes remain switched on and produce mRNA, which leads to the production of tryptophan synthetase. When tryptophan is present it combines with the repressor molecule, activating the repressor so that it does repress the operator gene, switching it off, and thus ceasing the production of tryptophan synthetase.

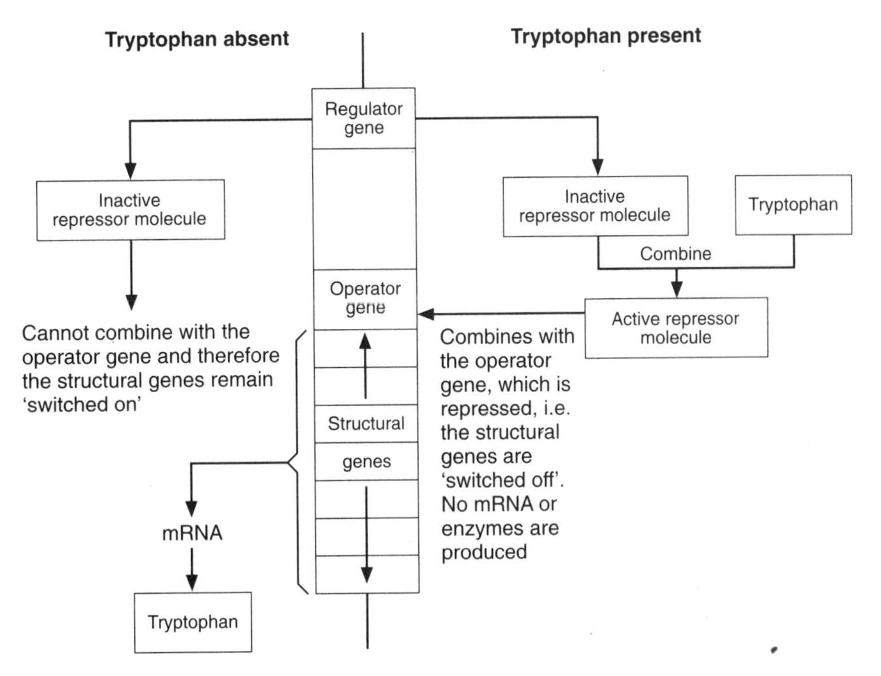

Fig. 3.8 Enzyme repression

GENETIC ENGINEERING

Such has been the development of science and technology that genes can now be manipulated and transferred from one organism to another. In this way, bacteria can be used as chemical factories for the production of substances such as hormones and antibiotics needed by mankind.

Recombinant DNA technology

This involves the manipulation of DNA and its transfer to another organism, where it is combined with that organism's own DNA. For example, the DNA portion responsible for the production of the human hormone insulin can be introduced into bacteria, which will then manufacture the hormone for use by diabetic patients. Recombinant DNA technology involves the following techniques.

1. **Splitting of DNA into smaller sections** – this involves the use of enzymes called **restriction endonucleases**, which act only between specific base sequences on the DNA.

2. **Copying the required DNA section** – this involves the use of an enzyme called **reverse transcriptase**, which can synthesize DNA from the relevant mRNA (i.e. the reverse of the normal process in which mRNA is made from DNA). If the mRNA which codes for a particular protein is collected, it can therefore be used to produce the DNA from which it came. This is called **copy DNA (cDNA)**.

   An alternative method uses **plasmids**. These are circular loops of DNA found in bacterial cells. The bacteria replicate these so that each cell has many copies of plasmid DNA. If a portion of human DNA is added to a plasmid, the bacterial cell will produce up to 200 identical copies of this human DNA.

3. **Joining together portions of DNA** – this involves the use of the enzyme **DNA ligase**, which carries out the linking of the DNA portions mentioned in the previous section.

The sequence of diagrams in Fig. 3.9 illustrates how these techniques can be used to produce human insulin.

Genetic engineering can be used to produce other hormones as well as insulin; human growth hormone and calcitonin are two examples. Other uses include transferring insecticidal properties from bacteria to plants such as potatoes and cotton, adding the genes for nitrogen-fixation to cereal crops and in the treatment of human diseases, such as thalassaemia, sickle cell anaemia and cystic fibrosis.

DNA fingerprinting

Everybody's DNA is unique to them in much the same way as fingerprints are. As every cell contains DNA, a few cells of blood, skin or sperm are sufficient to detect the individual from whom they came. The process entails:

1. separating the DNA from the sample cells;
2. cutting the DNA into sections by using restriction endonucleases;
3. separating the fragments by using electrophoresis;
4. transferring the fragments to a nylon membrane, by using a process called **southern blotting**;
5. attaching radioactive DNA probes to specific portions of the DNA fragments;
6. washing off any DNA not attached to the probes;
7. attaching the remaining DNA to X-ray film;
8. allowing the radioactive probes on the DNA to expose the film, which, when developed, reveals a unique pattern of dark and light bands. This is the genetic fingerprint.

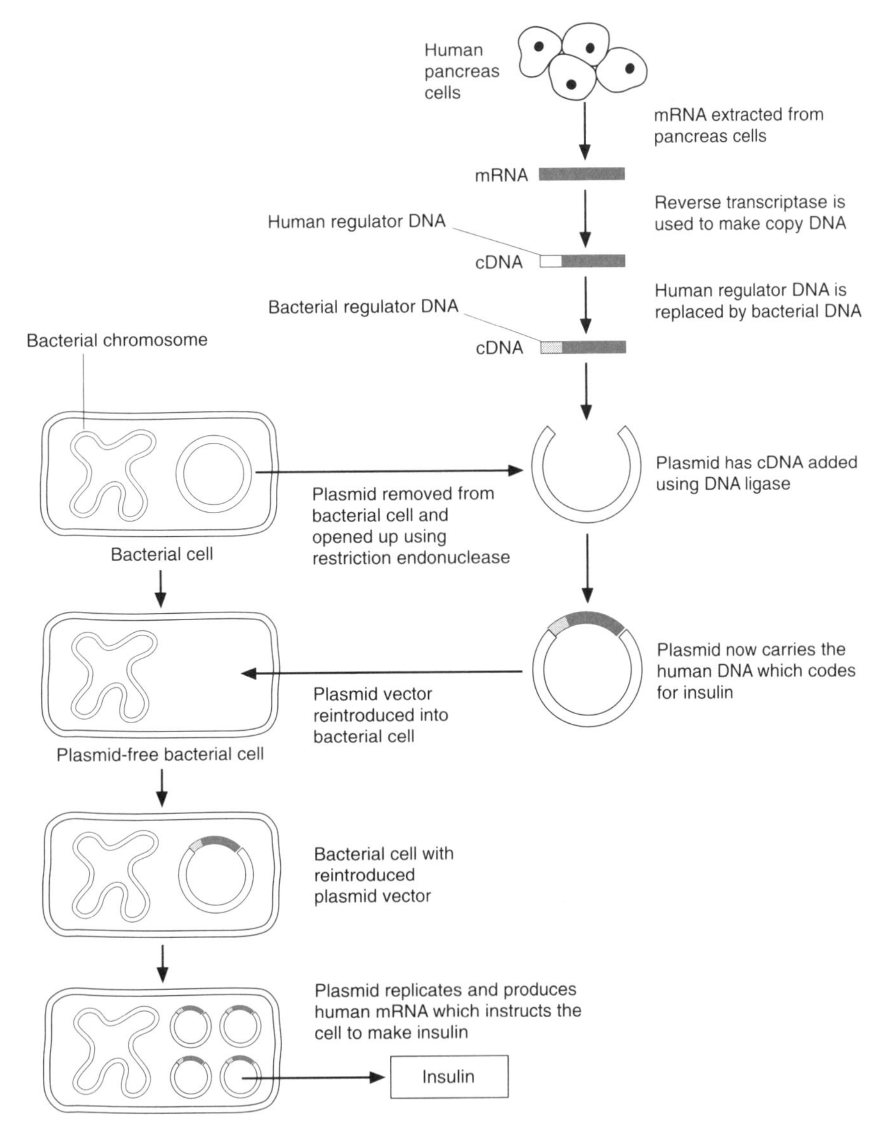

Fig. 3.9 Insulin production using recombinant DNA techniques

DNA fingerprinting can be used to help identify individuals who may have carried out crimes such as rape, settling paternity disputes, determining relationships for immigration purposes, detecting inherited diseases and monitoring bone marrow transplants. In Fig. 3.10, the dark DNA bands show that both children share some of their maternal and paternal DNA.

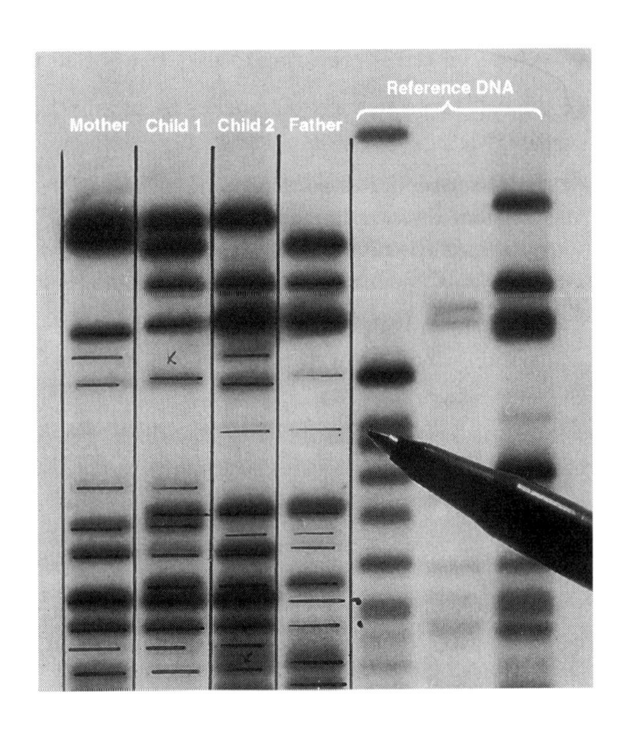

Fig. 3.10 DNA fingerprinting can be used to prove family relationships

3.2 CELL DIVISION

STRUCTURE OF CHROMOSOMES

Chromosomes are rod-like structures, consisting of nucleic acids and protein, located within the nucleus. During cell division chromosomes change their length as a result of coiling and uncoiling, dividing to form paired, joined **chromatids**. Each chromosome has somewhere along its length a well-defined region where the chromatids are particularly closely associated and which seems to be the point at which force is exerted in the separation of dividing chromosomes. This structure is called the **centromere**. Along the length of the chromosome there may be distinct constrictions and bumps, the pattern of which is quite constant for a particular chromosome from cell to cell. These 'bumps' are called **chromomeres** and they are probably caused by the coiling within the chromatids.

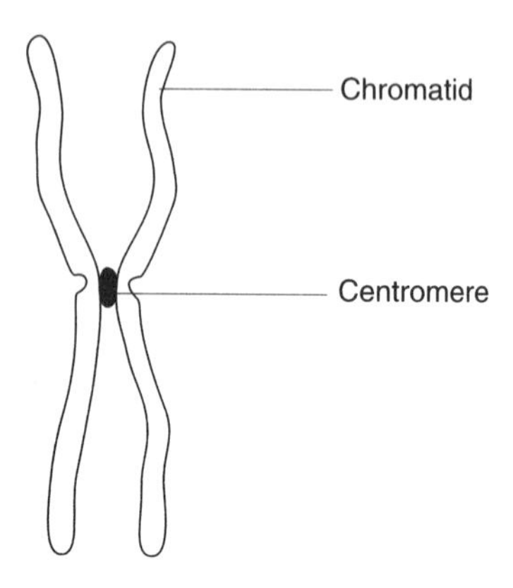

Fig. 3.11 Structure of a chromosome

The number of chromosomes per nucleus is normally constant for all the individuals of a species, e.g. man has 46, rat 42, garden pea 14 and tomato 24. The number of chromosomes characteristic of a species gives no indication of its level of organisation.

Chromosomes are present in pairs and therefore it is often convenient to speak of the chromosome number of a particular species in terms of the number of pairs (i.e. 23 for man). The members of each pair are alike but the different pairs are distinguishable. Every body (**somatic**) cell contains the characteristic number of chromosomes, but mature germ cells (**gametes**) contain only half the usual number, one member of each pair. The gametes are described as **haploid** in chromosome number and the somatic cells as **diploid**.

MITOSIS

In mitosis each chromosome duplicates itself and the duplicates are separated from each other at cell division, one going into the nucleus of one daughter cell and the second going into the other. The daughter cells are therefore identical with each other and with their parent cell in chromosome constitution.

When the cell is preparing to divide it is said to be in **interphase**. At this stage:

1. the cell forms new cell organelles to supply the daughter cells – **1st growth (G_1) phase**;
2. the DNA replicates so that there is sufficient for two daughter cells – **synthesis (S) phase**;
3. the cell builds up its store of energy to provide sufficient for cell division – **2nd growth (G_2) phase**.

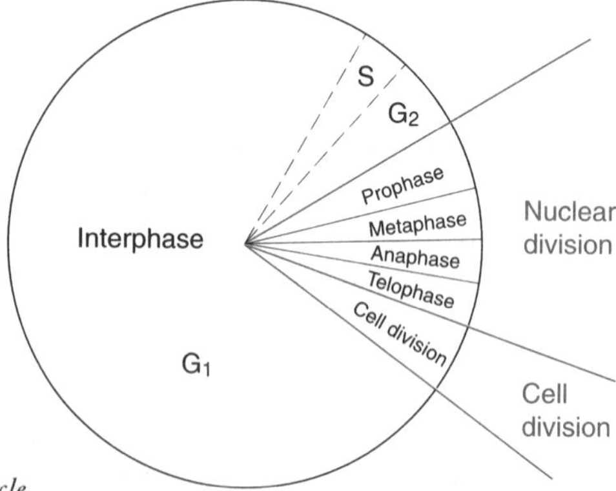

Fig. 3.12 The cell cycle

Nuclear division itself is a continuous process, but for ease of description four main stages are recognised: prophase, metaphase, anaphase and telophase.

The following series of diagrams represent mitosis in an animal cell and show only two different chromosomes.

(a) **Interphase**

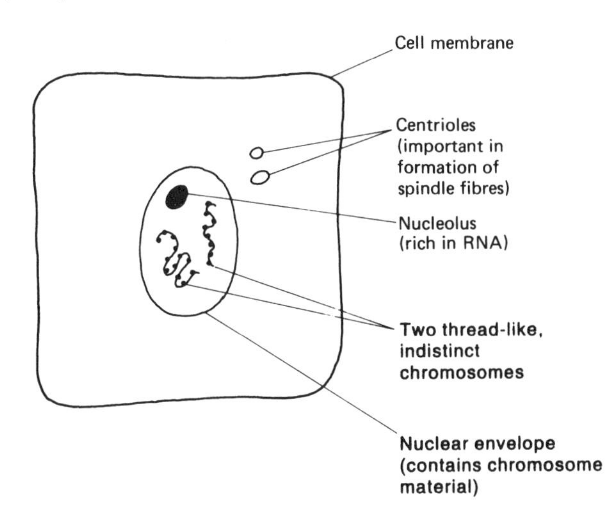

(b) **Early prophase**

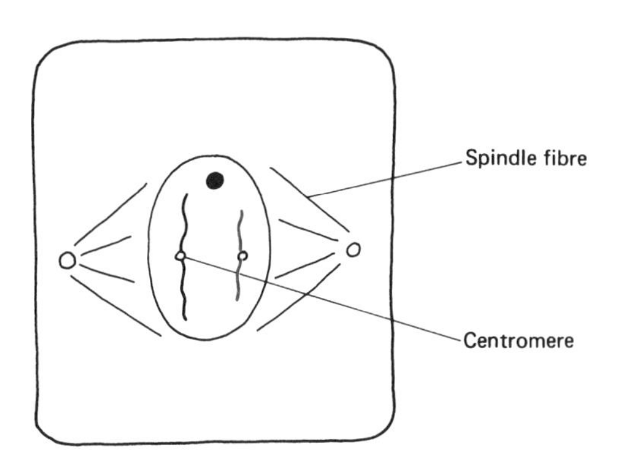

Chromosomes more distinct as they contract. Centromere visible. Spindle starts to form from centrioles lying at its poles. Nucleolus shrinks.

(c) **Late prophase**

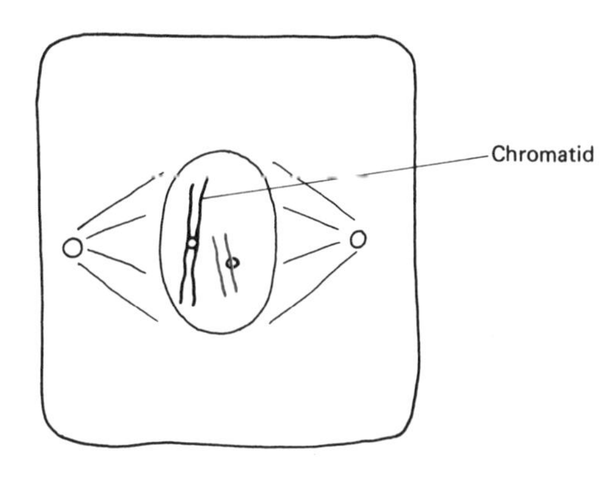

Chromatids visible as chromosomes shorten and thicken. Nucleolus disappeared. Nuclear envelope disappearing.

(d) **Early metaphase**

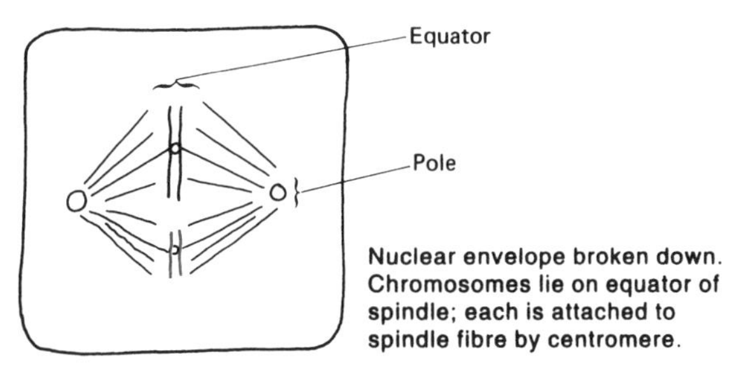

Nuclear envelope broken down. Chromosomes lie on equator of spindle; each is attached to spindle fibre by centromere.

Late metaphase: chromatids start to move apart.

(e) **Early anaphase**

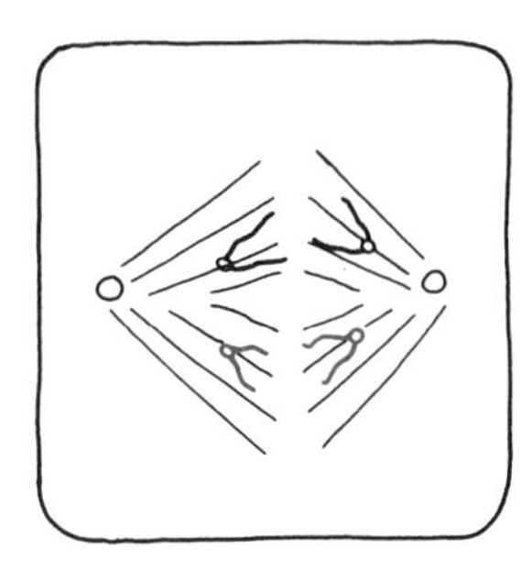

Chromatids move apart to opposite poles, probably by contraction of spindle fibres (a process requiring energy).

Late anaphase: chromatids reaching poles.

(f) **Early telophase**

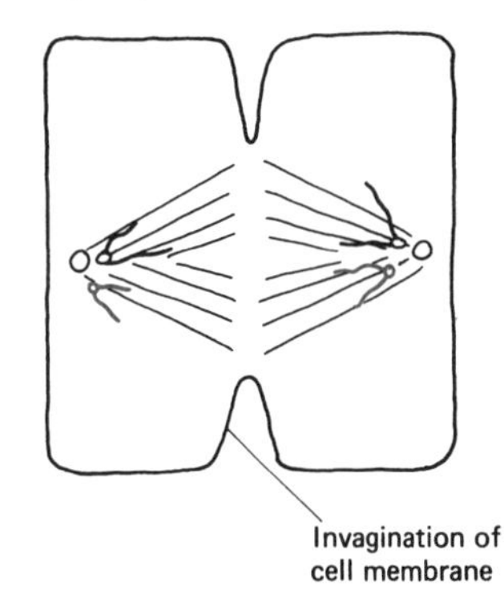

Chromatids assemble at poles and cell membrane invaginates (cell plate starts to form across cell in plants).

(g) **Late telophase**

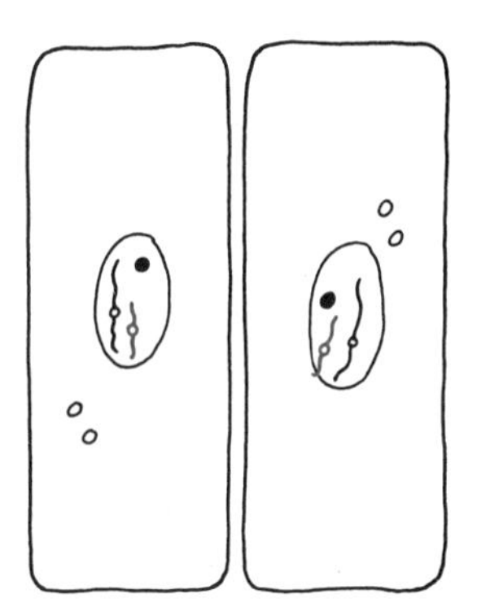

Two daughter cells formed. Spindle fibres degenerate. Nuclear envelope and nucleolus reform. Chromosomes regain thread-like form.

Fig. 3.13 Stages of mitosis

MEIOSIS

This results in the formation of haploid daughter cells since each receives only one of each type of chromosome instead of two. The same basic stages are recognised as in mitosis but they occur twice, i.e. first meiotic division followed by second meiotic division. The following series of diagrams is based on an animal cell containing two pairs of chromosomes.

(a) **Interphase**

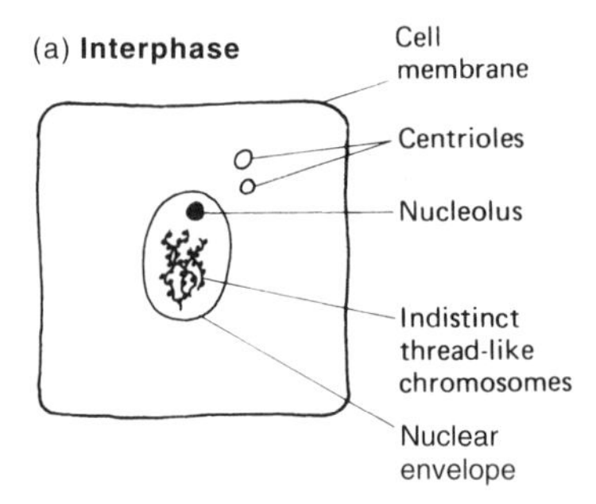

Five subdivisions of **prophase** are recognised: leptotene, zygotene, pachytene, diplotene and diakinesis

(b) **Prophase I (leptotene)**

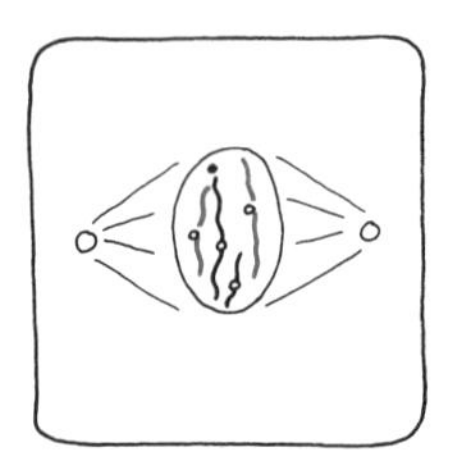

Chromosomes appear.
Spindle starts to form.

(c) **Prophase I (zygotene)**

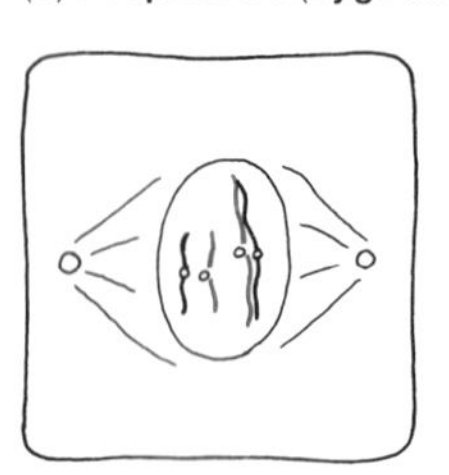

Nucleolus has disappeared. Homologous pairs of chromosomes (two chromosomes determining the same features) associate, forming a bivalent: a process known as synapsis.

(d) **Prophase I (pachytene)**

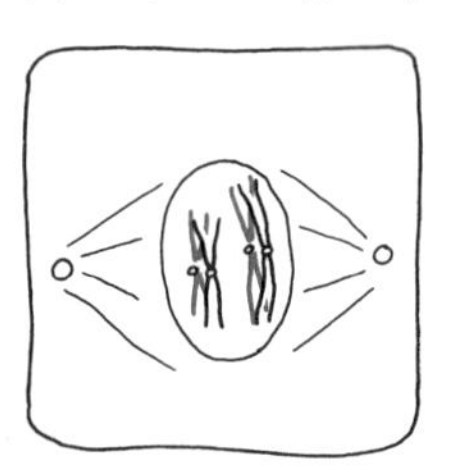

Chromatids become visible as they move apart from each other; they remain in contact at points called chiasmata.

(e) **Prophase I (diplotene)**

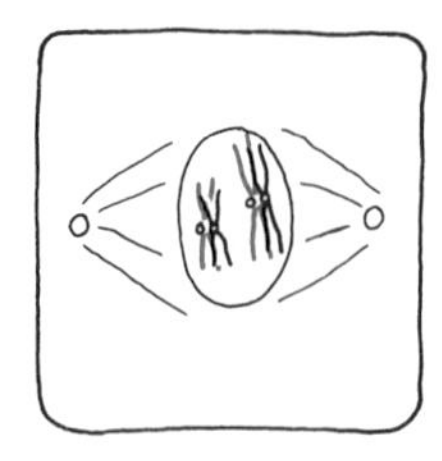

Chromatids continue to move apart as they shorten and thicken.

Diakinesis: Shortening and thickening continues; chiasmata move to ends; crossing over has occurred (see later); nuclear envelope breaks down.

(f) **Metaphase I**

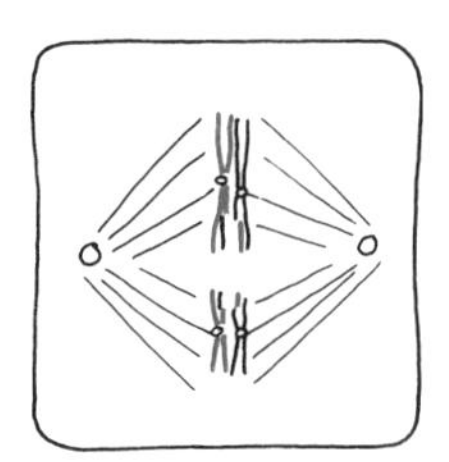

Homologous pairs of chromosomes align themselves on the equator of the spindle.

(g) **Anaphase I**

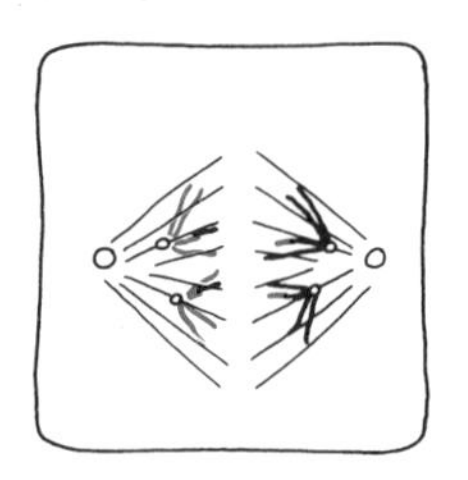

Homologous chromosomes move to opposite poles attached to spindle fibres by centromere (chromatids do *not* separate).

(h) **Telophase I**

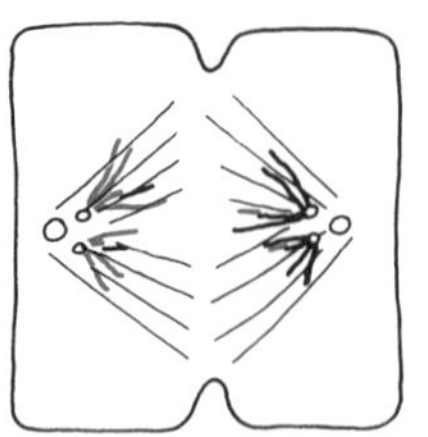

Chromosomes reach poles. Cell membrane invaginates.

There may be a short interphase or cells may move straight into the second meiotic division in which separation of the chromatids takes place.

(i) **Prophase II**

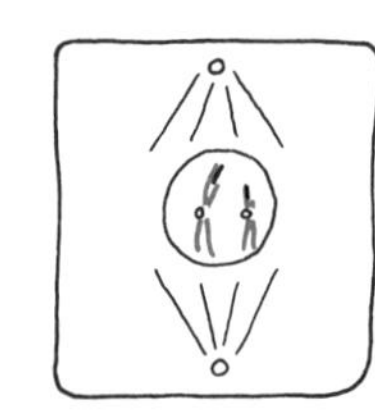
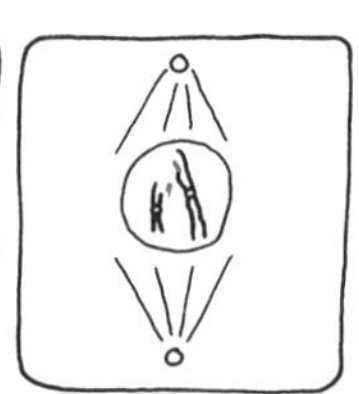

Spindles start to form, usually at right angles to one formed in meiosis I.

(j) **Metaphase II**

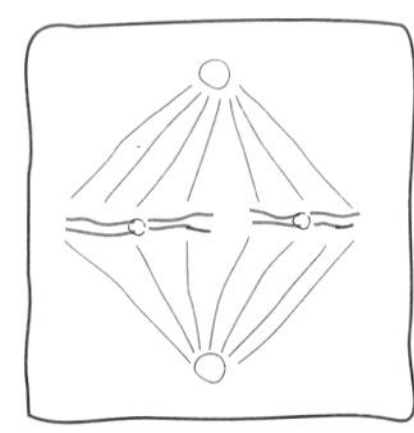
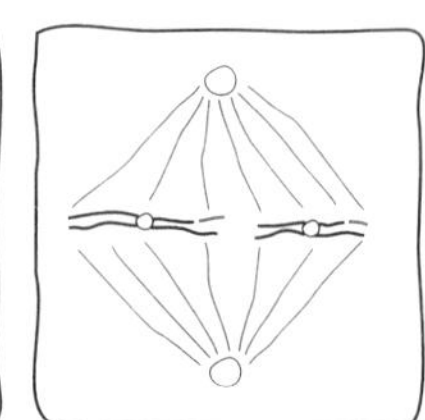

Chromosomes arrange themselves on the equator.

(k) **Anaphase II**

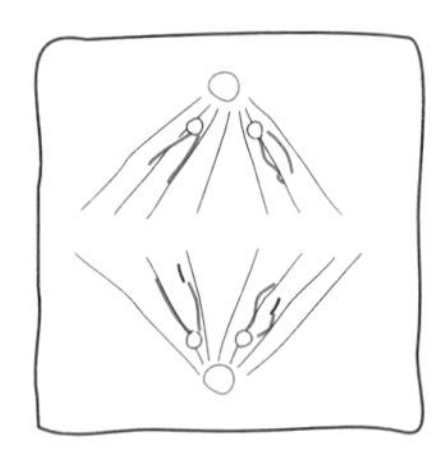
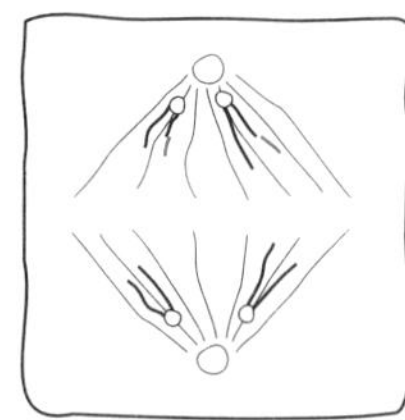

Chromatids pull apart and move to opposite poles.

(l) **Telophase II**

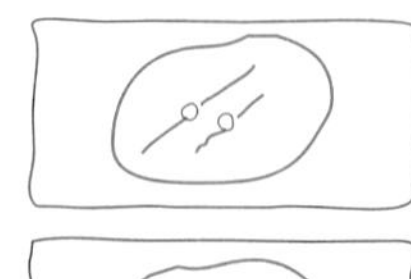
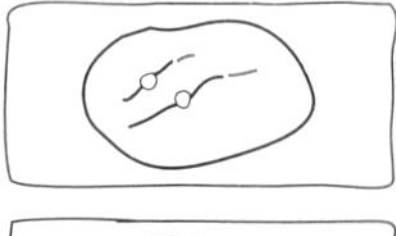
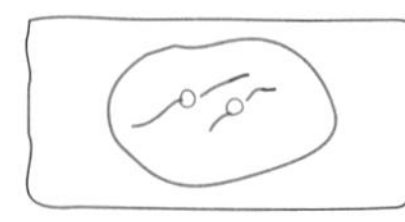
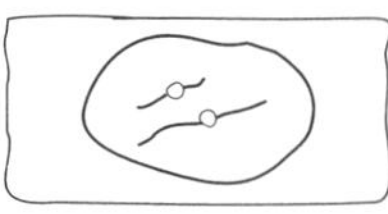

Cell membrane invaginates.
Nuclear envelope and nucleolus reform.
Two daughter cells formed, each with half the number of chromosomes present in original parent cell.

Fig. 3.14 Stages of meiosis

Details of crossing over

Chiasmata may form between any two of the four chromatids and there may be up to eight chiasmata in a bivalent.

Chiasmata have two functions:

1. to hold homologous chromosomes together while they move into position on the spindle prior to segregation;
2. crossing over (or exchange of genetic material) occurs at the chiasmata leading to increased variation, the raw material of evolution.

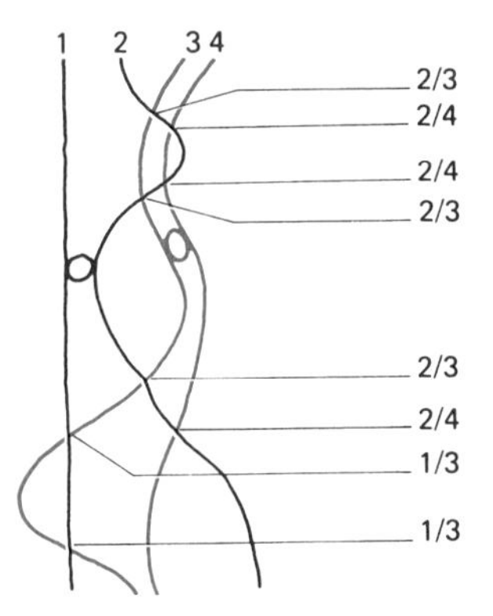

Fig. 3.15(a) Some possible chiasmata forming in a bivalent (the four chromatids are numbered for reference)

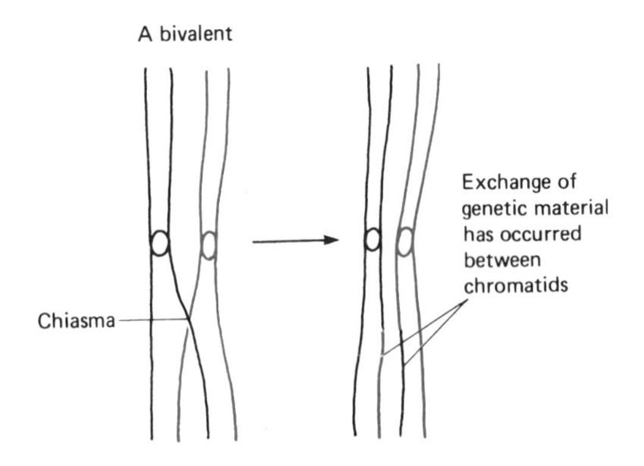

Fig. 3.15(b) Crossover showing exchange of genetic material

Significance of meiosis

1. Halving the chromosome number ensures that when gametes with the haploid number fuse to form a zygote the normal diploid number is restored.

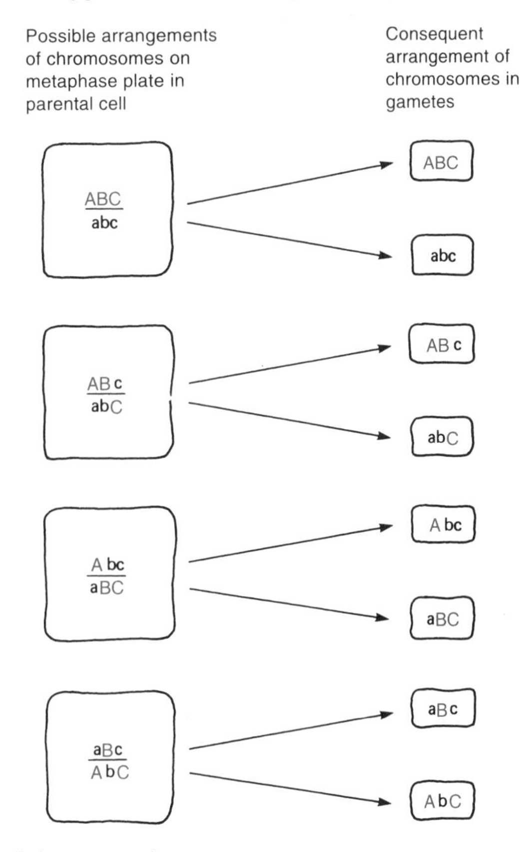

Fig. 3.16 Recombination of chromosomes in gametes

❷ Meiosis leads to increased variation because:

(a) when the haploid cells fuse at fertilisation there is recombination of parental genes;

(b) during metaphase I homologous chromosomes are together at the equator of the spindle, but they separate into daughter cells independently of each other;

(c) chiasmata and crossing over can separate and rearrange genes located on the same chromosome.

Table 3.3 *Differences between mitosis and meiosis*

Mitosis	Meiosis
One division of nucleus and one of chromosomes	Two divisions of nucleus and one of chromosomes
Chromosome number remains constant	Chromosome number halved
No association of homologous chromosomes	Homologous chromosomes associate in pairs
No chiasmata or crossing over	Chiasmata and crossing over occur
Two daughter cells formed	Four daughter cells formed (tetrad)
No variation (unless mutation occurs)	Variation due to exchange of genetic material
Chromosomes shorten and thicken	Chromosomes coil but remain longer than in mitosis
Chromosomes form a single line at the equator	Chromosomes form a double row at the equator
Chromatids move to opposite poles	Chromosomes move to opposite poles

3.3 HEREDITY AND GENETICS

You will be aware that organisms of the same species vary from each other within certain broad limits. Some characteristics in a population (e.g. height and weight) change gradually from one extreme to another with every conceivable intermediate represented. This is **continuous variation** and is a result of the environment acting on the organism's hereditary constitution. Other characteristics (e.g. eye colour and blood groups) fall into a few distinct groups with few, if any, representatives of intermediates. This is **discontinuous variation** and is the result of the organism's hereditary constitution alone (see 3.4).

In sexual reproduction a new individual results from the fusion of two gametes which between them contain all the necessary information for the development of a similar, but not identical, individual. If each gamete contributes information on every characteristic of the organism, it follows that at least two factors must control each characteristic in the offspring. These factors may or may not provide similar information on the character. If they do not, then either one or other factor manifests itself in the offspring, or an intermediate state between the two extremes becomes apparent. If a factor does not show itself in any one generation it nevertheless retains the capacity to do so in a later one.

Organisms are recognisably similar to their parents (**heredity**); however, they are not identical (**variation**).

Genotype, which is the genetic make-up of organisms, often determines limits, e.g. maximum height.

Phenotype is the actual appearance of an organism. It is caused by interaction between genotype and environment.

MENDELIAN INHERITANCE

The Austrian monk Gregor Mendel (1822–1884) observed clearly defined characters in the garden pea. Initially he studied only one pair of contrasting characters (**monohybrid inheritance**). He isolated plants that had 'pure bred' for several generations, artificially pollinated them and observed and counted the offspring (**first filial** or **F_1 generation**). He then crossed the F_1 plants with each other to give an F_2 generation. The characteristic apparent in the F_1 was termed **dominant** and the opposing character which disappeared in the F_1 and reappeared in the F_2 was termed **recessive**. Mendel knew nothing of meiosis, genes or chromosomes. He spoke of a pair of factors (now called **alleles**) determining a character. For example, the gene for height is composed of the allele for tall and the allele for short. It is usual to denote the dominant allele with a capital letter (upper case) and the recessive with a lower case letter.

Let T = tall; t = short

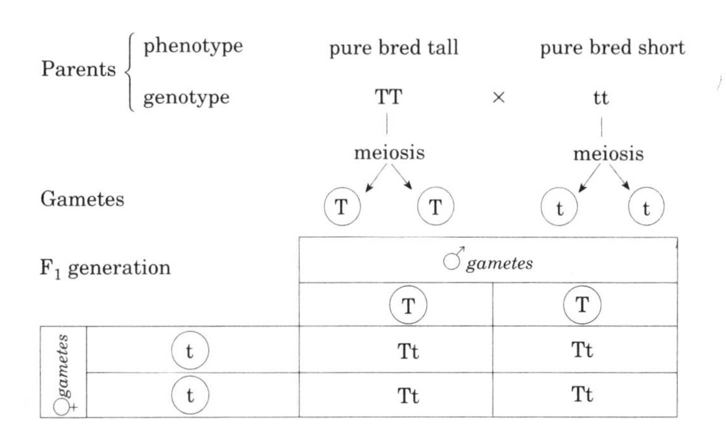

All the offspring have the genotype Tt and the phenotype tall. Now cross two F_1 plants, i.e. Tt x Tt.

F_2 generation

	♂ gametes: T	♂ gametes: t
♀ gametes: T	TT	Tt
♀ gametes: t	Tt	tt

TT = homozygous dominant, phenotype is tall

Tt = heterozygous, phenotype is tall

tt = homozygous recessive, phenotype is short

Therefore the ratio of phenotypes is 3 tall : 1 short

This 3 : 1 ratio will only be apparent if samples are large enough. From these crosses Mendel formulated the following law.

The Law of Segregation An organism's characteristics are determined by internal factors which occur in pairs. Only one of a pair can be represented in a single gamete of the organism.

Further instructions on how to represent genetic crosses are given in Table 3.4.

Table 3.4 *Representation of genetic crosses*

Instruction*	Reason/notes	Example– tall plant and short plant
Choose a single letter to represent each characteristic.	This is an easy form of shorthand. In some conventional genetic crosses, e.g. in *Drosophila*, there are set symbols, some of which use two letters.	
Choose the first letter of one of the contrasting features.	When more than one character is considered at one time, such a logical choice means it is easy to identify which letter refers to which character.	Choose either T or S.
If possible, choose a letter in which the higher and lower case forms differ in shape as well as size.	If the higher and lower case forms differ it is almost impossible to confuse them regardless of their size.	Choose T because the higher case form T differs in shape from the lower case form t whereas S and s differ only in size, and are more likely to be confused.
Let the higher case letter represent the dominant feature and the lower case letter the recessive one. Never use two different letters where one character is dominant. Always state clearly what feature each symbol represents.	The dominant and recessive features can easily be identified. Do *not* use two different letters as this indicates incomplete dominance or codominance.	Let T = tall t = short Do *not* use T for tall and S for short.
Represent the parents with the appropriate pairs of letters. Label them clearly as 'parents' and state their phenotypes.	This makes it clear to the reader what the symbols refer to.	Parents { Tall plant TT × Short plant tt
State the gametes produced by each parent. Label them clearly and encircle them. Indicate that meiosis has occurred.	This explains why the gametes only possess one of the two parental factors. Encircling them reinforces the idea that they are separate.	meiosis meiosis Gametes (T) (t)
Use a type of chequerboard or matrix, called a **Punnett square**, to show the results of the random crossing of the gametes. Label male and female gametes even though this may not affect the results.	This method is less liable to error than drawing lines between the gametes and the offspring. Labelling the sexes is a good habit to acquire – it has considerable relevance in certain types of crosses, e.g. sex-linked crosses.	♂ gametes: (T) (T) ♀ gametes (t): Tt Tt ♀ gametes (t): Tt Tt
State the phenotype of each different genotype and indicate the numbers of each type. Always put the higher case (dominant) letter first when writing out the genotype.	Always putting the dominant feature first can reduce errors in cases where it is not possible to avoid using symbols with higher and lower case letters of the same shape.	All offspring have the genotype Tt and the phenotype tall.

* **Note** Always carry out these instructions in their entirety. Once you have practised a number of crosses, it is all too easy to miss out stages or explanations. Not only does this lead to errors, but often makes your explanations impossible for others to follow. You may understand what you are doing, but if the reader cannot follow it, it isn't much use, and neither will it bring full credit in an examination.

TEST CROSS (BACK CROSS)

An organism with a dominant phenotype may be homozygous or heterozygous. In order to find the genotype of an organism it is crossed with a homozygous recessive individual of the same species. The following ratios are expected.

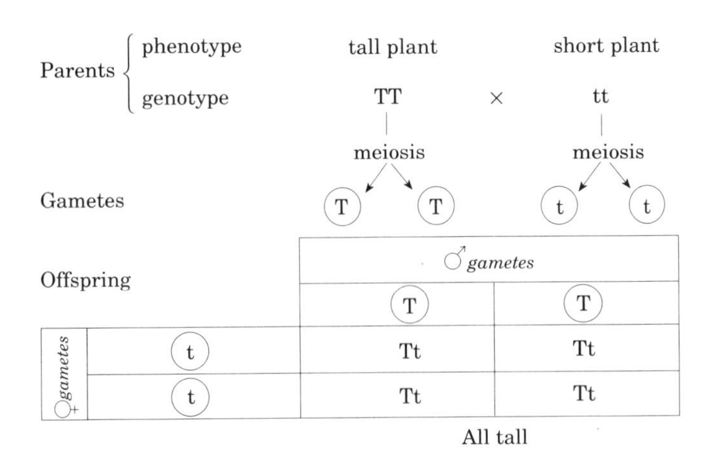

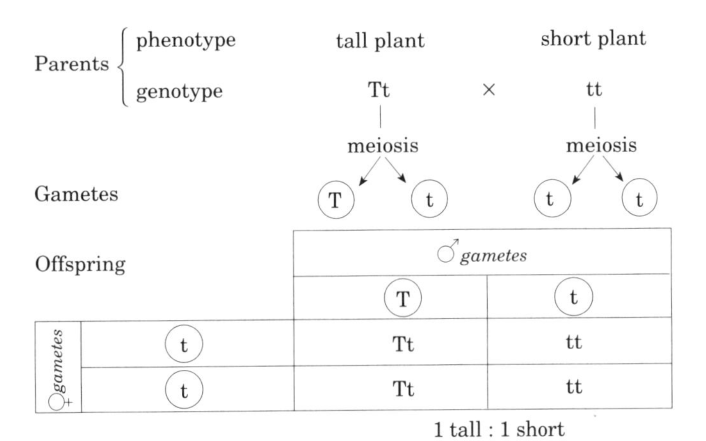

Dihybrid inheritance

This is inheritance of two pairs of characteristics.
Let T = tall; t = short; C = coloured; c = white

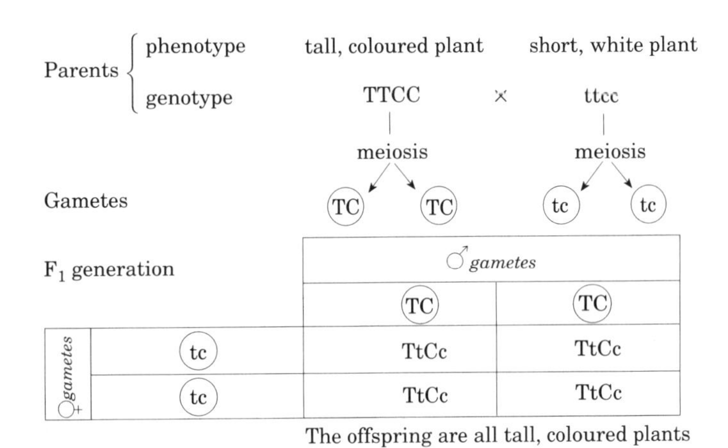

Then an F_1 **intercross** (i.e. a cross between two of the F_1 offspring).

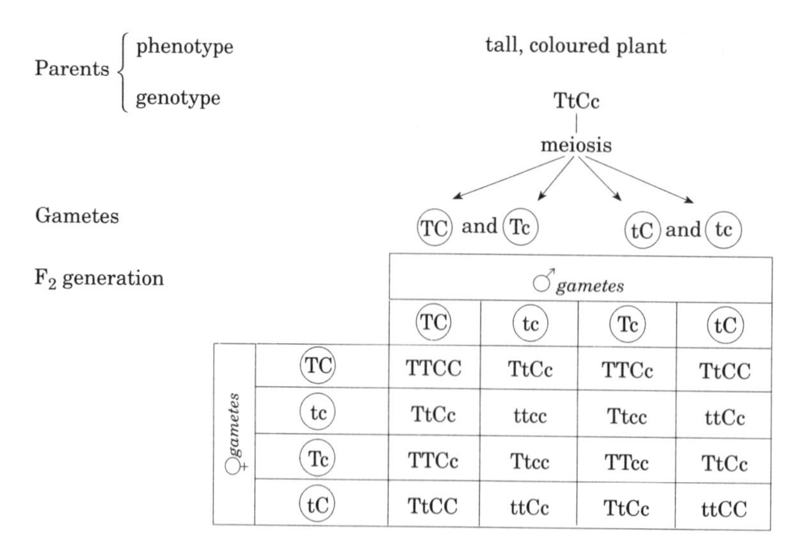

	TC	tc	Tc	tC
TC	TTCC	TtCc	TTCc	TtCC
tc	TtCc	ttcc	Ttcc	ttCc
Tc	TTCc	Ttcc	TTcc	TtCc
tC	TtCC	ttCc	TtCc	ttCC

Therefore in the F_2 generation there are 9 tall, coloured
3 tall, white
3 short, coloured
1 short, white

From these results Mendel formulated the following law.

The Law of Independent Assortment Each of a pair of contrasted characters may be combined with either of another pair.

In modern terms: each member of an allelic pair may combine randomly with either of another pair.

GENETICS OF SEX

In humans the female sex chromosomes are XX; therefore, all female gametes contain an X chromosome (**homogametic**). The male sex chromosomes are XY; therefore, the gametes may contain either an X or Y chromosome (**heterogametic**).

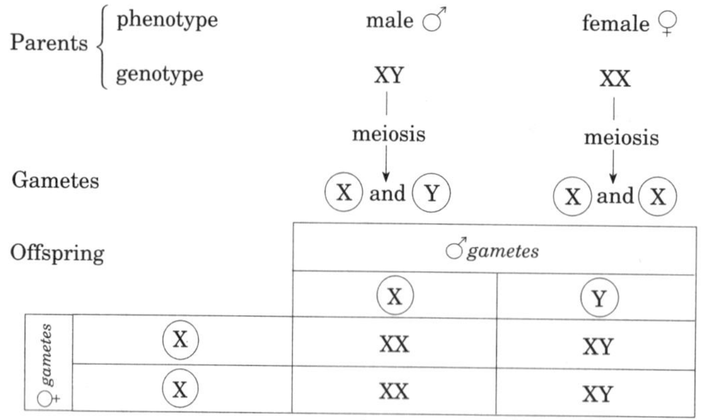

	X	Y
X	XX	XY
X	XX	XY

Sex ratio 1 female : 1 male

In birds the female is the heterogametic sex.

In *Drosophila* the female is XX and the male XY. However, unlike many species, the Y chromosome is not shorter than the X, but it is a different shape.

Sex linkage

Two common examples of recessive alleles linked to the X chromosome are red/green colour blindness and haemophilia.

Let X^h represent the X chromosome carrying the allele for haemophilia.

Let X^H represent the X chromosome carrying the normal allele.

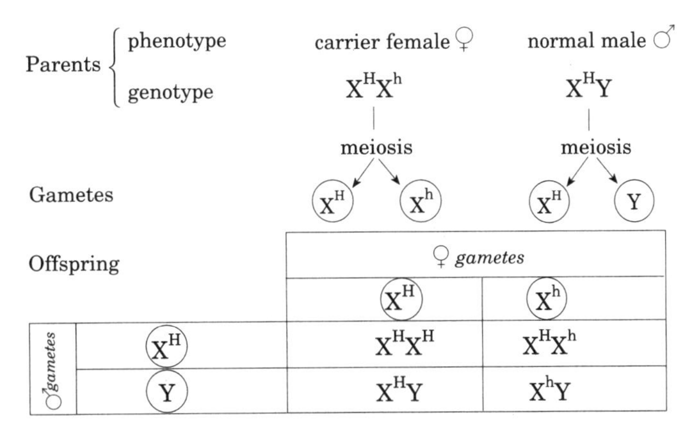

This cross gives rise to 1 carrier female (X^HX^h)
1 normal male (X^HY)
1 haemophiliac male (X^hY)
1 normal female X^HX^H

Autosomal linkage

It may have occurred to you that, with only 23 pairs of chromosomes to determine all the different characteristics in a human, each chromosome must possess numerous alleles. All alleles on the same chromosome are said to be linked; they move together from generation to generation. Linkage is the association of two or more alleles so that they tend to be passed from generation to generation as an inseparable unit and fail to show independent assortment.

Crossing over and recombination

In practice the situation is not so simple because chiasmata formation and crossing over between homologous pairs during the first prophase of meiosis (see 3.2) mean that linked alleles can be separated.

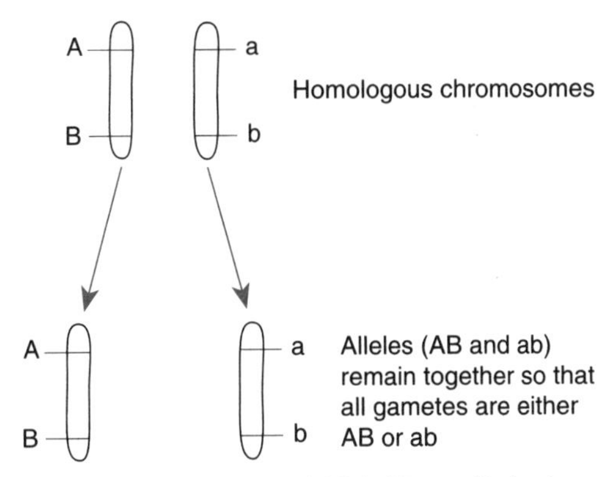

Gametes produced if A/B are linked

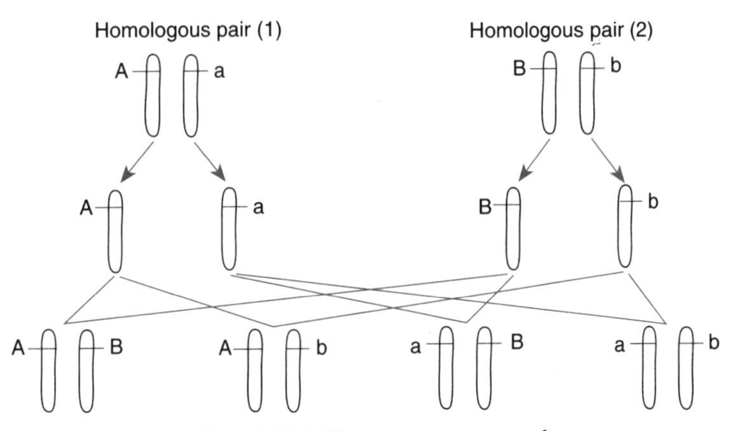

Gametes produced if A/B are on separate chromosomes

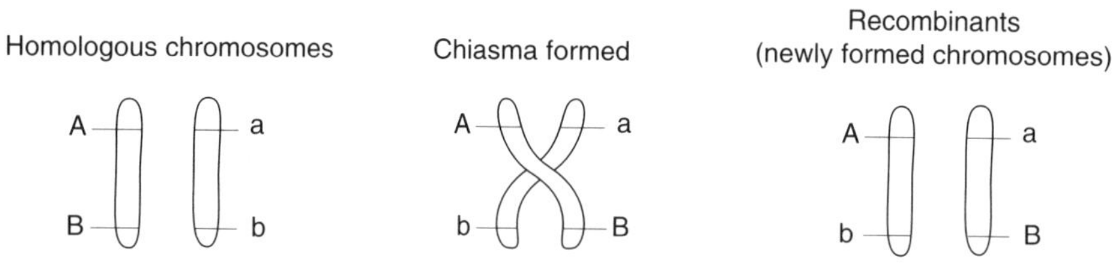

Homologous chromosomes — Chiasma formed — Recombinants (newly formed chromosomes)

A a B b — A a b B — A a b B

Recombination of linked alleles due to crossover

If crossing over always occurred in the above case then all the gametes would be recombinants, i.e. Ab and aB. None would be AB and ab. This is most unlikely. If crossover occurred in 50% of cases then half the gametes would be Ab and aB (recombinants), and the remaining half would be the original AB and ab. With crossover in 25% of cases only 25% would be recombinants, etc.

How can recombinants be detected?

The test cross is used. Take an organism of genotype AaBb. If crossed with the homozygous recessive aabb the following offspring could arise:

	Gametes of heterozygous parent			
	AB	Ab	aB	ab
Gametes of homozygous parent ab	AaBb	Aabb	aaBb	aabb
Number of offspring in typical sample	125	45	40	115
Phenotype	both dominant features visible	one dominant feature (A) visible	one dominant feature (B) visible	no dominant feature visible

The offspring can be determined phenotypically.

If the alleles for characters A and B were on separate homologous pairs the offspring ratio should be 1:1:1:1, but it clearly is not here. There are three possible explanations:

(i) A/a is not segregating in a 1:1 ratio as expected. To test this add up all the offspring arising from a gamete containing a and those from a gamete with A.
a gamete aB(40) + gamete ab(115) = 155
A gamete AB(125) + gamete Ab(45) = 170
As these figures are approximately equal, this is not the explanation.

(ii) B/b are not segregating in a 1:1 ratio as expected. To test this follow the above procedure.
b gamete Ab(45) + gamete ab(115) = 160
B gamete AB(125) + gamete aB(40) = 165
These figures are also approximately equal; therefore this is not the explanation.

(iii) Alleles are linked. To test this add the offspring from the homozygous gametes (AB and ab), i.e. 125 + 115 = 240, and compare with the offspring from the heterozygous gametes (Ab and aB), i.e. 45 + 40 = 85. These are clearly not in the ratio of 1:1; therefore the alleles are linked and Ab and aB are the recombinants.

CODOMINANCE

When homozygous red and white antirrhinums are crossed the F_1 does not produce the expected result of all the dominant type. Instead all the F_1 are pink. When these are selfed the F_2 has 2 pink, 1 red and 1 white. This is called **codominance**.

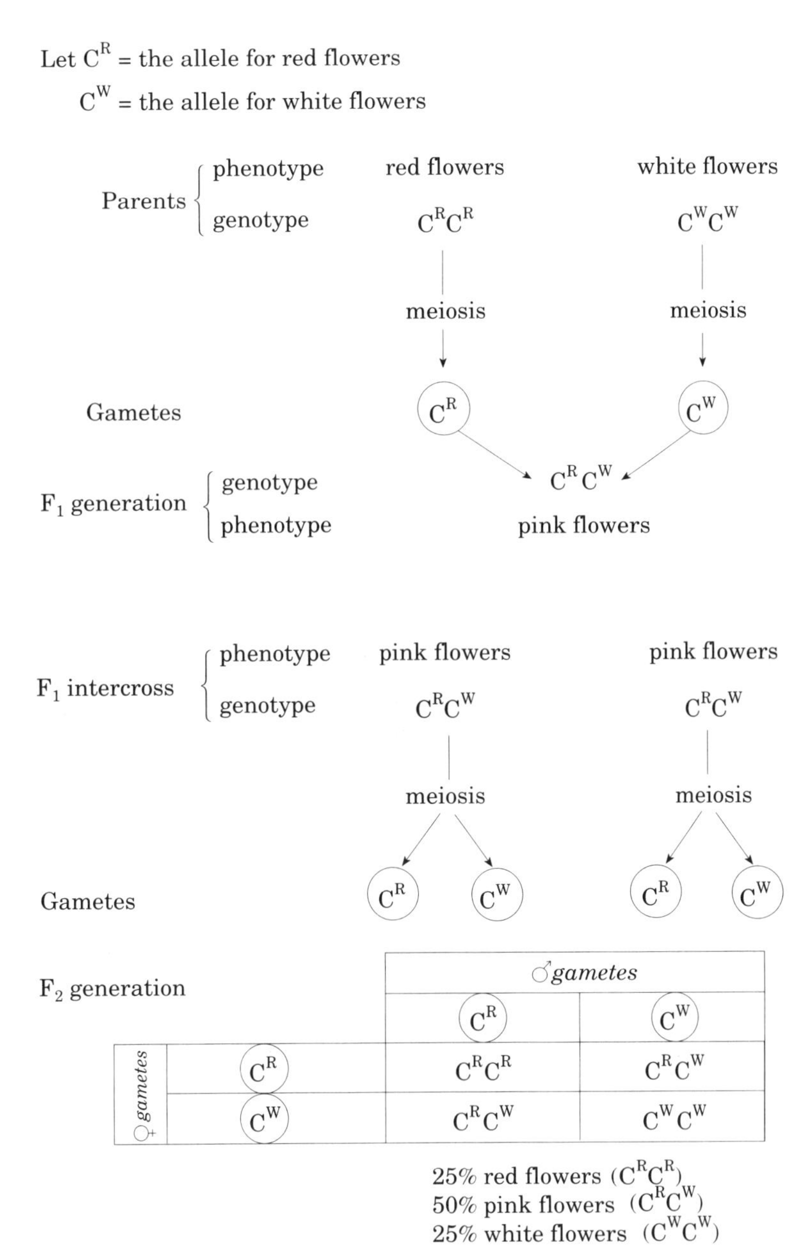

MULTIPLE ALLELES

Sometimes more than two alleles control a particular characteristic. Only two alleles can occupy a locus on a pair of homologous chromosomes at any time. The ABO blood group system is controlled by three alleles:

allele I^A = production of antigen A on erythrocyte
allele I^B = production of antigen B on erythrocyte
allele I^O = production of no antigens

Therefore the possible combinations are I^AI^A; I^AI^O; I^AI^B; I^BI^B; I^BI^O; I^OI^O. The alleles I^A and I^B show equal dominance but both are dominant to I^O.

Blood group	Possible genotype
A	I^AI^A or I^AI^O
B	I^BI^B or I^BI^O
AB	I^AI^B
O	I^OI^O

Transmission of the alleles is in the normal Mendelian fashion.

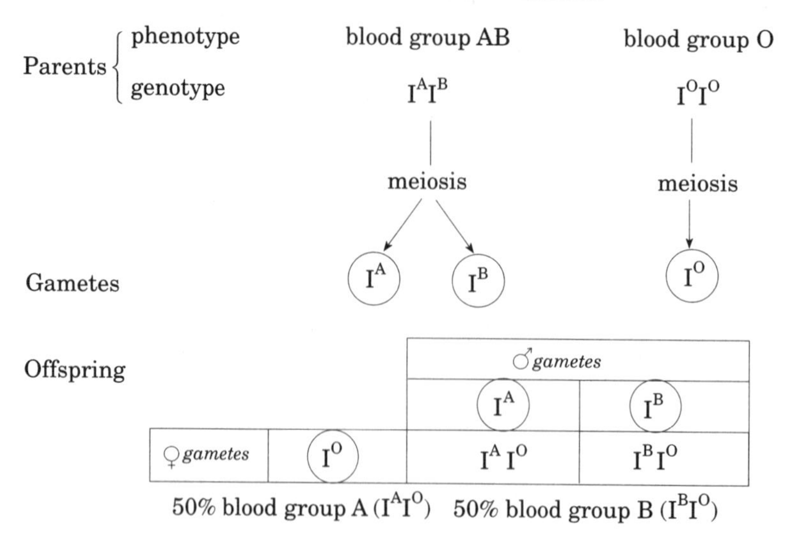

3.4 GENETIC VARIATION AND EVOLUTION

There are two types of variation.

1. **Continuous variation** If you were to measure the heights of a group of humans and plot the number of people at each height you would most likely end up with a graph similar to that shown in Fig. 3.17. This type of graph is referred to as a normal distribution curve. Characteristics which display continuous variation are controlled by many genes and the character is therefore called a **polygenic character**.
2. **Discontinuous variation** Some features of organisms in a population do not display a gradual change from one extreme to the other but rather belong to distinct forms. An example is the human ABO blood group system which has four discrete types: A, B, AB and O. Features exhibiting discontinuous variation are determined by a single gene.

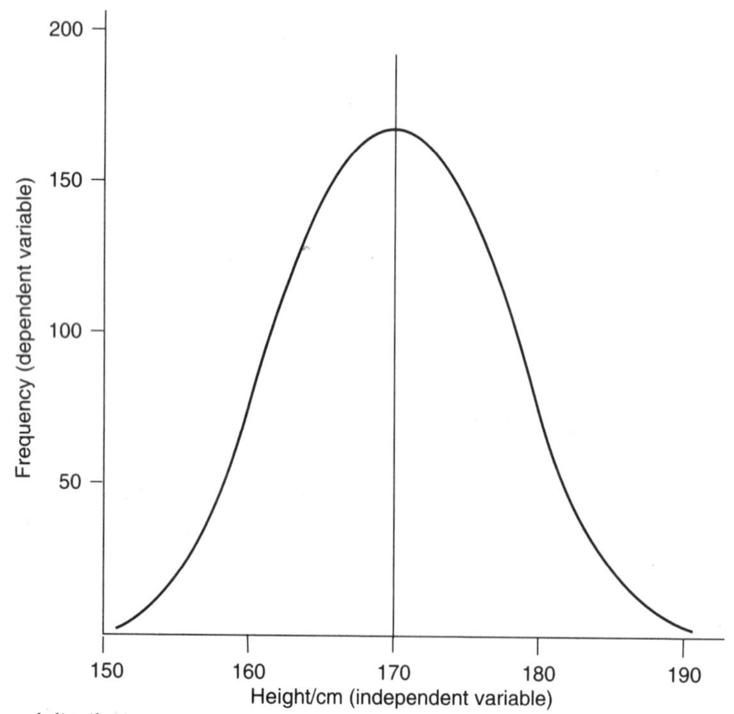

Fig. 3.17 Normal distribution curve

It is obvious as you look around you that not only is there considerable variety between different species of organisms, but there is also much variation between individuals of a single species. How then does such variety arise?

THE INFLUENCE OF THE ENVIRONMENT

Even if two individuals are genetically identical, they may display differences as a result of the environmental conditions to which they are exposed. Two genetically identical plants which are separated, with one grown in the light and the other in the dark, will quickly acquire a very different appearance.

GENETIC EFFECTS

Gene recombination

Sexual reproduction produces variety in three ways:

1. Where cross-fertilisation occurs the genes from two different parents are mixed.
2. During metaphase I of meiosis the random distribution of chromosomes on the equator and subsequent segregation lead to further mixing of genes.
3. Crossing over between homologous chromosomes during prophase I of meiosis recombines linked genes.

Mutations

A mutation is a change in the structure or amount of DNA in an organism. The rate of mutation for a single gene is constant, but all mutate very rarely. This is perhaps fortunate as most mutations are harmful, often lethal. Nevertheless they are the raw material upon which evolution thrives. Only mutations which arise in the formation of gametes are inherited.

Gene (point) mutation

These arise as the result of an alteration to the amount of DNA at a single locus. Any change to one or more nucleotides on the DNA results in a different sequence of bases. As you saw in 3.1, this sequence determines the order of amino acids in the polypeptide the DNA codes for. The resulting protein may therefore have a different composition and structure. As the function of many proteins (e.g. enzymes) depends crucially on their precise shape and composition, such an alteration may have major, even fatal, consequences for the organism.

An example of a gene mutation is **sickle cell anaemia**. The DNA molecule which codes for one of the polypeptide chains of the haemoglobin molecule has an alteration to a base on a single nucleotide. This leads to a sequence of events which results in abnormal red blood cells. The events are summarised in Fig. 3.18(a).

There are a number of forms of gene mutations:

1. **Substitution** One or more nucleotides are replaced with others possessing different bases (e.g. sickle cell example in Fig. 3.18(a)).
2. **Addition** One or more additional nucleotides become incorporated into the DNA molecule.
3. **Deletion** One or more nucleotides are lost from the DNA molecule.
4. **Duplication** A portion of the nucleotide sequence is repeated within the DNA molecule.
5. **Inversion** A portion of the nucleotide sequence is reversed within the DNA molecule.

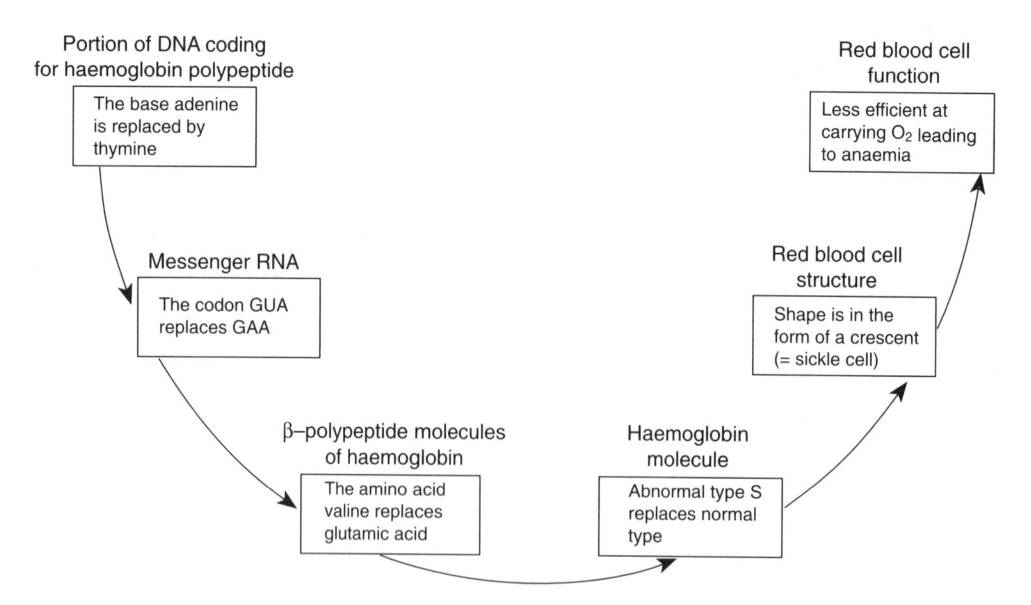

Fig. 3.18(a) A gene mutation – the sequence of events which leads to the development of sickle cell anaemia

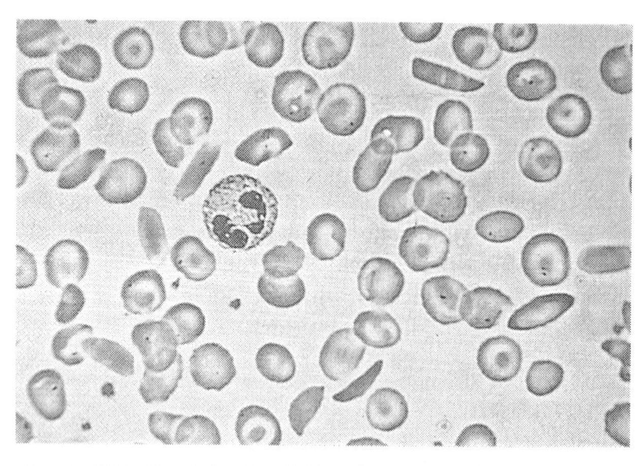

Fig. 3.18(b) Blood showing sickle cells

Apart from sickle cell anaemia, other examples of gene mutations include cystic fibrosis and Huntington's chorea. You can find out more about these diseases from textbooks or the appropriate leaflets obtainable from health education centres.

Chromosome mutations

Numerical changes Occasionally chromosomes fail to segregate during anaphase of meiosis resulting in a gamete having the diploid rather than the haploid number of chromosomes, i.e. in humans this would mean that the sperm and/or egg would have 46 rather than 23 chromosomes. When this diploid gamete fuses with a normal haploid one the resultant zygote will have three sets of chromosomes, i.e. it is triploid (3n). It is possible to obtain organisms with four or more sets of chromosomes. Any condition which results in complete sets of chromosomes being duplicated is referred to as **polyploidy**. Polyploids may be sterile – in the case of a triploid, for example, there will always be an odd number of chromosomes and these cannot divide equally when the cell divides meiotically. Where polyploidy occurs within the same species it is known as **autopolyploidy**. When it arises as a result of the combination of chromosomes of two individuals from different species it is known as **allopolyploidy**. Many commercial varieties of plants are polyploids, e.g. wheat, banana and tomatoes.

It may be a single chromosome rather than a whole set which fails to segregate during anaphase of meiosis, a condition called **nondisjunction**. One of the resultant gametes therefore has an additional chromosome, the other lacks one. For example, in humans one gamete would contain 24 chromosomes and the other 22, rather than the normal 23 each. On fusion with a normal gamete the resultant zygote will either possess one less than the diploid number (2n-1, i.e. 45 chromosomes in humans) or one extra (2n+1, i.e. 47 chromosomes in humans).

An example of **nondisjunction** in humans is **Down's syndrome** where the 21st chromosome fails to segregate and those zygotes with the additional chromosome develop. The resultant offspring has 47 chromosomes and survives with varing degrees of disability and with a shorter life expectancy. Other examples which you might like to research in textbooks are Klinefelter's and Turner's syndromes, both of which affect the sex chromosomes.

Structural changes We saw in 3.2 that during prophase I of meiosis chromosomes may break and exchange portions with their homologous partner (crossing over). Occasionally this process leads to an alteration in the sequence of genes on the chromosome. The various types of structural changes are illustrated in Fig. 3.19.

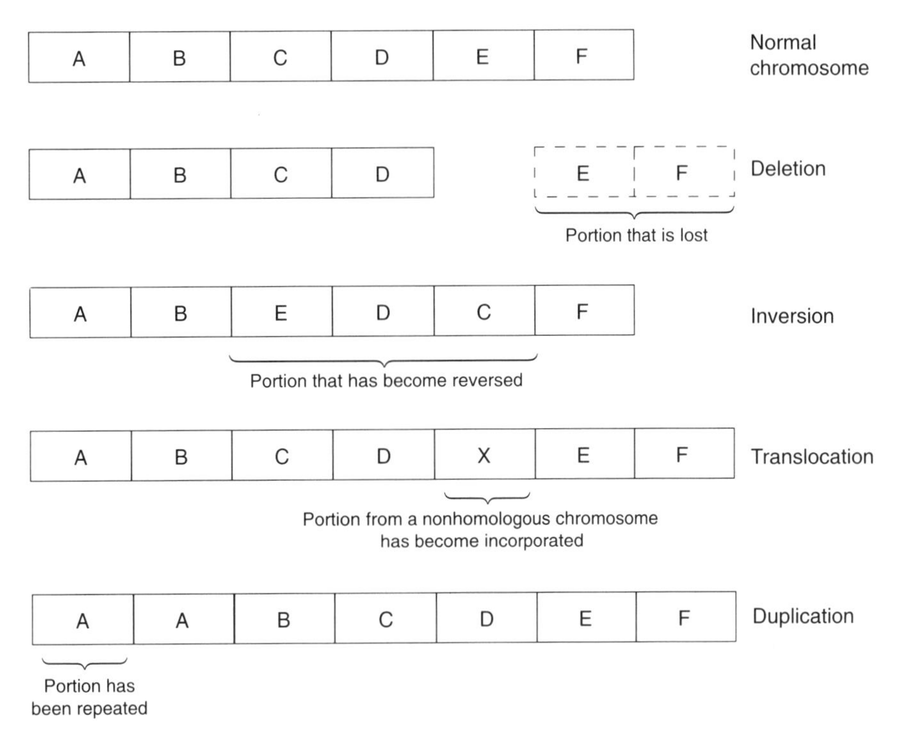

Fig. 3.19 Examples of chromosome mutation

Causes of mutations

All genes appear to mutate naturally at a constant rate – in animals typically one or two mutations for each 100 000 genes in each generation. Certain agents increase this natural rate: high-energy radiation (X-rays, ultraviolet light, α and β particles and neutrons) and certain chemicals (formaldehyde, mustard gas). Agents which increase mutation rates are called **mutagens**.

Genetic counselling and screening

Genetic counselling involves researching the family history of inherited disease in order to advise parents of the probability of their children suffering an inherited disease. On the basis of this advice, the parents can choose whether or not to have children. One disease which is the subject of genetic counselling is cystic fibrosis.

Genetic screening can be used to determine whether a fetus possesses a genetic defect. In a process called **amniocentesis,** a little of the amniotic fluid is removed using a hypodermic syringe. This fluid contains fetal skin cells whose chromosomes can be examined for such genetic abnormalities as Down's Syndrome.

EVOLUTION

Evolution is the process by which new species arise as the result of changes to existing species over a period of time.

Evolution through natural selection (Charles Darwin and Alfred Wallace)

Darwin became the naturalist on HMS Beagle which sailed in 1832 to South America and Australasia. During the voyage he was influenced by his knowledge of fossils and noted their similarity to present-day forms. He also noticed the differences in South American animals and plants, which could be related to differences in their environment. In the Galapagos Islands he observed that, while the finches there were unique, they had a general resemblance to finches which existed on the mainland of South America. He considered that originally a few finches had strayed from the mainland to these islands and, as they bred, they produced new types with differences that allowed them to fare better in the new conditions. After the voyage he spent over 20 years developing his views on evolution.

It is problematical whether he would have published his findings had not his theory been anticipated by Alfred Wallace. Wallace sent his theory of how evolution might have come about to Darwin, who found it was, in essence, the same as his own. Darwin was advised to read a joint paper on the subject to the Linnaean Society. This he did in 1858 and the following year published his book *On the Origin of Species by Means of Natural Selection and the Preservation of Favoured Races in the Struggle for Life.*

The Darwinian Theory of Natural Selection

Darwin's theory is concerned with three observable facts plus two deductions.

Over-reproduction Malthus attempted to show that all organisms tend to increase in a geometric ratio. Evidence for the enormous reproductive capacity includes:

- The very slow-breeding elephant: if it were to produce six young in a lifetime, and if their descendants continued to breed at this rate, in 750 years there would be 19 million elephants.
- Many fish are capable of laying millions of eggs, e.g. female cod lays about 2–3 million eggs per year.
- Flowering plants generally produce many hundreds of seeds.

Relative constancy of the numbers of species Darwin's second fact was that despite this tendency towards a geometric increase in numbers, the numbers of a given species tended to remain constant. Hence, many plants and animals must be destroyed at some stage of their lives.

Struggle for existence From these two facts, Darwin arrived at the first deduction. It is obvious that if the parents produce vast numbers of offspring or over many years are capable of doing so, and yet the number of a given species remains constant over many years, then there must be some sort of struggle between the organisms.

Variation among the offspring The third observable fact was that among the offspring of any two parents (animals and plants), there are usually slight and perhaps hardly noticeable differences. Generally, no two offspring are identical.

Natural selection/survival of the fittest This second deduction arises from the third fact. If a variation confers upon the individual some greater ability to withstand the hazards and competition of the struggle for existence, then that organism will stand a better chance of survival, and so live to breed. Darwin anticipated inheritance by noting that **like produces like**, i.e. that the advantageous variation will be inherited, resulting in more offspring with this variation.

It is important to qualify this: the inheritance of one small variation will not by itself produce a new species. However, the production of variations in a particular direction over many generations, and their inheritance, will gradually lead to the evolution of a new species.

The process of natural selection

We have seen that natural selection is the mechanism by which organisms which are better adapted to their environment are more likely to survive and so more likely to pass on their characteristics to succeeding generations. Every organism is therefore subjected to selection according to the environmental conditions that exist at the time. The environment in effect exerts a **selection pressure**.

There are three types of selection.

1. **Directional selection** As environmental conditions gradually change, so there is a selection pressure which results in those organisms best suited to the new conditions surviving and breeding. The species therefore shows slow adaptive change (see Fig. 3.20 (a)).
2. **Stabilising selection** Where there is little environmental change there is a selection pressure which favours the survival of those individuals which do not show adaptations to extreme conditions, i.e. those in the middle of the range of types. This type of selection tends to make members of a species more similar by eliminating the extreme types (see Fig. 3.20 (b)).
3. **Disruptive selection** If environmental conditions take a number of distinct forms (e.g. hot summers and cold winters), then selection may favour the formation of two discrete forms of a species, each adapted to one set of conditions. In our example one form might adapt to hot summers and hibernate in winter, the other might adapt to the cold winter, aestivating during the hotter summer (see Fig. 3.20 (c)). Where interbreeding continues but there is more than one form of the species, the condition is known as **polymorphism**. One example of polymorphism, that of the peppered moth (*Biston betularia*), is described on page 118.

In a population of a particular mammal fur length shows continuous variation.

1 When the average environmental temperature is 15°C, the optimum fur length is 2.0 cm. This then represents the mean fur length of the population.

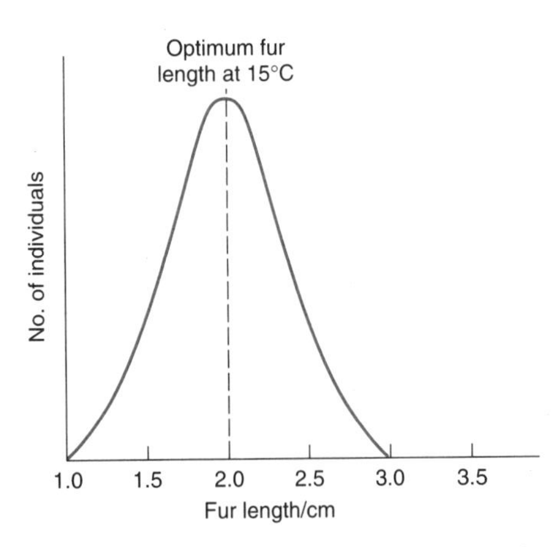

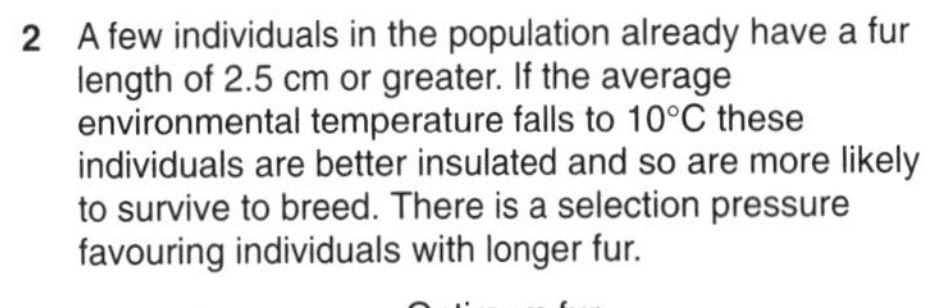
2 A few individuals in the population already have a fur length of 2.5 cm or greater. If the average environmental temperature falls to 10°C these individuals are better insulated and so are more likely to survive to breed. There is a selection pressure favouring individuals with longer fur.

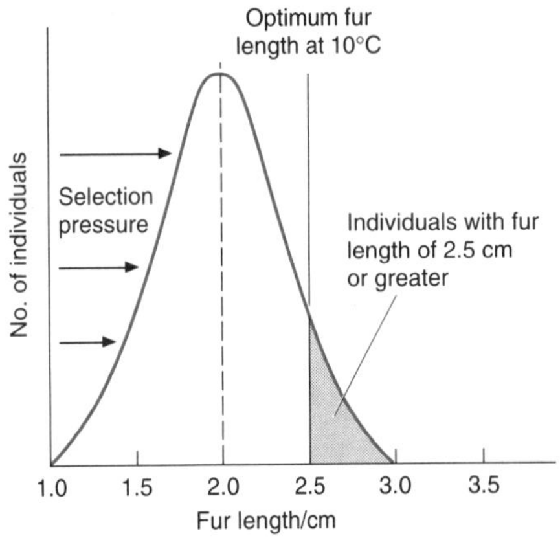

3 The selection pressure causes a shift in the mean fur length towards longer fur over a number of generations. The selection pressure continues.

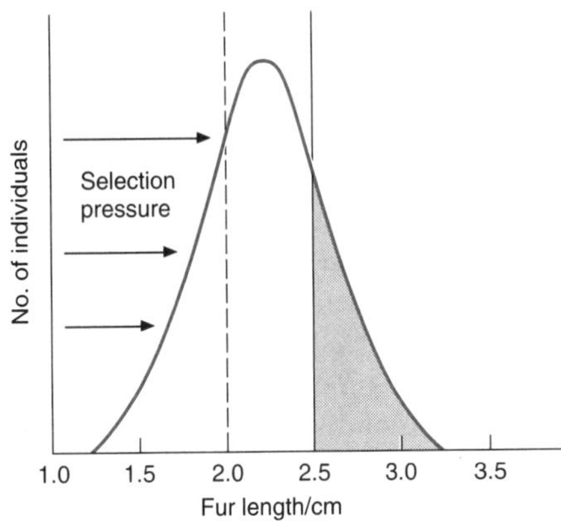

4 Over further generations the shift in the mean fur length continues until it reaches 2.5 cm – the optimum length for the prevailing average environmental temperature of 10°C. The selection pressure now ceases.

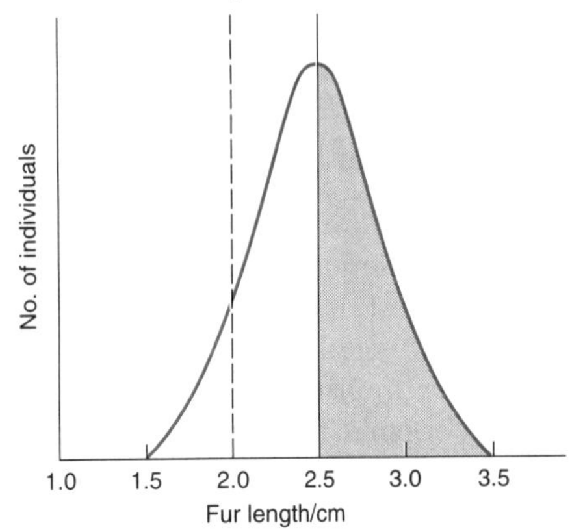

Fig. 3.20 (a) Directional selection

1 Initially there is a wide range of fur length about the mean of 2.0 cm. The fur lengths of individuals in the shaded areas are maintained by rapid breeding in years when the average temperature is much warmer or colder than normal.

2 When the average environmental temperature is consistently around 15°C with little variation, individuals with very long or very short hair are eliminated from the population over a number of generations.

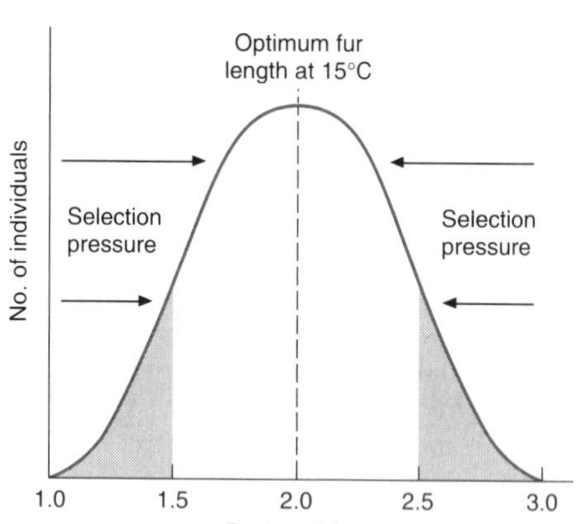

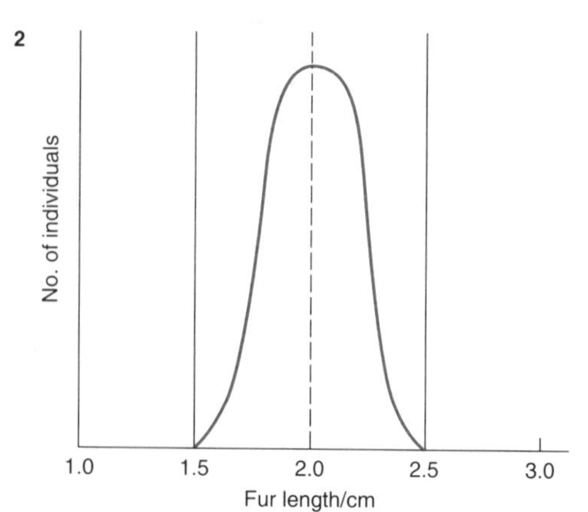

Fig. 3.20 (b) Stabilizing selection

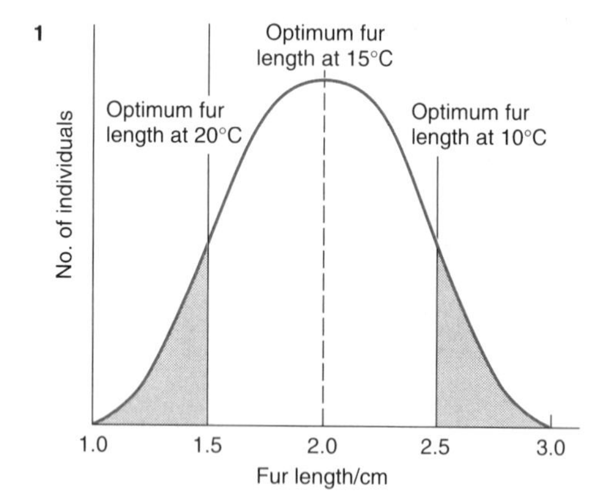

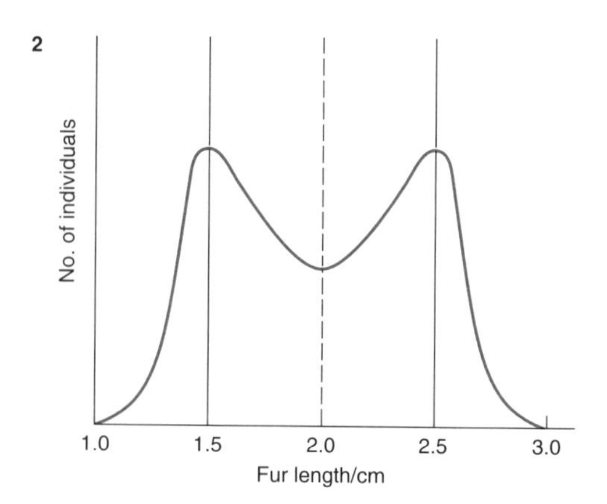

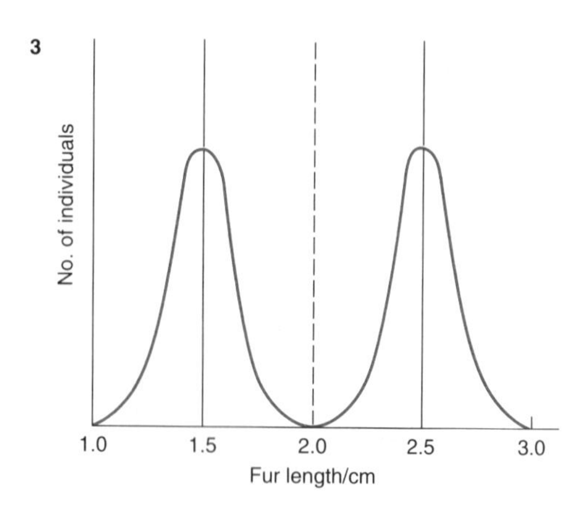

1 When there is a wide range of temperatures throughout the year, there is continuous variation in fur length around a mean of 2.0 cm.

2 Where the summer temperature is static around 20°C and the winter temperature is static around 10°C, individuals with two distinct fur lengths predominate: 1.5 cm types which are active in summer and 2.5 cm types which are active in winter.

3 After many generations two distinct sub-populations are formed.

Fig. 3.20 (c) Disruptive selection

Examples of selection pressures

Industrial melanism Industrial melanism in moths is an example of genetic selection induced by pollution. It was recognised in soot-polluted Manchester in 1850, when the black, melanic form of the peppered moth, *Biston betularia*, was first seen. By 1900, it had almost replaced the typical, mottled form. The increase from 1% to 99% took 50 generations (the moth has an annual life cycle). During the second half of the nineteenth century, the rest of Britain began to receive the melanic (black) form as a result of dispersal from Manchester and natural selection in other cities. The most substantial selective pressure for melanic forms is visual predation by moth-eating birds, who remove the moths from trees and other vertical surfaces when they rest there during daylight. Where surfaces are blackened by soot the melanic form is better camouflaged and predation is largely restricted to the mottled form. In areas not polluted by soot the reverse is true and the mottled form predominates. There are more then 150 species of moths known to exhibit industrial melanism.

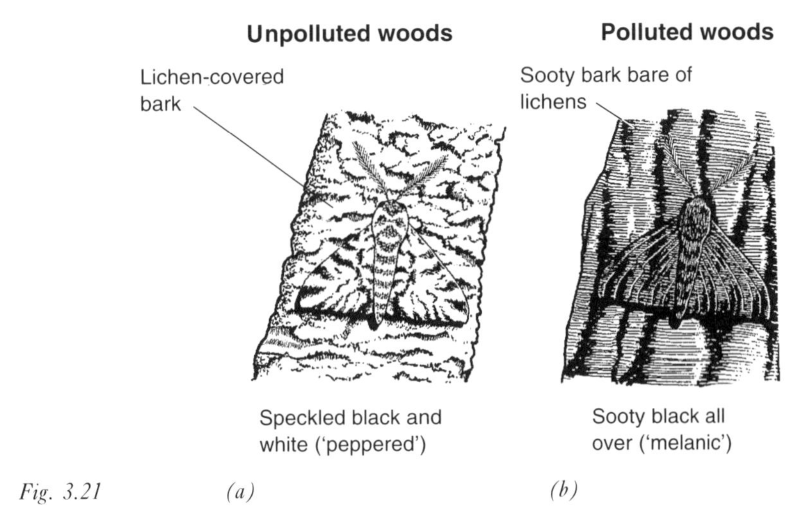

Fig. 3.21

Metal tolerance in plants At least 21 species of plants have been recognised as having metal-tolerant strains. Species of *Festuca* and *Agrostis* have become genetically adapted to high concentrations of zinc and lead and are tolerant on sites with a high concentration of the metals, but may not survive where the concentration is low. Where several metals are present, such as nickel and copper, plants may be tolerant to both but tolerance to one metal is genetically independent of another.

Insecticides The widespread use of insecticides has provided a selection pressure which has resulted in adaptive changes within a very short time. It has taken only a few years, and in some cases only months, to produce strains of insects which have become completely resistant to many insecticides. At least 225 resistant species of insects have been recognised. Many of these show resistance to DDT, dieldrin, aldrin, lindane, malathion and fenthion. The housefly, *Musca domestica*, has developed strains in some parts of the world which are resistant to every insecticide which can be 'safely' used, with the exception of pyrethrum compounds. In some types of resistance, enzyme induction has been responsible for the ability of the insect to break down the insecticides, i.e. mutant strains 'switch on' genes which help control the synthesis of oxidoreductase enzymes to render the insecticides harmless. Many species of insects which are vectors of disease have become resistant in this way, among them the yellow fever mosquito *Aedes aegypti* and the malaria-carrying mosquitoes *Anopheles gambiae* and *A. culicifacies*.

Antibiotics The resistance of bacteria to antibiotics has arisen by the selection of resistant mutants. As with types of insecticide resistance, some resistance to antibiotics has evolved by 'switching on' genes to control the production of antibiotic-splitting enzymes. All naturally occurring penicillin-resistant strains of *Staphylococcus*, as well as penicillin-resistant strains of many other species, owe their resistance to their ability to produce penicillin B lactamase. Resistance to antibiotics has already destroyed the usefulness of several drugs. The first signs of staphylococcal resistance appeared soon after penicillin became extensively used in hospitals.

Table 3.5 *Incidence of penicillin-resistant infection at a general hospital*

Date	Total patients	Patients with penicillin-resistant strains
Apr–Nov 1946	99	15
Feb–June 1947	100	38
Feb–June 1948	100	59

By 1950, the majority of staphylococcal infections in all British general hospitals were penicillin resistant. Resistance has now appeared in certain staphylococci to all major antibiotics, often as a triple resistance to penicillin, tetracycline and streptomycin.

Adaptive immunity in mammals

Microorganisms and insects reproduce at such an alarming rate that it is relatively easy to visualise how they can adapt faster than man can combat them. Prolifically breeding mammals such as the brown rat and the rabbit have also evolved resistant strains capable of withstanding warfarin and myxomatosis virus, respectively.

Warfarin was developed as a rat poison. It contains di-coumarol which interferes with normal clotting mechanisms in the blood. In 1960, strains were recognised in Shropshire which were immune to warfarin. The incidence of resistant rats in parts of Wales had risen to 50% by 1970 because of the intensive selective pressure of the widespread use of warfarin. Their resistance is due to a single mutant gene and individuals homozygous for this gene are weaker than normal warfarin-sensitive rats.

Myxomatosis resistance in rabbits became widespread in the mid-1950s. Prior to this, the virus was responsible for up to 90% mortality of rabbits in certain areas. The genetic response to this selective pressure was the production of mutants which spent a greater proportion of time above the ground, like hares. Normal rabbits live in crowded warrens underground where the vector of the virus, the rabbit flea, can spread rapidly throughout a population. Those that spent most time above ground were favoured, whereas previously they were selected against because of predation. Even though the rabbits were not physiologically resistant, their altered behaviour offered them protection. In the Australian population, however, a genetically resistant strain emerged after the initial epidemics. Most British rabbits are now genetically resistant to myxomatosis.

Sickle cell anaemia

This is caused by a mutation which results in the incorporation of an incorrect amino acid at one point in the protein chains of the haemoglobin molecule. The mutated gene is known as Haemoglobin S or HbS and is recessive. Thus only those persons having the sickling gene from both parents (i.e. homozygous for the gene) suffer acutely from the disease. The disease is an often fatal form of anaemia which is relatively common in West Africa. Whenever the blood cells of a victim encounter a low level of oxygen, as in the venous blood of tissues, they are liable to collapse to a sickle shape and may form blockages and other complications in blood vessels. Sufferers have only a 20% chance of surviving to maturity as compared with nonsufferers. Heterozygous individuals also suffer sickling when the oxygen tension falls below a critical level. Given this information, we would not expect the mutation to be favoured. Indeed, such harmful mutations are frequently eliminated. This happens rapidly where it is a dominant gene, but more slowly where it is a recessive, as selection acts most often on a mutation when it appears in the phenotype. New HbS genes arise spontaneously, so they will always be present in a population, but a percentage of them should be eliminated at each generation.

However, observations do not bear out this hypothesis. There are large areas of Africa and Asia where the gene occurs, usually in a single dose, in 15–20% of the population. There are even communities within these regions with frequencies of 40%. A heterozygote can be recognised because the blood cells show some sickling when the person is artificially exposed to low oxygen pressure in the laboratory. Clearly, the gene is not being removed from the population. We can only conclude that in some way HbS confers an advantage over the normal Hb gene, enough to redress the losses of genes at each premature death from anaemia. The answer lies in resistance to malaria. Children with HbS have been shown to have a 25% better chance of surviving malarial attacks than those with normal genes. Selection is not acting to remove the genes causing sickle cell anaemia. They are removed when homozygous but this selection is outweighed by their selective advantage when heterozygous, due to better resistance on the part of heterozygotes to malaria.

Chapter roundup

When you consider that all the genetic information needed to produce the whole of mankind (more than 5 000 000 000 individuals) was contained within the 5 000 000 000 sperm and ova from which they were derived, and that these together would occupy a five litre vessel, it is obvious that the genetic code is an extremely complex and highly condensed information source.

We have seen in this chapter how exactly this genetic information is stored, how it is translated into cells and organisms, and how it is passed to succeeding generations. We have also seen how a fundamental problem for living organisms is overcome – namely how to maintain genetic stability without sacrificing the necessity for genetic change. It follows that if an organism has survived to the point where it can reproduce, it must be successful. It is logical, therefore, that the offspring should be as near identical as possible to their parents if they too are to survive. At the same time the environment is constantly changing and organisms must adapt to meet these changes. Offspring therefore need to be different from their parents in order that there is sufficient variety amongst individuals of a species to ensure that some, at least, will possess the necessary characteristics to suit the changed environment. The remarkable chemical stability of DNA ensures that offspring are almost always recognisably similar to their parents. At the same time meiosis allows the DNA of a species to be constantly mixed into new combinations of characters. Mutations produce sudden, often dramatic, changes of features and add to the variety of individuals without which evolutionary change would be impossible.

In the next chapter we will look at the various strategies used by organisms to try to ensure that it is *their* DNA which is passed to succeeding generations.

Illustrative questions and worked answers

1 (a) Biochemical analysis of a sample of DNA showed that 33% of the nitrogenous bases were guanine. Calculate the percentage of the bases in the sample which would be adenine. Explain how you arrived at your answer.

(b) What name is given to the triplet of three bases which designates individual amino acids?

(c) If the triplet of mRNA bases which designates the amino acid lysine is AAG (where A = adenine and G = guanine), what is the complementary triplet of three bases on the tRNA molecule? Give a key for the letters that you use.

Time allowed 9 minutes

Tutorial notes

Questions relating to DNA frequently test your ability to translate the sequence of bases on DNA to those on mRNA and/or tRNA. When doing this you not only need to remember the complementary base pairs on DNA (namely adenine and thymine, cytosine and guanine) but also that thymine is replaced by uracil in RNA. The numerical part of the question also requires you to know the complementary base pairs in DNA and therefore deduce that the amount of each partner in a pair must be equal.

Suggested answer

(a) 33% of the bases on the sample DNA are guanine. Guanine pairs with cytosine and this must also therefore comprise 33% of the bases. Between them these two bases account for 66% of the total. The other two bases, adenine and thymine, must therefore make up the remainder, namely 100% – 66% = 34%. As these two bases are also in equal amounts, then each comprises 34/2 = 17% of the total. Adenine therefore makes up 17% of the total.

(b) The triplet of three bases is called a codon.

(c) If the mRNA triplet is AAG then the tRNA triplet is UUC where U = uracil and C = cytosine.

2 Mice from a group known to be heterozygous for genes A and B were mated with a group known to be homozygous recessive for both genes. The resultant offspring are shown in the table below.

Genotypes	Numbers
AaBb	151
aabb	146
Aabb	53
aaBb	59

(a) If the genes A and B were independently inherited, in what ratio would you have expected the four genotypes to have occurred?

(b) Do the actual results obtained confirm your predictions?

(c) Could the actual results be explained if
  (i) the two genes were linked?
  (ii) the two genes were not linked but selection operated against those offspring which were heterozygous for one of their two genes?

Explain your answers fully.

Time allowed 15 minutes

Tutorial notes

In order to answer this question you will need to know and understand Mendel's laws and the concept of crossing over during meiosis. Above all else, it is essential to lay out genetic questions in the standard format and to include every stage. It is all too easy to skip certain steps because you are clear in your own mind what is happening. The danger is that the reader may not be able to follow your reasoning in the absence of full explanations and valuable marks will be lost. Remember that the examiner cannot do the work for you and, while they may be fairly certain what you are *trying* to say, they can only give marks for what you *actually* write. For this reason, however tedious it might seem, always state clearly 'parents','gametes', 'meiosis', 'F_1 offspring', etc. Never cut corners!

Suggested answer

(a) One group of parents is heterozygous, i.e. AaBb. The other group is homozygous recessive for both genes, i.e. aabb.
If the two genes are inherited independently the following results should occur:

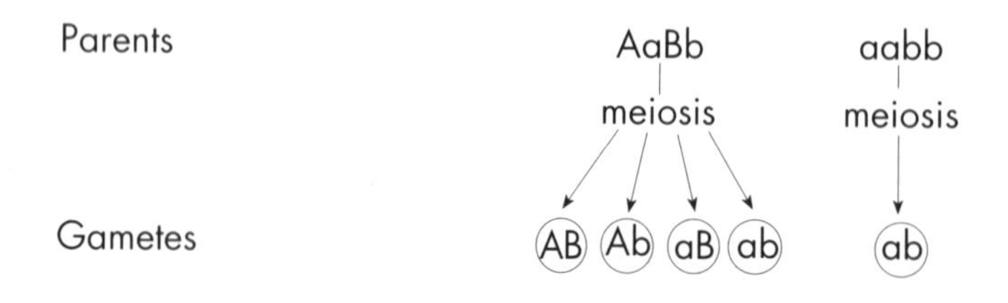

Offspring

Gametes	ab	Predicted ratio	Actual results
AB	AaBb	1	151
Ab	Aabb	1	53
aB	aaBb	1	59
ab	aabb	1	146

(b) Clearly the results do not confirm the prediction.

(c) (i) If the two genes are linked then the heterozygous parent would only produce two types of gamete, AB and ab, in which case the results would be:

Gametes	(ab)
(AB)	AaBb
(Ab)	Aabb

This only gives two types of offspring rather than the four actually produced. It is therefore *not* the explanation.

(ii) If selection were operating against offspring that were heterozygous for one of their genes, then those offspring with one pair of genes in the heterozygous state (i.e. Aabb and aaBb) should be present in smaller numbers than expected. This is the case and so could explain the result. In practice such selection normally operates where two enzymes (produced by two genes) are needed for a vital metabolic process. If these two enzymes are only produced when their respective dominant genes are present, then at least one A and one B gene are required in the genotype. Aabb and aaBb lack one dominant gene, in each case, and so the metabolic pathway is affected and the organisms could be at a disadvantage. In such an example the genotype aabb might be expected to be similarly disadvantaged and selected against. The results given do not, however, confirm this, which casts doubt on (ii) as a likely explanation. Nevertheless, in the absence of detailed information on the mechanism and nature of genes A and B, it must remain a possibility.

3 What do you understand by a species? Outline the ways in which new species arise.
Time allowed 35 minutes

Tutorial notes

This is an open-ended essay type question and to answer it effectively will require you to have a critical appreciation of the concept of a species. As with most questions relating to evolution, a wide knowledge of the topic gained from background reading is essential to ensure that sufficient supporting detail and actual examples are included. Examiners are unlikely to have an objective marking plan for such a question. It is more likely that they will make marks available for certain areas and will allocate them according to whether the required point is made, apparently understood and supported by evidence and examples.

In the first part of the question a definition of a species is not sufficient; for full marks there will need to be discussion of the difficulties in defining a species and examples of where any alternative definitions given do not suit a particular case. In both parts try to be as specific as possible referring to particular geographical and climatic conditions, named habitats and, where possible, actual animal and plant examples. In the second part refer to variation, natural selection and adaptation.

Suggested answer

The smallest group that a biologist distinguishes is called a species. It is not possible to find a completely satisfactory definition of a species. While the many members of a given species resemble one another more closely than do the members of any other species, the most widely used and satisfactory definition of a species is not based upon the criterion of appearance at all. It depends on the closeness of their relationship with one another. A species can be described as a population of related organisms, individuals of which are able to mate (if reproduction is sexual) and produce viable offspring which can interbreed. The major drawback in using this concept of a species is apparent when one considers those organisms which now only exist as fossils. It is obviously impossible to tell whether Fossil X can interbreed with Fossil Y. So we can never be sure whether Fossil X is a different species from Fossil Y. A more recent definition of a species is 'a recognisable lineage, a collection of organisms that shares a unique evolutionary history and is held together by the cohesive forces of reproduction, development and ecology'.

The second part of the question relies on the much broader concept of evolution of a species. Geographical isolation of populations and the effects of natural selection on their gene pools will be the major theme of the answer. The way in which variation, natural selection and adaptation lead to evolution of a new species is summarised by this diagram.

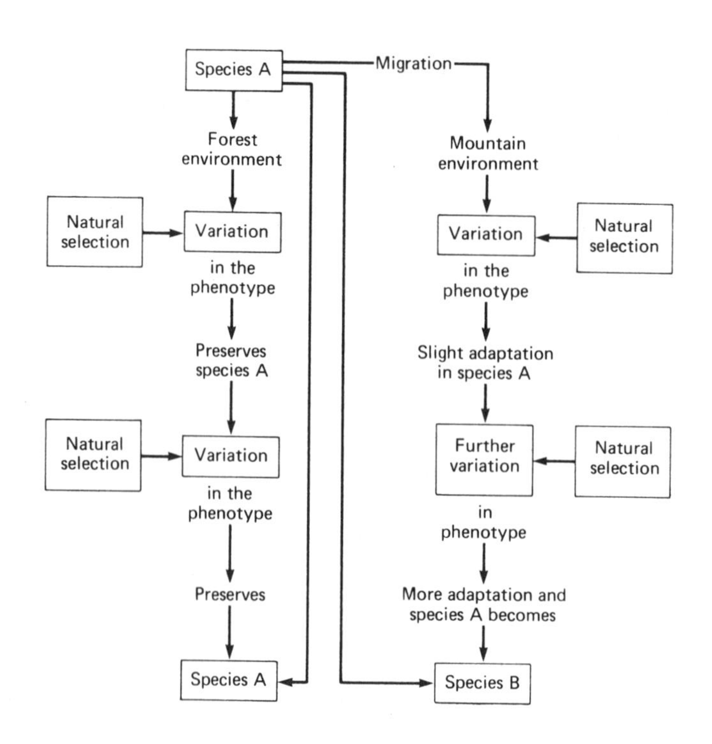

Question bank

1 (a) How is sex genetically determined in humans and birds? (4)

(b) A woman has four sons, one of which is haemophiliac and the remainder are normal. What are the genotypes of the woman and her husband? (8)

Is it possible for them to have a haemophiliac daughter? Give an explanation for your answer. (5)

(c) Why are there more colour-blind individuals than haemophiliacs in the population when both conditions are inherited in the same fashion? (3)

Time allowed 30 minutes

Pitfalls

In part (a) you must appreciate that sex determination is different in birds and mammals. In both cases, however, it is effectively the presence or absence of the Y chromosome which determines the sex.

Genetic questions such as (b) give you the opportunity to obtain high marks because there is usually a definite right answer. Equally, they can prove disastrous when misunderstood and/or poorly explained. Many candidates have embarked on a genetic explanation only to have to abandon their efforts having wasted precious time. You should therefore read the question carefully and plan a sketch answer to be certain you are on the right lines. The layout should follow the set format (see Table 3.4) and each stage should be logically explained in words, flow diagrams or Punnett squares. A clearly annotated key should be provided for all symbols used.

Note in (c) that three marks are available, indicating that three points need to be made.

Points

(a) Sex is determined by chromosomes not alleles. In humans the genetic composition XX produces a female, while in birds a male results. Likewise XY in humans is male, but is a female in birds.

(b) The allele for haemophilia is: (i) recessive, (ii) sex linked and (iii) carried on the X chromosome.

The sons are XY and the Y chromosome could only have been contributed by the father; the mother provides the X chromosome. As one son has haemophilia and the allele is carried on the X chromosome, the mother must have one X chromosome which bears the haemophiliac allele. As her other sons are normal the other X chromosome does not carry the allele (i.e. mother is X^HX^h). The genotype of the father could be either X^HY if he is normal or X^hY if he is haemophiliac. If he is the latter, then a haemophiliac daughter is possible. If he is the former, this would not be possible. The following genetic diagrams summarise the information.

Let X^h represent the chromosome carrying the haemophilia allele.
Let X^H represent the chromosome carrying the normal allele.
If the father is normal his genotype is X^HY.
From the above we know the mother to be X^HX^h.

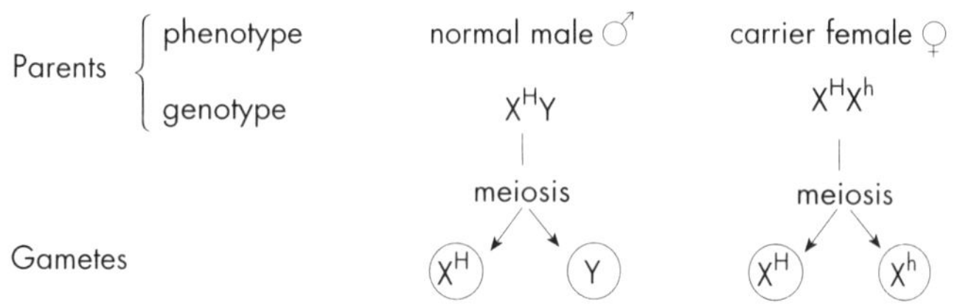

Parents	phenotype	normal male ♂	carrier female ♀
	genotype	X^HY	X^HX^h
		meiosis	meiosis
Gametes		X^H, Y	X^H, X^h

Offspring

	♀ gametes	
♂ gametes	X^H	X^h
X^H	X^HX^H	X^HX^h
Y	X^HY	X^hY

None of the offspring have the genotype X^hX^h and therefore none are haemophiliac daughters.

If the father is haemophiliac his genotype is X^hY and the gametes he produces are X^h and Y. With the mother's genotype unchanged the cross is:

Offspring

	♀ gametes	
♂ gametes	X^H	X^h
X^h	X^HX^h	X^hX^h
Y	X^HY	X^hY

One in four offspring (half the daughters) are haemophiliac (X^hX^h).

(c) The colour-blind allele occurs more frequently in the population than the haemophiliac one because:

1. haemophilia is potentially lethal and therefore is selected against, sufferers being less likely to survive to sexual maturity and so produce offspring;
2. haemophiliacs may choose not to have children knowing the risk that their offspring may suffer the disease;
3. haemophiliac females rarely survive to child-bearing age and hence only carriers pass on the disease. With colour blindness both sufferers and carriers can pass on the condition.

2 The following graph shows the emergence of bacterial strains which are resistant to an antibiotic used. The figures relate to one hospital.

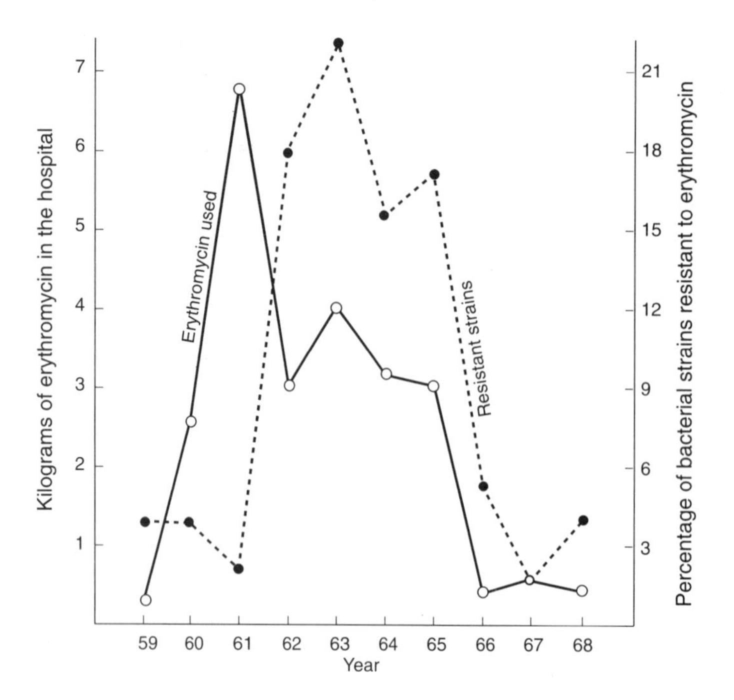

(a) Briefly explain how each of the following could contribute to the rapid emergence of resistant strains.
   (i) DNA in a bacterium is haploid. (2)
   (ii) Bacteria usually reproduce asexually by fission. (2)
   (iii) Fission may take place as often as every 30 minutes. (2)

(b) The graph may be used to demonstrate evolution in action. What selection pressure occurs and what is the result of such pressure? (2)

Time allowed 10 minutes

Pitfalls

You should study the graph with care and appreciate that the number of resistant strains reflects the amount of antibiotic used after a time lag of two years. The key word is 'rapid' and your answers must explain the rapidity of development of resistant strains.

Suggested answer

(a) (i) As the DNA is haploid, all alleles are effectively dominant. A recessive mutation in a diploid organism may be masked by its dominant partner and thus take many generations to build up in the population before a double recessive allows it to express itself. In this case, however, with no second allele, a recessive mutation for antibiotic resistance expresses itself immediately.
(ii) Asexual fission means that the mutation (resistance to erythromycin) is passed on to all cells arising from the mutant parent cell. There is no mixing of genetic material (and possible masking of the condition) as might arise with sexual reproduction.
(iii) The rapidity of cell division means that millions of copies of cells resistant to erythromycin can be made in a single day.

(b) The selection pressure referred to is the antibiotic erythromycin. It will destroy all susceptible (i.e. normal) strains of the bacteria, but not the mutants. These resistant mutants therefore survive to reproduce their own kind and have the potential to further mutate into more resistant forms. The greater the quantity of antibiotic used, the greater is this selection pressure and the more rapidly resistance develops.

CHAPTER 4

REPRODUCTION AND DEVELOPMENT

Units in this chapter

Chapter objectives

No living organism survives forever and therefore, if a species is to continue to exist, its members must reproduce their own kind. In this chapter we will look at the strategies that different species have adopted to ensure their survival. These include various types of asexual reproduction, with its rapid production of numerous but usually identical offspring, and sexual reproduction – often a more protracted process yielding fewer offspring, but with the advantage of producing the variety so essential to a species' evolution and survival in a changing world. We will investigate the many types of life cycle which have developed to help organisms adapt to a particular mode of life, such as the use of larval stages for dispersal or exploitation of a food source. In addition, we will explore the structural, physiological and behavioural processes involved in reproduction and the development of individuals into sexually mature adults ready to play their own part in the next reproductive cycle.

4.1 REPRODUCTIVE STRATEGIES

There are basically two forms of reproduction – asexual and sexual. Table 4.1 is a list of differences between the two.

ASEXUAL REPRODUCTION

This involves only one organism and the individuals produced are genetically identical, i.e. belong to a **clone**. Asexual reproduction ensures rapid multiplication and is especially useful where an organism needs to colonise a localised area in a short time. There are five major types of asexual reproduction.

Table 4.1 *Differences between asexual and sexual reproduction*

Asexual	Sexual
No mixing of genetic material; therefore less variation in offspring; therefore less evolutionary potential	Genetic mixing; therefore increased variation and great evolutionary potential
No gametes	Gametes
Usually more offspring	Fewer offspring
One parent	Usually two parents
Rapid; therefore takes advantage of favourable conditions	Longer process
May be resistant phases, e.g. spores, perennating organs	Rarely special resistant phase; may overcome difficult periods by delayed implantation (e.g. bats) or prolonged gestation (e.g. beavers)
The structures produced in asexual reproduction are more often used as a means of dispersal in animals than in plants	The structures produced are more often used as a means of dispersal in plants than in animals

Fission

This is typical of bacteria and protoctista. The cell divides into two or more equal parts. In **binary fission** (splitting in two) growth of the population is exponential, i.e. one cell divides into two, two into four, four into eight, etc. This is often a very rapid form of reproduction, e.g. many bacteria divide every 20 minutes. In many parasitic protoctista **multiple fission** occurs giving an even greater rate of reproduction, e.g. in the malarial parasite *Plasmodium*, where the nucleus divides into thousands of parts. This splitting is called **schizogamy** and each resultant cell is a schizont.

Budding

The parent produces an outgrowth, or bud, which detaches to become a separate individual, e.g. in *Saccharomyces* (yeast), *Hydra*, *Obelia*.

Fragmentation

This only occurs in simple organisms where the tissue is relatively undifferentiated (e.g. sponges and filamentous green algae). If small parts break off an individual, they will form whole new organisms.

Sporulation

This is the formation of small unicellular bodies which detach from the parent and, given suitable conditions, grow into new organisms. Spores are formed by bacteria, protozoa and many lower plants. Spore-bearing structures vary but the spores are generally small, light, easily dispersed, have resistant walls and are produced in vast numbers.

Vegetative propagation

Part of a plant becomes detached and grows into a new plant, e.g. the leaves of the African violet. Some organs of vegetative propagation are also known as **perennating organs** since they enable the plants to survive adverse conditions, such as the development of a stem into a corm (e.g. crocus) or stem tuber (e.g. potato); the swelling of a root in carrot or dahlia; and the swollen buds of bulbs such as onion.

SEXUAL REPRODUCTION

This involves the fusion of specialised haploid cells called **gametes**, which generally arise from two individuals. Usually the gametes differ in structure, size and behaviour (heterogametes), one being small, highly motile and produced in large numbers, the other being larger, non-motile and produced in small numbers. Some lower organisms, e.g. certain algae and fungi, produce identical gametes (isogametes). Fusion of gametes is called **syngamy**.

Parthenogenesis

This is the development of a new organism from an unfertilised female gamete. The parents are always diploid and, where the gamete is produced by mitosis as in aphids, the offspring are also diploid. Where the gamete is the result of meiosis, e.g. in honeybees, the offspring are haploid. This results in a male bee, or drone. Female honeybees develop from fertilised gametes and are therefore always diploid.

4.2 REPRODUCTION IN MAMMALS

Reproduction is concerned with the perpetuation of a species. Any mechanisms that increase the chance of fertilisation and survival of the offspring will be of evolutionary advantage. Mammals owe much of their evolutionary success to the development of such mechanisms, which include:

1. development of secondary sex characteristics to allow sexually mature individuals to recognise and mate with each other;
2. seasonal breeding cycles that restrict copulation to times that will ensure birth during seasons most favourable to the survival of offspring;
3. female receptiveness to the male only when ovulation is taking place, or even ovulation being stimulated by the act of copulation;
4. internal fertilisation bringing sperm and egg close together within the relative safety and stability of the female genital tract;
5. internal development of the embryo in stable and protected conditions;
6. the placenta acting largely as a barrier to harmful substances and an exchange mechanism for beneficial ones;
7. suckling as a means of providing the newly born with a fairly secure source of food ideally suited to its early development;
8. parental care allowing development of the young in controlled and protected conditions with the maximum use of learned behaviour, which has the advantage of being adaptable to meet varying circumstances.

Mammals, as with most animals, have the capacity to move from place to place. As we saw in 3.4, genetic variety, and hence evolutionary potential, is produced when two different genotypes are combined during sexual reproduction. The more an organism can move around, the greater the number of potential mates it can come into contact with, and hence the greater the number of genotypes that can be combined. This leads to a greater degree of outbreeding and hence more variety of offspring.

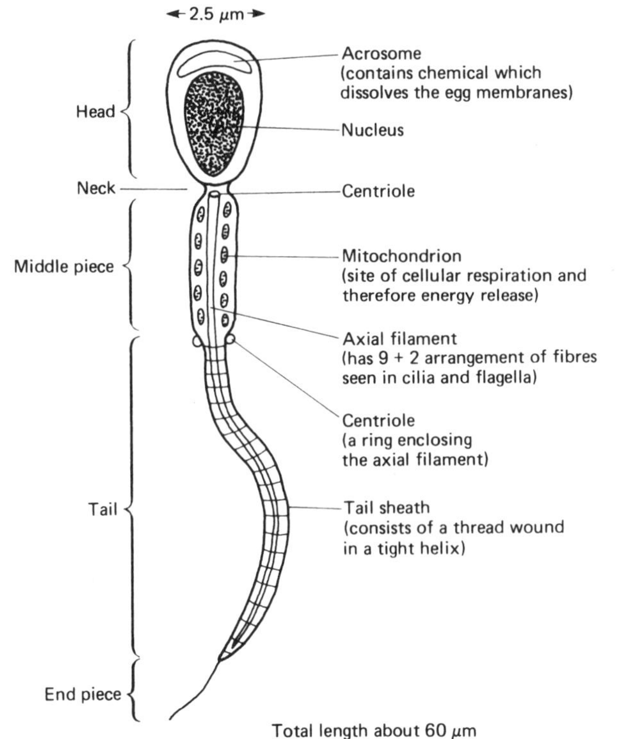

Fig. 4.1 Human spermatozoan

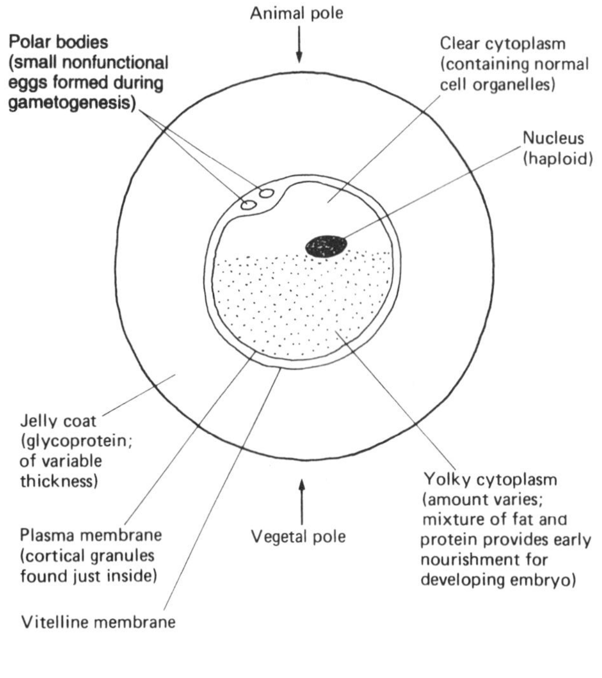

Fig. 4.2 Generalised egg cell (ovum)

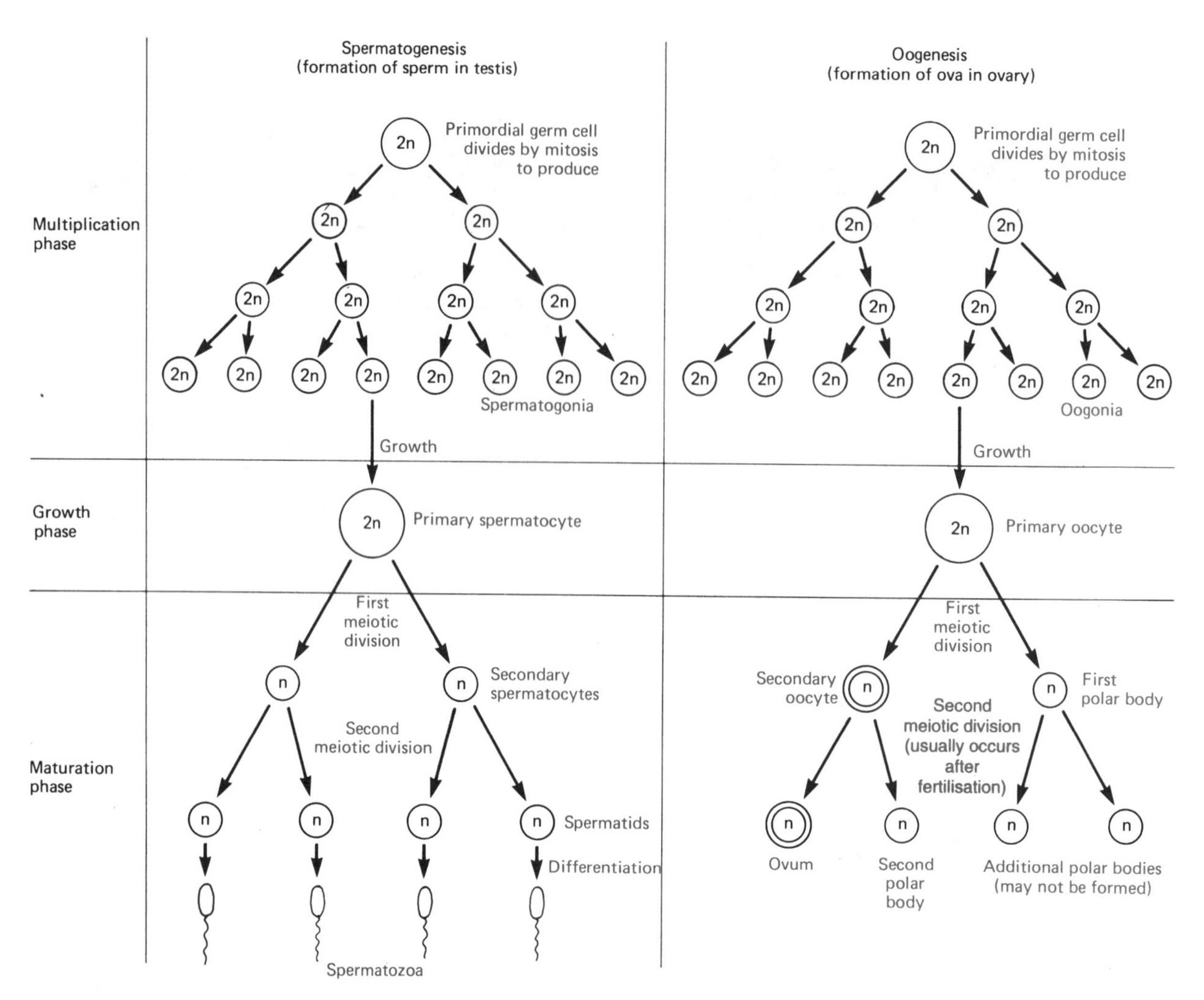

Fig. 4.3 Gametogenesis

In mammals, the male gamete, called the **sperm** (see Fig. 4.1), is small, highly motile and produced in large numbers. In contrast, the female gamete, the **ovum** (see Fig. 4.2), is larger, non-motile and produced in much smaller numbers. The sperm are produced in the **testes** by a process called **spermatogenesis**; the ova are produced in the **ovaries** by **oogenesis**. Both processes are collectively called **gametogenesis** (see Fig. 4.3).

Male human reproductive system

Sperm are produced in the testes which are suspended by a spermatic cord which comprises the vas deferens (sperm duct), an artery and vein, lymph vessels and nerves all bound together by connective tissue. The testes contain over 1 km of tiny tubes called **seminiferous tubules,** the walls of which divide to produce cells which ultimately mature into sperm. These tubules combine until finally they merge into a 6 m long tube called the **epididymis**, which stores the sperm. The epididymis leads into another muscular tube, the **vas deferens**, which carries sperm into the urethra and hence to the outside of the body through the **penis**. Secretions from the **seminal vesicles**, **Cowper's glands** and the **prostate gland** are added to the sperm to provide nutrients and an optimum pH. The resultant fluid is called **semen**.

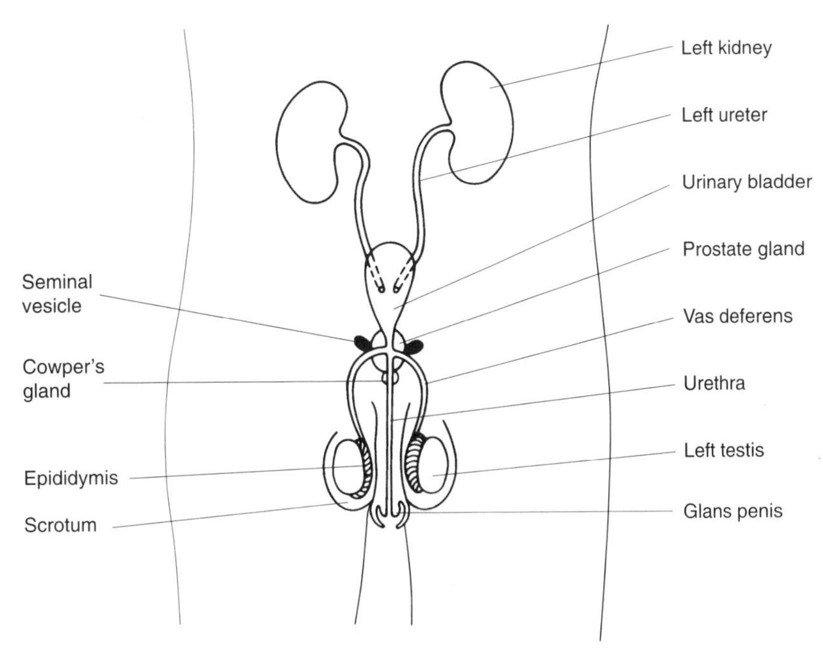

Fig. 4.4 Front view of the male human reproductive and urinary systems

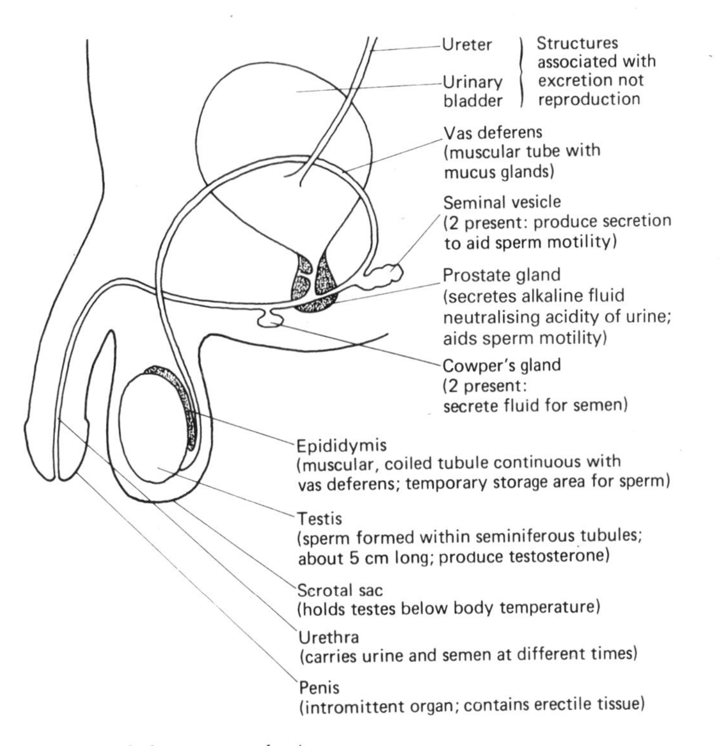

Fig. 4.5 Side view of the male human reproductive system

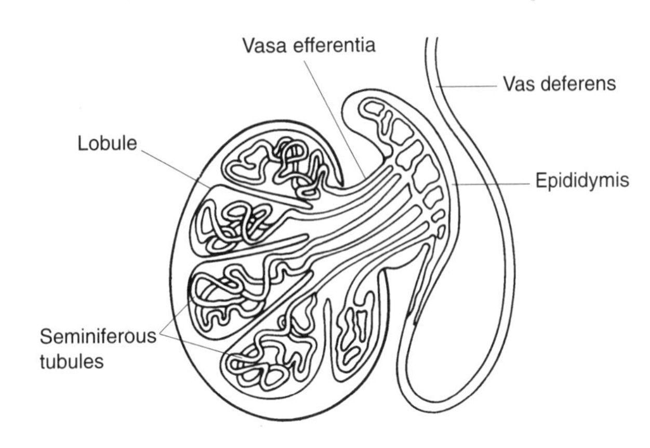

Fig. 4.6(a) LS testis

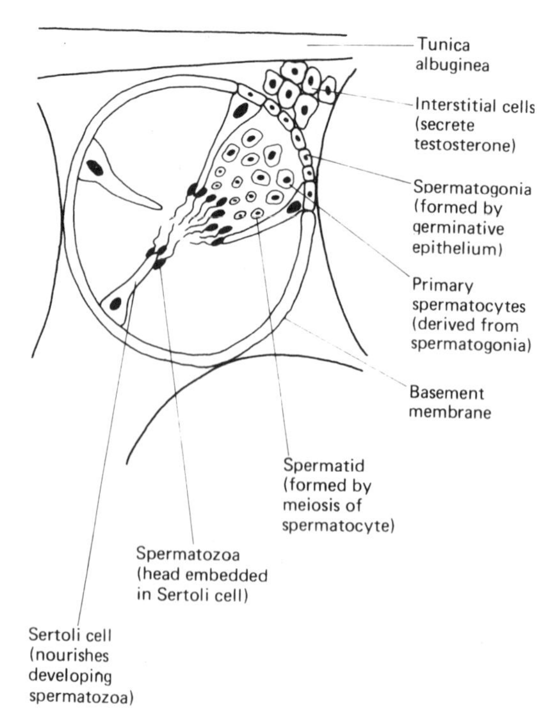

Fig. 4.6(b) TS seminiferous tubule × 500 (each testis contains about 1000)

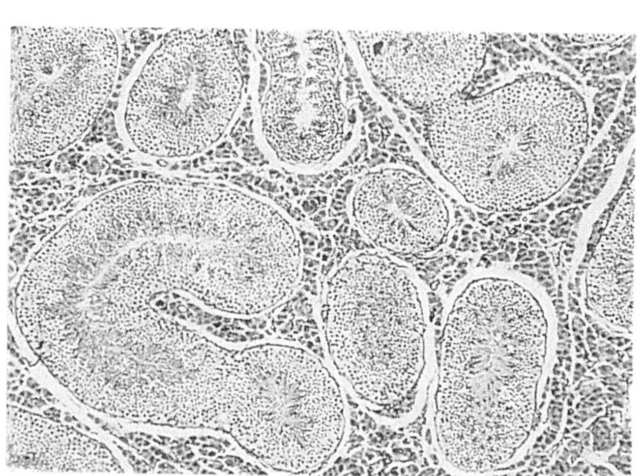

Fig. 4.6(c) TS seminiferous tubules, magnification 60 ×

The female human reproductive system

The ovaries contain around 400 000 ova at birth, which mature in turn and one is normally released from alternate ovaries each month after puberty. They pass into the funnel-shaped opening of the **oviduct (Fallopian tube)**. The oviduct is a thin muscular tube lined internally with cilia, which help move the ova down towards the **uterus**, a pear-shaped organ with a wall of unstriated muscle. The inner lining of the uterus is a soft mucus membrane called the **endometrium**. The uterus opens through a ring of muscle called the **cervix** into a muscular tube lined with striated epithelium – the **vagina**. This in turn opens to the outside through the external genitalia called the **vulva**.

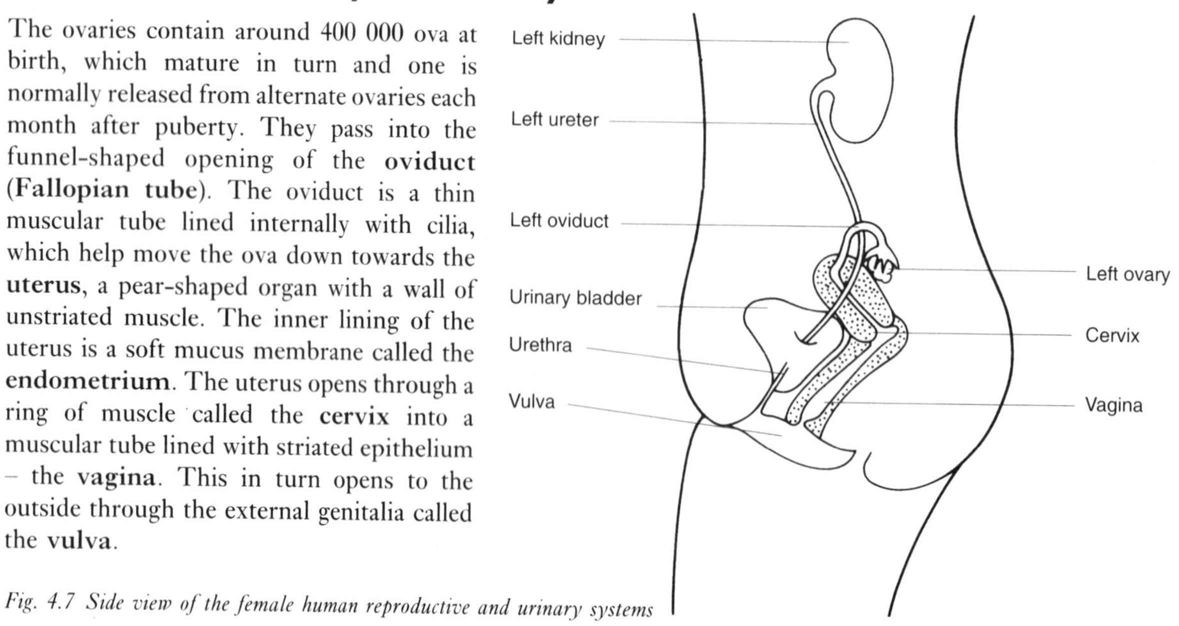

Fig. 4.7 Side view of the female human reproductive and urinary systems

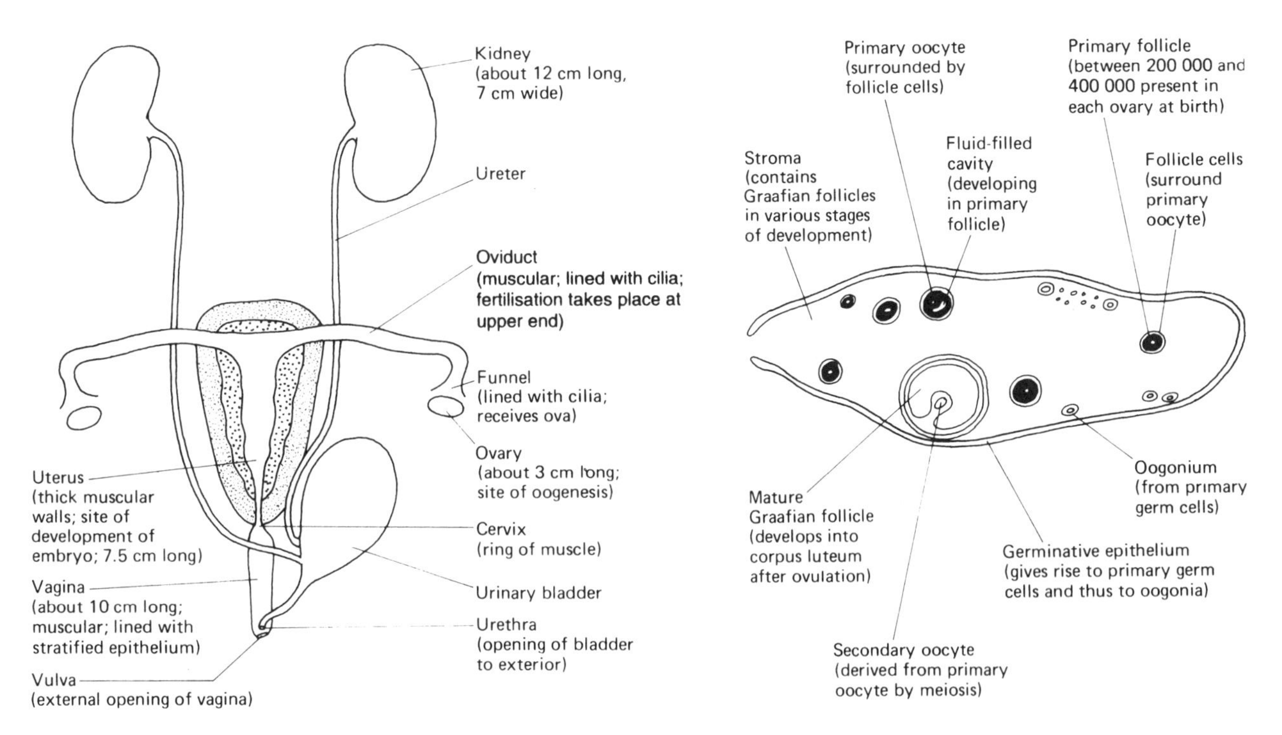

Fig. 4.8 Front view of female human reproductive and urinary systems

Fig. 4.9 LS ovary

Courtship and mating

The rituals which precede mating in many species are often long and elaborate. They serve the purpose of ensuring that both partners are sexually mature and ready for mating, as well as establishing a pair-bond, which is necessary if both are to share in the rearing of the young.

In humans sexual arousal can be stimulated in a wide variety of ways and results in an increased blood supply to the genital regions. In females this leads to the swelling of the labia and clitoris and the secretion of fluid by the vaginal wall which assists penetration of the penis. In males the penis becomes erect so that it can be introduced into the vagina. Movement of the penis within the vagina leads to the reflex contraction of muscles in the epididymis and vas deferens which causes the sperm to be moved by peristalsis into the urethra. At the same time secretions from the seminal vesicles and the prostate and Cowper's glands are released and mixed with the sperm. The semen so produced is ejaculated from the penis by powerful contractions of the muscles of the urethra. Each ejaculate of around 3 cm^3 of semen contains some 500 million sperm.

Fertilisation

The ejaculated sperm released at the top of the vagina swim through the uterus and into the oviducts. Apart from the lashings of their own tails, muscular contractions of the uterus and oviduct help move the sperm towards the ovaries. If an egg has been released from an ovary it normally meets, and is fertilised by, the sperm in the top third of the oviduct some three days after being released. Only a few hundred of the initial 500 million sperm reach the ovum and only one actually fertilises it.

The successful sperm releases a protease enzyme from the acrosome, which digests the vitelline membrane around the ovum. The acrosome now turns itself inside out to produce a filament which can pierce the newly softened region of the vitelline membrane. The penetration of this membrane causes it to thicken, forming a **fertilisation membrane** which prevents other sperm penetrating it. The head and middle piece of the sperm enter the ovum while the tail is discarded. The addition of the haploid number of chromosomes in the sperm to those of the ovum results in a new diploid cell called a **zygote**.

Infertility

Couples sometimes experience difficulty conceiving a baby and the reasons include:

1. an irregular menstrual cycle, which makes the timing of sex to coincide with the female's fertile period difficult;

2. intercourse occurring only at times when an ovum is not available for fertilisation;
3. blocked oviducts preventing the sperm and ovum meeting;
4. non-production of ova;
5. the absence of sperm in the ejaculate;
6. inability to erect the penis (impotence).

Development of the zygote

The zygote moves down the oviduct to the uterus, a journey which takes about one week in humans. By the time it has reached the uterus it has divided mitotically to form a hollow ball of cells – a **blastocyst**. This becomes implanted in the lining of the uterus and the outer layer of cells forms trophoblastic villi which project into the uterine wall. As the embryo develops it becomes surrounded by a series of extra-embryonic membranes: the **amnion**, which encloses a fluid-filled cavity that eventually fills the entire uterus, and the **allantois** and **chorion**, which unite to form the allanto-chorion, which develops into the placenta. The only connection between the embryo and the wall of the uterus is the stalk of the allantois which becomes the **umbilical cord**. This contains the umbilical arteries and vein conveying fetal blood to and from the placenta.

The placenta

An exchange of materials takes place between the maternal blood spaces in the uterine lining and the fetal capillaries in the chorionic villi.

The functions of the placenta are as follows.

1. Oxygen, water, soluble food and salts pass from the maternal to fetal blood.
2. Carbon dioxide and nitrogenous waste pass from the fetal blood to maternal blood for removal by the mother.
3. The mixing of fetal and maternal bloods is prevented since the fetus may inherit the father's blood group which, if incompatible with the mother's, would damage the fetus if mixing occurred.
4. During pregnancy the placenta progressively takes over the role of producing the hormones which prevent ovulation and menstruation, e.g. progesterone.
5. It allows maternal and fetal blood pressures to differ from one another.
6. It prevents the passage of some pathogens from the mother to the fetus, although some viruses and bacteria and many antibodies can cross between the two.
7. While allowing some hormones across it, the placenta prevents the passage of those maternal hormones which could adversely affect fetal development.

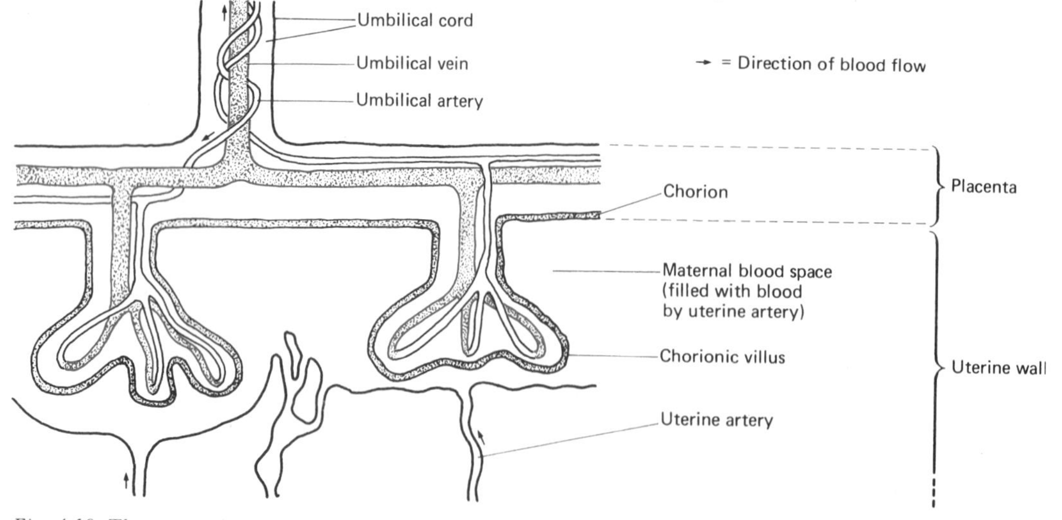

Fig. 4.10 The mammalian placenta

Birth

During pregnancy the quantity of different hormones produced changes; the amount of oestrogen increases while that of progesterone decreases. This helps to trigger the birth process. A hormone called **oxytocin** is produced from the posterior lobe of the pituitary gland and this causes contractions of the uterus which lead to the expulsion of the fetus, followed a little later by the placenta, which at this stage is called the **afterbirth**.

Parental care

In mammals parental care is extensive and includes the production of milk. After the birth the anterior lobe of the pituitary gland produces the hormone **prolactin**, which causes the mammary glands to produce milk, and suckling by the baby causes its release from the nipple. The milk first secreted has a mild laxative effect and is known as **colostrum**. Apart from being a nutritious and balanced mixture of foods, the milk contains antibodies which provide the young with some initial immunity to certain diseases.

Parental care may include the provision of a protective nest and some training in the ways of living. The extent of this further care varies between species.

Female sexual cycle (oestrous or menstrual cycle)

In humans this is a 28-day cycle controlled by hormones secreted by the pituitary gland and the ovary. The main hormones involved and their functions are as follows.

1. Follicle stimulating hormone (FSH)
   (a) causes Graafian follicles to develop in the ovary;
   (b) stimulates the tissues of the ovary to produce oestrogen.
2. Oestrogen
   (a) repairs the uterine wall following menstruation;
   (b) builds up in concentration during the first two weeks of the menstrual cycle until it stimulates the pituitary gland to produce luteinising hormone.
3. Luteinising hormone (LH)
   (a) brings about ovulation;
   (b) causes a Graafian follicle to develop into a corpus luteum which produces progesterone.
4. Progesterone
   (a) inhibits FSH production and therefore stops further follicles developing (a fact made use of in the development of the contraceptive pill);
   (b) causes development of the uterine lining prior to implantation.

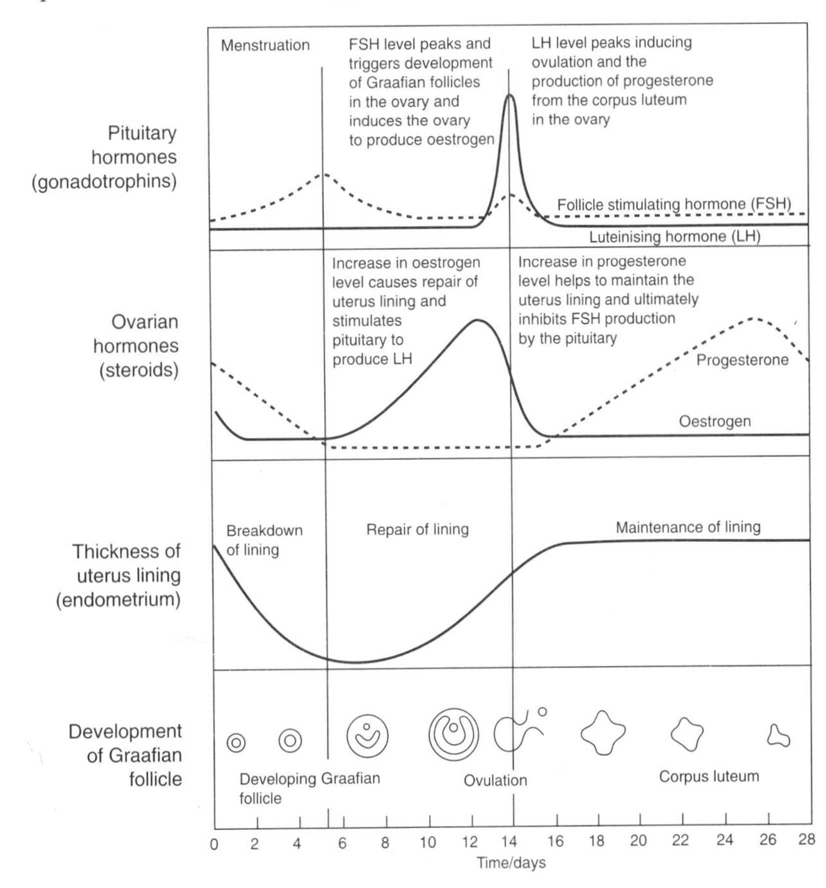

Fig. 4.11 Human menstrual cycle

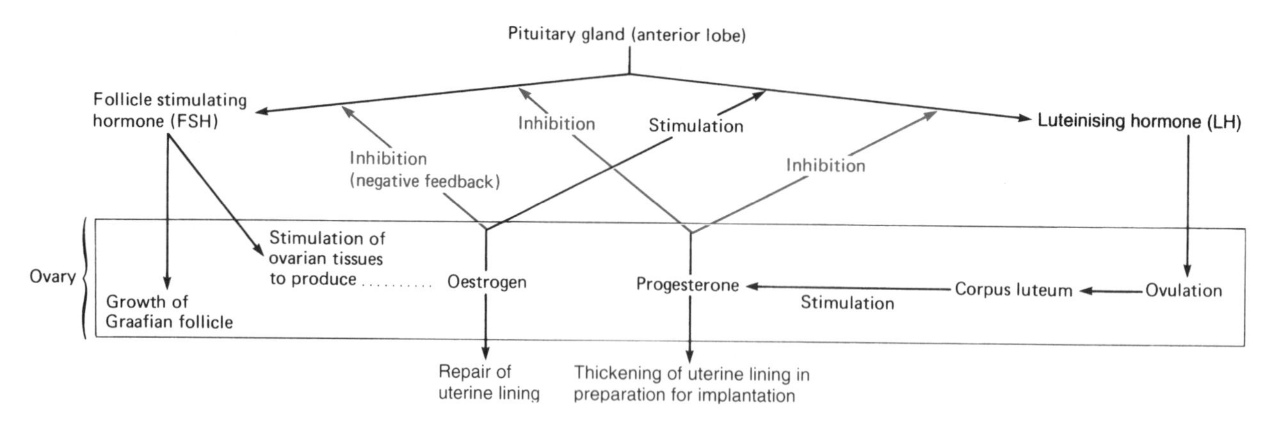

Fig. 4.12 Hormonal control of the menstrual cycle

During pregnancy, progesterone, produced initially by the corpus luteum and later by the placenta, is present in high concentrations. Therefore, LH production and hence ovulation are inhibited. As FSH production is also inhibited, ripening of follicles and production of oestrogen cease.

Male sexual cycle

There is no cycle in male humans, but the gonads are regulated by hormones identical to those of the female. FSH, which is generally called interstitial cell stimulating hormone (ICSH) in the male, promotes spermatogenesis. LH, also called ICSH, causes the interstitial cells between the seminiferous tubules to secrete androgens, which are hormones that stimulate the development of male secondary sex characteristics.

Breeding cycles in vertebrates

If the cycle is seasonal, environmental changes trigger the endocrine system at the most advantageous time of the year for breeding. Whereas in the stickleback *Gasterosteus aculeatus* the length of day is the most important environmental trigger and temperature plays a subsidiary role, in the minnow *Couesius plumbeus* the cycle is most affected by temperature.

The breeding of birds is nearly always seasonal. In temperate latitudes the increase in the length of day in spring acts through the pituitary gland to cause ripening of the gonads. Changes in behaviour associated with the development of the gonads vary with species. Other factors such as the degree of activity and food eaten may be important, especially near the equator where there is little seasonal variation.

If the cycle is more or less continuous, as in most endotherms and some tropical ectotherms, the controls are usually within the organism and rely on a feedback mechanism. Usually the female will only receive the male during a strict period of oestrus (or heat) but in humans receptivity may occur throughout the cycle. In some animals (e.g. rabbit and ferret) coitus induces ovulation and fertilisation usually takes place within a few hours. In certain bats copulation occurs in autumn and fertilisation is delayed until spring. Implantation usually occurs about one week after fertilisation, as in humans and rabbits. However, in some animals (e.g. pig, cat and dog) implantation is delayed for about two weeks and in badgers mating takes place in the summer but implantation is delayed until late in the winter.

4.3 REPRODUCTION IN FLOWERING PLANTS

Flowering plants have successfully colonised land by reducing the water-dependent gametophyte generation to the point where it is entirely protected within the sporophyte generation. You should bear in mind that, while the flower is the main organ of reproduction in angiosperms, leaves, stems and roots may all play a reproductive role.

A typical flower is, in effect, composed of four sets of modified leaves. The outer set comprises the **sepals**, which are collectively called the **calyx**. These protect the flower when in bud, although they often remain photosynthetic and so contribute to the food production of the plant. The next layer of leaves is the **petals**, collectively called the **corolla**. These are brightly coloured in insect-pollinated plants, where they are used to attract the insects.

Next, moving into the centre of the flower, come the **stamens**, known collectively as the **androecium**. Each stamen comprises a long **filament** at the end of which are the **anthers**. The anthers produce pollen grains, which contain the male gamete.

Finally in the centre of the flower are the **carpels**, collectively called the **gynoecium**. These have an ovary at their base containing one or more **ovules**, each of which encloses the female gamete. Extending from the carpel may be a slender stalk, the **style**, at the end of which is the sticky **stigma**, upon which pollen grains need to be deposited if fertilisation is to occur.

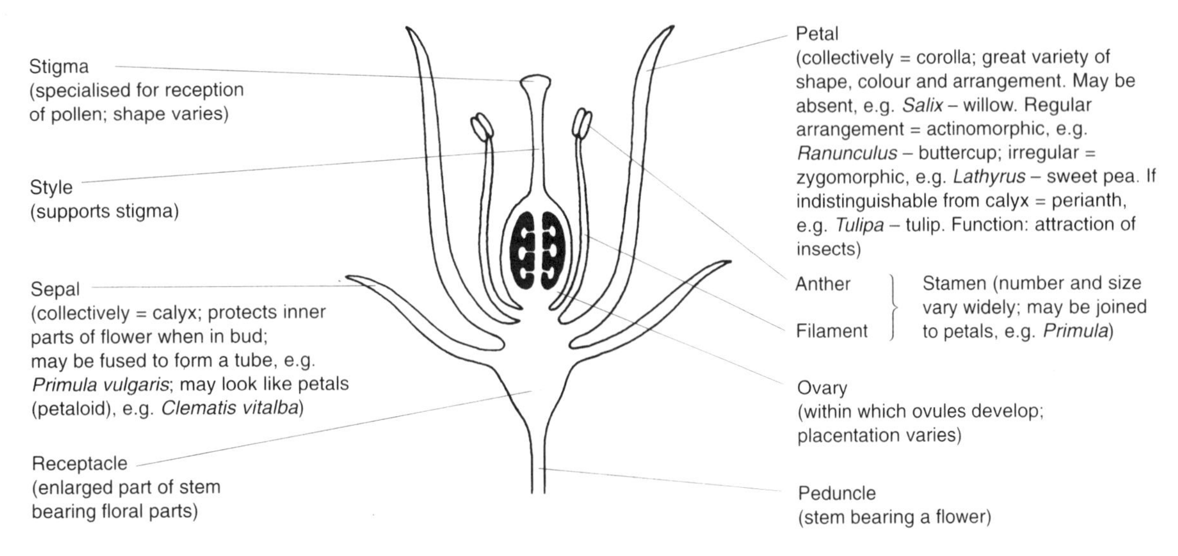

Fig. 4.13 LS generalised flower

Production of pollen

The centres of the pollen sacs contain **pollen mother cells**, each of which divides meiotically to give a tetrad of pollen grains. These four cells later separate. Each young grain has one nucleus but this divides into two. One of these, surrounded by denser cytoplasm, forms the **generative cell** and this later gives rise to the male gametes; the other forms the **tube nucleus**. The pollen grain has a double wall, a delicate inner one of cellulose (the intine) and an outer one of variable thickness (the exine). Pollen grains vary in shape and in size from 3–300 μm. They are liberated by the longitudinal rupture of the anther lobe together with the breakdown of the wall between each pair of sacs.

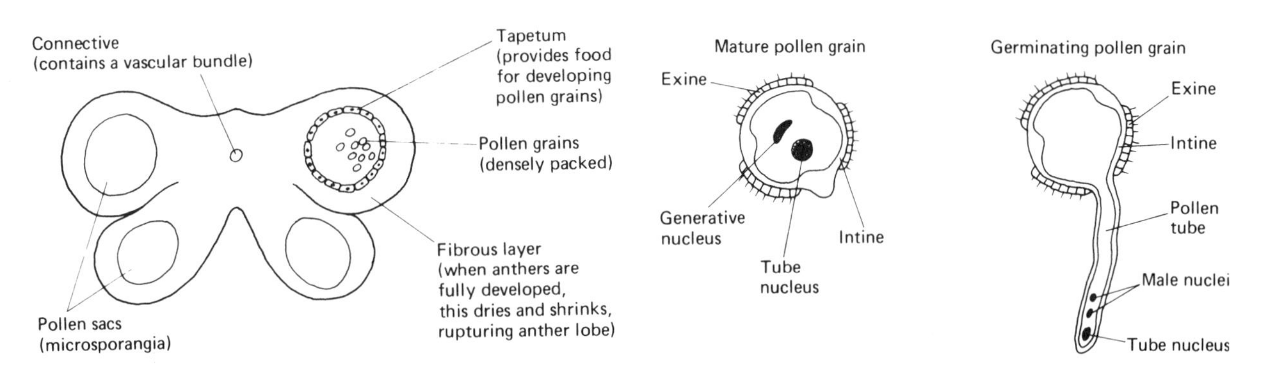

Fig. 4.14(a) TS anther

Fig. 4.14(b) Germination of pollen grains

Structure and development of an ovule

One cell of the nucellus becomes enlarged and is known as the **embryo sac mother cell**. This divides meiotically to give four cells, three of which are crushed as the remaining one enlarges to form the embryo sac. The single nucleus within the embryo sac divides mitotically and the two nuclei move to opposite ends of the sac. Each of the two nuclei divides mitotically twice so that there are four haploid nuclei at each end of the sac. One nucleus from each end moves to the centre and these fuse to form the **primary endosperm nucleus** (central fusion nucleus). The three remaining at the micropyle end form the **egg apparatus** and the three at the other end form the **antipodal cells**.

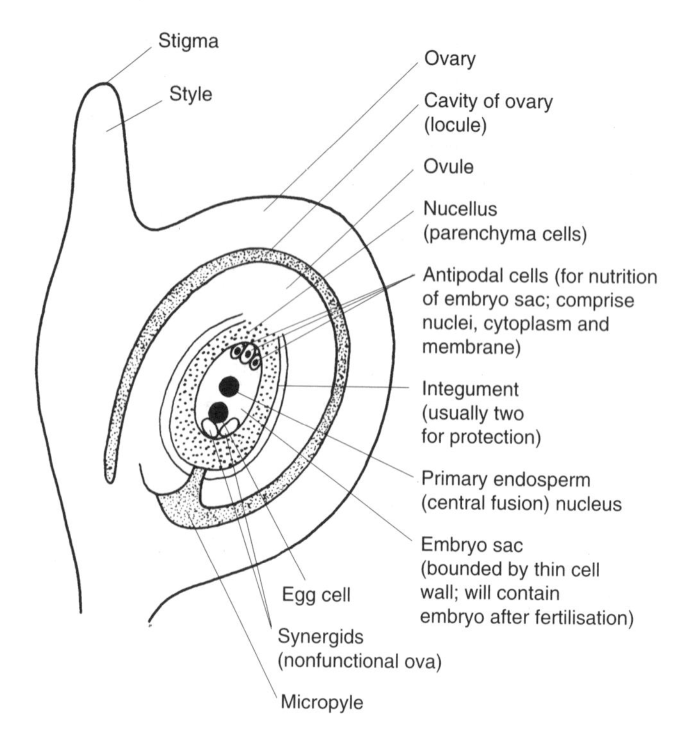

Fig. 4.15 LS generalised carpel

Pollination

This is the transfer of pollen from the anther to the stigma.

Self-pollination This is the transfer of pollen from the anther to the stigma of the same flower, or a different flower on the same plant. Many plants are designed to self-pollinate should cross-pollination fail, e.g. in the Compositae as the flower ages the stigmas curl to touch the anthers. A few species of plants produce flowers which become self-pollinated while still in the bud stage (cleistogamy), but this usually only occurs in a small proportion of the flowers, e.g. *Viola* (violet), where open flowers may also be cross-pollinated by insects.

Cross-pollination This is the transfer of pollen from the anther to the stigma of a flower on a different plant of the same species. Although some plants regularly show self-pollination there is a general tendency towards cross-pollination, which reduces inbreeding and increases the variability of a population.

The two main agents of cross-pollination are wind and animals (mainly insects).

Table 4.2 *Differences between wind- and insect-pollinated flowers*

Wind	Insects
Enormous amount of pollen produced because there is great wastage	Less pollen produced because mechanically more precise
Pollen small, light and smooth	Pollen larger and heavier with projections
Often found in plants which occur in groups	Often found in plants which are more or less solitary
Often found in unisexual flowers (usually with an excess of male flowers)	Mostly in bisexual (hermaphrodite) flowers
Flowers dull, scentless and nectarless	Bright, scented flowers with nectar, therefore attractive to insects
Stigmas long and protrude above petals	Stigmas often deep in corolla
Stigmas often feathery or adhesive	Stigmas often small
Stamens long and protrude above petals	Stamens may be within corolla tube
Examples: grasses (e.g. *Festuca*), hazel (*Corylus*), willow (*Salix*), plantain (*Plantago*)	Examples: snapdragon (*Antirrhinum*), foxglove (*Digitalis*), buttercup (*Ranunculus*), clover (*Trifolium*)

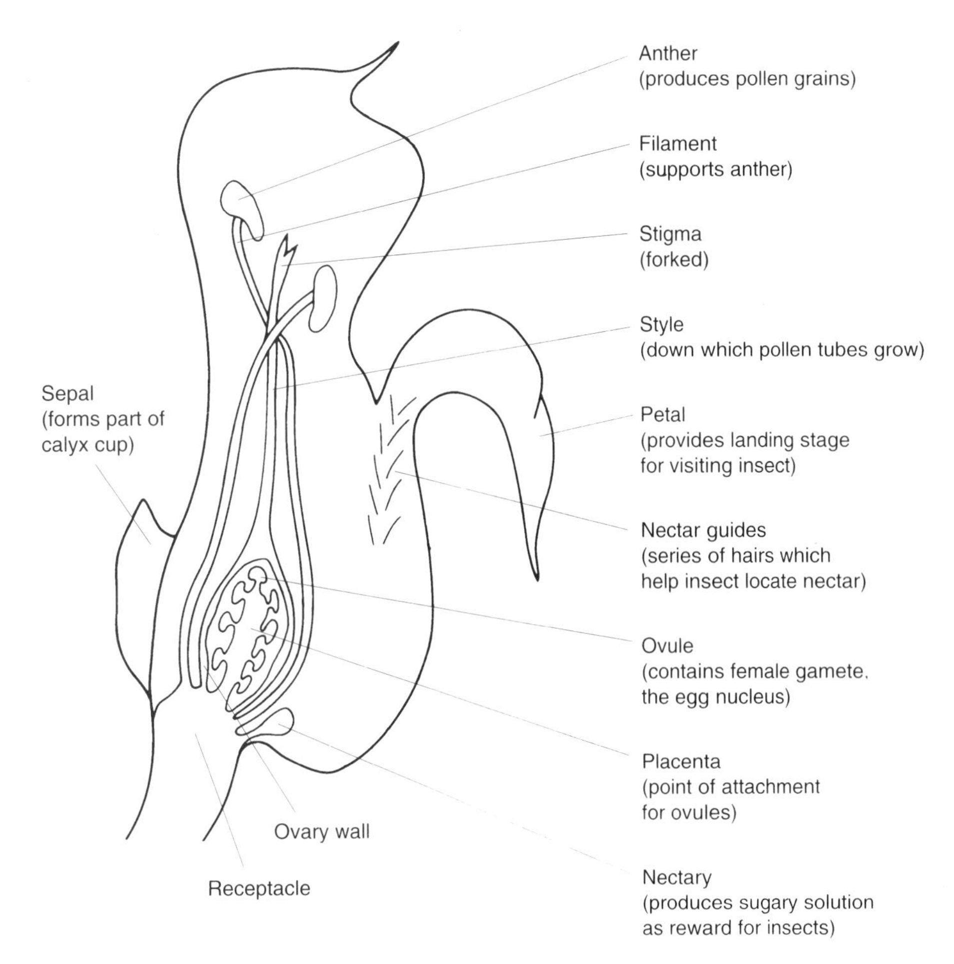

Fig. 4.16 Antirrhinum (half-flower) – an example of an insect-pollinated flower

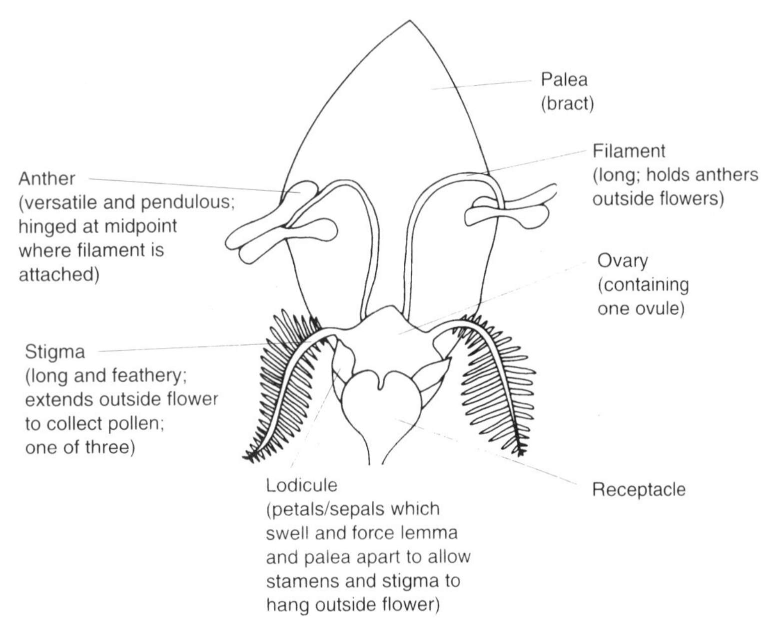

Fig. 4.17 Rye grass (with front bract removed) – an example of a wind-pollinated flower

Prevention of inbreeding

While some plants regularly self-pollinate, most avoid doing so because such inbreeding reduces variability and leads to loss of vigour. Mechanisms used by plants to prevent self-pollination include:

1. Being **dioecious**, a condition in which all the flowers on a plant are male or all are female. Provided the species is totally dioecious, self-pollination is impossible. *Fraxinus* (ash) is predominantly dioecious.
2. Being **monoecious** by having separate male and female flowers, i.e. the plant is both male and female, but individual flowers are one sex or the other, never both. This does not prevent self-pollination but makes it less likely. *Zea* (maize) is an example of a monoecious plant.
3. The structure of the flower makes self-pollination unlikely. For example, in the iris a flap protects the stigma from coming into contact with its own pollen as an insect leaves the flower.
4. The anthers and stigma mature at different times. For example, in *Lamium* (dead nettle) the anthers ripen first and their pollen has long been dispersed by the time the stigma is receptive to pollen.
5. Sometimes self-pollination occurs, but the pollen tube does not develop fully unless it is of a different genetic composition (i.e. from a different plant).

Fertilisation

As the female gamete in angiosperms is protected within the carpel, the male gamete can only reach it via the pollen tube, which provides a channel of entry and protection for the male nuclei. The pollen grain germinates within a few minutes of landing on the stigma and the pollen tube pushes between the loosely packed cells of the style. The entire contents of the pollen grain move into the tube, the tube (vegetative) nucleus moving first. Callose plugs block the older, empty parts of the tube as it grows. The pollen tube may secrete pectases to soften the middle lamellae of the cells of the style and the growth towards the micropyle is thought to be chemotropically controlled. The normal mode of entry to the ovule is through the micropyle.

The pollen tube contains three nuclei, the tube nucleus and two male gametes derived from the generative nucleus, which keep near the tip as growth proceeds. When the pollen tube penetrates the embryo sac, one male nucleus fuses with the egg cell (oosphere) and the other with the primary endosperm nucleus. Since this second fusion involves three nuclei (the primary endosperm nucleus was derived from two polar nuclei), it is called triple fusion. The fertilised oosphere gives rise to the embryo and the fertilised primary endosperm nucleus to the endosperm. In non-endospermic seeds the endosperm is used up to form the cotyledons before the seed is ripe, but in endospermic seeds nuclear and cellular divisions of the fertilised primary endosperm nucleus give rise to an extensive endosperm.

Seeds and fruits

The development of the fertilised ovule produces a seed and the ovary as a whole develops into the fruit.

The mature seed is covered by a hard leathery testa, which is the product of one or both integuments and may have a scar, or hilum, on it which marks the point of attachment to the ovary wall. The embryo within the testa consists of one or two cotyledons (monocotyledon or dicotyledon) which may or may not store food. The cotyledons are attached to a central axis differentiated into a **plumule** (shoot) at one end and a **radicle** (root) at the other.

Examples of types of seeds:

- dicotyledon, non-endospermic: broad bean, pea, sunflower
- dicotyledon, endospermic: castor oil
- monocotyledon, non-endospermic: water plantain
- monocotyledon, endospermic: maize, onion

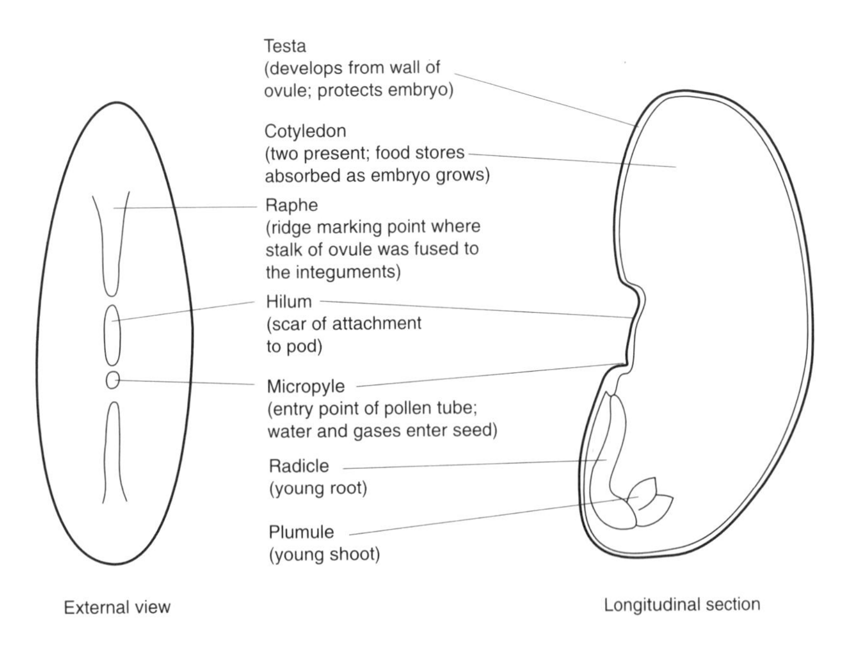

Fig. 4.18 Broad bean seed

Dormancy

If seeds germinated at the earliest time that conditions permitted it, many would begin growth before the onset of winter only to find themselves killed by an unfavourable climate because they had not developed fully enough to survive. For this reason many seeds undergo a period of dormancy before germinating. Sometimes a particular set of conditions needs to be satisfied before the seeds begin growth. Examples include:

1. a period of sustained cold;
2. a set period of time to allow certain very slow chemical processes to occur;
3. a certain amount of light above a given intensity;
4. partial breakdown of the seed coat (testa), e.g. by the digestive processes of animals;.
5. the heat of a flash fire.

Germination

A seed begins to develop after a period of dormancy and is dependent on the food reserves in the cotyledon until the first true leaves develop. Germination requires an adequate supply of oxygen and water and a suitable temperature. As the first stage of germination is the absorption of water and since testas are frequently impermeable to water, germination may be delayed until the testa is broken or decayed. The water renews the physiological activities of the endosperm and embryonic tissues. The enzymes in the seeds are activated and the food is changed from an insoluble to a soluble form. There are two main types of germination:

1. **hypogeal**, in which the cotyledons remain below the ground, e.g. broad bean;
2. **epigeal**, in which the cotyledons emerge above the ground, e.g. runner bean.

As the seed develops the embryonic shoot (**plumule**) and the embryonic root (**radicle**) emerge from the testa. The plumule grows upwards, often in a hooked position to help protect the growing point, while the radicle grows downwards. If growth is mainly in the region below the cotyledons (hypocotyl) then the cotyledons are carried above the soil surface as growth proceeds. This is called **epigeal germination**. If growth is mainly in the region above the cotyledons (epicotyl) then they remain below ground. This is called **hypogeal germination**.

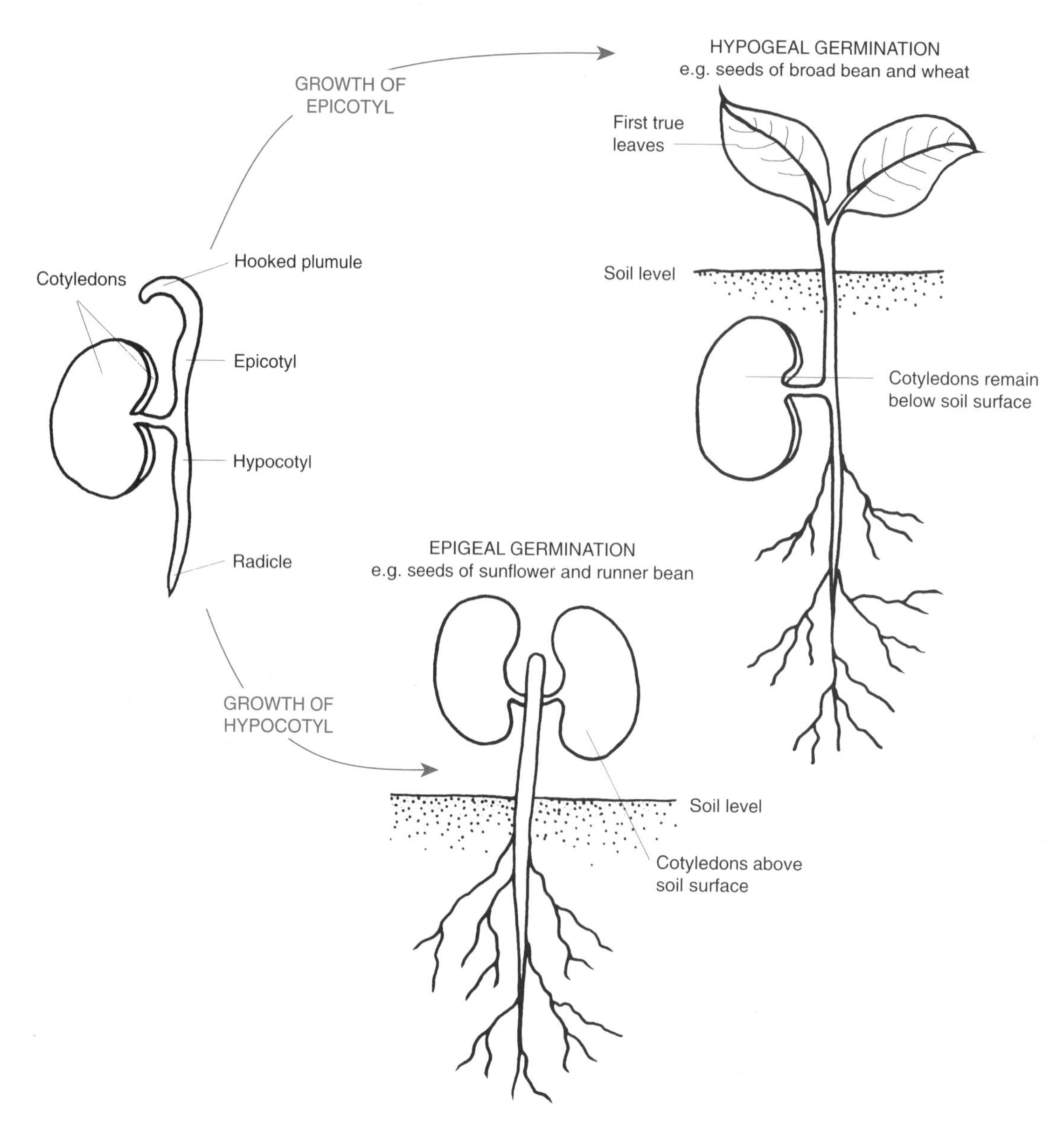

Fig. 4.19 Differences between hypogeal and epigeal germination

4.4 GROWTH AND DEVELOPMENT

There is a limit to the size of any individual cell, partly due to the distance over which a nucleus can exert its controlling influence. Therefore single-celled organisms will absorb and assimilate materials until they reach a certain size, at which they will divide to give two separate individuals. The multicellular condition allows organisms to attain greater size. This brings its own problems but these are largely outweighed by such advantages as:

1. increased scope for differentiation, specialisation of function and hence greater efficiency;
2. the ability to separate more easily processes requiring different conditions, e.g. different pH values;
3. increased ability to store materials;
4. the ability to replace damaged cells from those remaining;

5. greater competitive advantage, e.g. large plants have better access to light than small ones;
6. some security from predators that find a large organism more difficult to capture and ingest.

All multicellular organisms originate as a single cell, the zygote, and undergo three phases of development:

1. Growth: an irreversible increase in mass.
2. Differentiation: the development of differing cell structures and functions.
3. Morphogenesis: the development of the overall form of organs and hence the organism.

In addition to the importance of growth and development in the early stages of an organism's life, these events continue to manifest themselves later in such processes as regeneration, repair and gametogenesis.

GROWTH

Growth is an irreversible increase in size during development. It usually occurs in three distinct phases: cell division, cell assimilation and cell expansion.

Measurement of growth

Using a single parameter for the measurement of growth may not take into account growth in all directions, e.g. measuring an increase in length does not take into account changes in girth. Changes in volume are difficult to measure in irregular organisms and changes in fresh weight may be complicated by temporary changes, e.g. drinking water. Dry weight gives a less misleading picture but, since the measurement of this involves killing the organism, numerous similar organisms are required. A further complication is that an overall measurement of growth does not take into account the fact that different organs may have their own peculiar growth rates (**allometric growth**).

Growth curves

Fig. 4.20 presents a variety of growth curves in plants and animals.

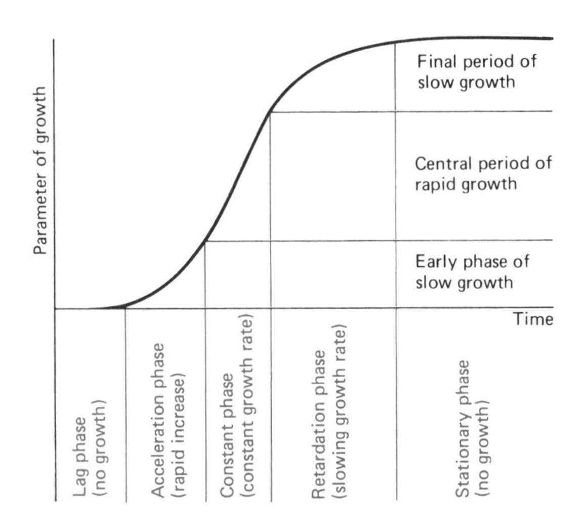

This sigmoid (S-shaped) curve can be applied to a population, an individual or an organ of an individual, although the pattern may be modified.

Fig. 4.20(a) General growth curve

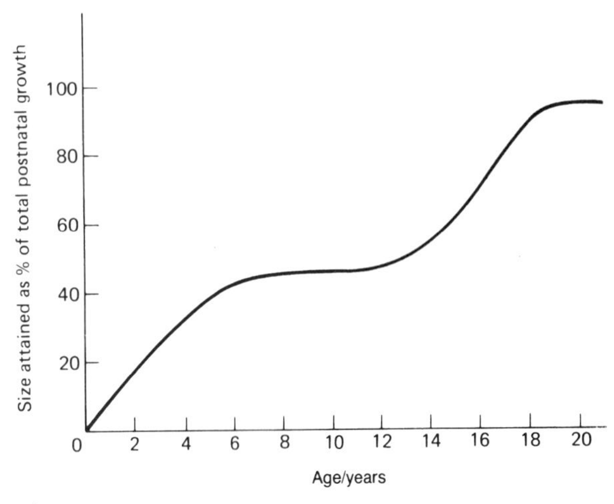

This resembles two flattened sigmoid curves on top of each other. The first represents the early growth phase in a child, the second the later growth phase in adolescence.

Fig. 4.20(b) Human growth curve

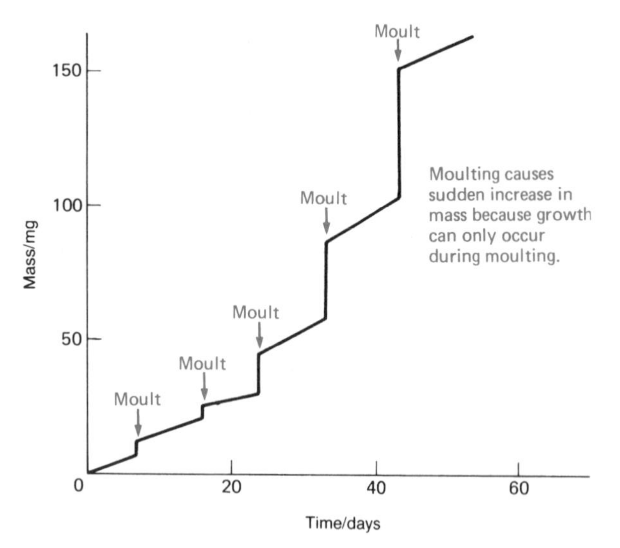

This is known as intermittent growth.

Fig. 4.20(c) Growth curve of an arthropod, e.g. Notonecta glauca

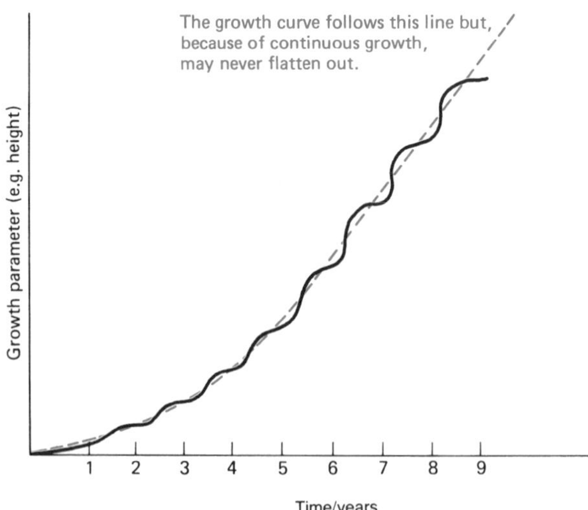

This is known as continuous growth.

Fig. 4.20(d) Growth curve of a perennial plant in temperate regions

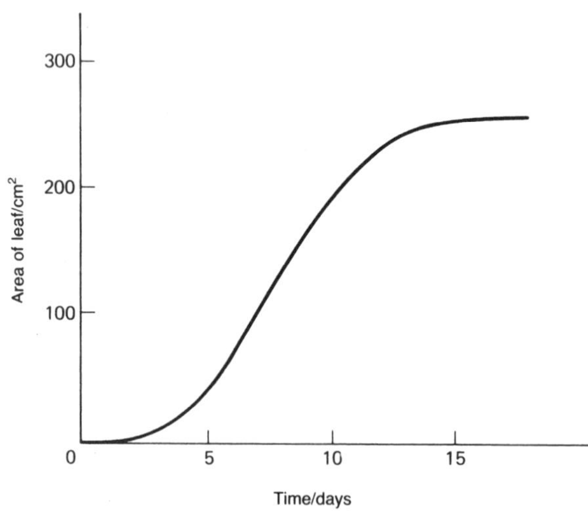

In plants growth of an organ often reflects the growth of the whole organism.

Fig. 4.20(e) Growth in area of cucumber leaf

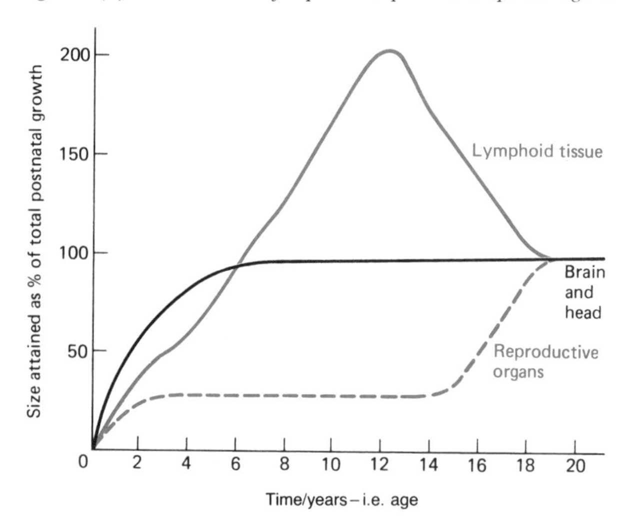

In animals growth of organs is often allometric.

Fig. 4.20(f) Growth curves of some human organs

Growth rate

A graph of growth rate is drawn by plotting growth increments (i.e. increase in growth over successive periods of time) against time (see Fig. 4.21).

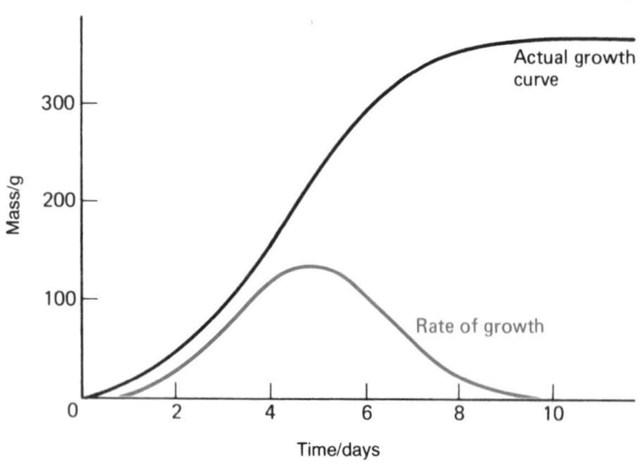

The growth rate curve of most organisms is bell-shaped.

Fig. 4.21(a) Relationship between actual growth and rate of growth

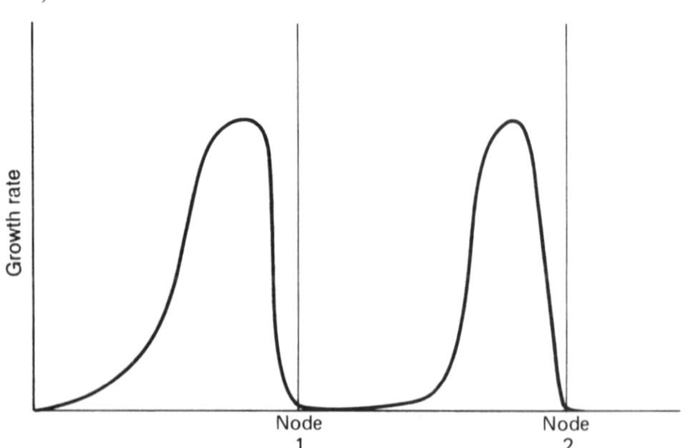

In *Tradescantia* maximum growth occurs at the base of each internode; there is no growth at the nodes.

Fig. 4.21(b) Growth rate of Tradescantia

The rate of growth in plants and animals is affected by:

1. **Genotype:** dominant and recessive alleles determine such things as protein synthesis, metabolism and size.
2. **Hormones:** auxins and gibberellins in plants and thyroxine and somatotrophin in animals.
3. **Nutrition:** plants require carbon dioxide, light and water; animals need proteins, fats, carbohydrates, vitamins, mineral ions and water.
4. **Environment:** e.g. light for vitamin D synthesis in some animals and for photosynthesis in plants; effect of length of day on meristematic activity; temperature has different effects on ectothermic and endothermic animals; thermoperiodicity in plants influences flowering, etc.

Growth may be limited by such negative factors as disease, pollution and parasites. These are the most important influences on growth, but some organisms may be affected by stress, atmospheric pressure and gravity.

DEVELOPMENT IN PLANTS

In the early stages of development, cell division occurs throughout the embryo, but as it develops into an independent plant the addition of new cells is restricted to certain parts – **meristems**. The presence of meristems, whose activity permits growth throughout the life of the organism, distinguishes plants from animals.

Meristems may be apical, i.e. located at the tips of main and lateral roots and shoots, or lateral, arranged parallel to the organ in which they occur, e.g. vascular cambium and cork cambium (phellogen).

Typically, meristematic cells have large nuclei, compact cytoplasm and small vacuoles. When conditions are favourable these cells divide by mitosis, then elongate and differentiate.

DEVELOPMENT IN ANIMALS

In contrast to plants, development takes place all over the body in animals. Embryonic development is triggered by fertilisation and the multiplication of cells mostly ceases after the organism reaches adult size; the number of organs remains constant.

Cleavage

Following fertilisation the nucleus of the zygote divides mitotically, each division being accompanied by cleavage of the cytoplasm to form separate cells, or blastomeres. These divisions give rise to an embryonic structure called a blastula, whose form depends on the amount of yolk in the egg, which varies for different animals.

If the egg contains relatively little yolk (i.e. is microlecithal), as in mammals, and the segmentation is holoblastic (cleavage of all the cytoplasm occurs), the blastomeres are more or less equal in size, although those at the vegetal pole may be slightly larger and fewer due to the accumulation of yolk (see Fig. 4.22).

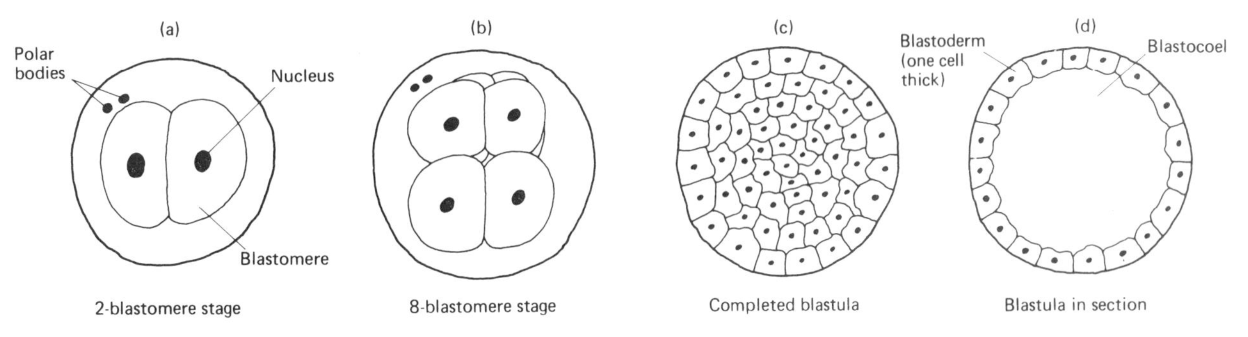

Fig. 4.22 Cleavage in Amphioxus *(a chordate)*

Chapter roundup

We have seen in this chapter that while individual organisms must die, the reproductive process ensures that the species as a whole survives. Reproduction is, however, more than a mechanism to retain a stable species; on the contrary, the process of sexual reproduction is designed to incorporate variety. This variety is achieved by:

1. the combining of two, often very different, parental genotypes;
2. the process of meiosis (see 3.2) which
   (a) randomly segregates the chromosomes on the metaphase plate;
   (b) permits crossing over during prophase I;
3. mutations;
4. the influence of the environment;

These processes produce the almost infinite variety upon which evolution flourishes. We have also seen that while reproduction confers many advantages, there is, in many organisms, a place for the asexual process. Although the scope for variation is less (only mutation and environmental effects), asexual reproduction allows the rapid build-up of individuals which possess a suitable genotype. This allows the exploitation of a favourable habitat by just a single individual.

We have seen that dispersal is necessary to:

1. prevent overcrowding and hence stop excessive competition for food, light, water, mates, etc., as well as reducing the likelihood of epidemic attack;
2. increase variety by reducing the risk of backcrossing and ensuring a greater mix of the genes within the gene pool;
3. exploit the variety which reproduction creates by spreading offspring into new and varied habitats where some may possess features which allow them to survive and breed successfully.

Obviously, as most animals can move themselves from place to place, dispersal is less of a problem than it is for the immobile plants, which have therefore evolved more elaborate means of dispersing their fruits and seeds.

Illustrative questions and worked answers

1 The graph below shows the thickness of the uterus lining throughout the menstrual cycle in an adult human female.

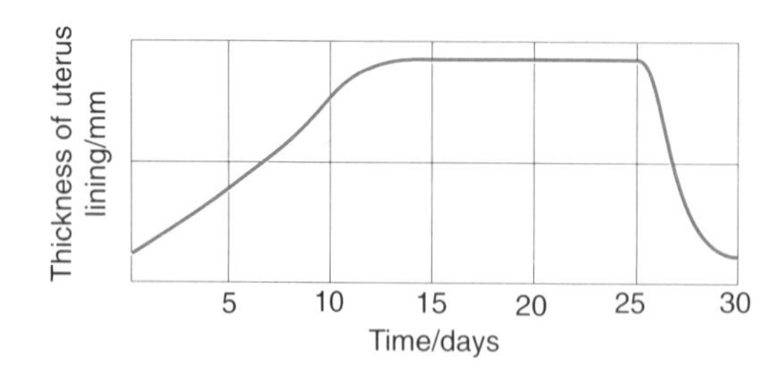

(a) From the graph state the day
   (i) ovulation is most likely to happen
   (ii) assuming sperm are present, when fertilisation is most likely to occur
   (iii) the corpus luteum begins to break down
   (iv) menstruation begins

(b) (i) Describe how the menstrual cycle is controlled by hormones, stating for each hormone where it is produced and exactly what it does.
(ii) Give two medical uses of the hormones you have mentioned.

Time allowed 40 minutes

Tutorial note

In this highly structured form of essay you should read the complete question before embarking on any part. You should then decide to which part of the question each piece of information is most relevant. If you dash into a detailed answer for part (a) it is very possible that you might include detail which is needed in part (b)(i) and which would then need to be repeated. As the same statement is highly unlikely to bring marks twice within the same question you will have wasted precious time. The moral, as always, is 'read the *whole* question carefully and then plan your answer thoroughly'.

As with many questions in biology the answer can be obtained in two ways. The first method involves learning all the facts and memorising information, including diagrams, which might arise on an examination paper. This is time consuming, demands an excellent memory, and may fail when an original and unexpected question is set. The second method also involves thorough learning – this is inescapable at A level as many candidates have found to their cost – but is selective. From a relatively few fundamental facts, you can deduce many others. This method requires less learning but the ability to reason logically is essential; it is also very flexible being equally applicable to the unexpected question as to the predictable one.

Part (a) is an example. You could have learnt the relevant days on which ovulation, menstruation, etc. normally occur and so answer effectively. If, however, the timings shown on the graph were atypical you could come unstuck with this memorised answer. On the other hand, if you had learnt and understood why there are changes in the thickness of the uterine lining you could deduce your own answer.

In part (i), if you appreciate that the uterus lining is thickening in readiness to receive and nourish a newly fertilised ovum, it follows that the ovum should arrive in the uterus when the lining is fully thickened. The journey from the ovary to the uterus may take up to a week and therefore release of the ovum 5 or 6 days before the lining reaches maximum thickness will ensure that it reaches the uterus when it is fully thickened. Later dates of release will also ensure this, but clearly early implantation is preferable in case menstruation occurs early. On the graph the probable days are 8–15 with 9 or 10 as the most likely. You should note the words 'from the graph' and not use memorised dates since these do not correspond with the ones in this question. For instance, day 14 is usually cited as the date for ovulation, because most cycles use the start of the menstrual flow as day 1. This graph uses the end of the menstrual flow as day 1 and so is displaced by about 5 days compared with the usual graphs. If you use memory rather than your ability to interpret the graph given you will fare badly.

For (ii) you need to know that an ovum is usually fertilised 2 or 3 days after release, making the most likely time for fertilisation between days 11 and 13 on the graph.

The corpus luteum breaks down shortly before menstruation. It is then easy to see that, since menstruation involves the breakdown of the uterus lining and it begins on day 25, the corpus luteum must begin to break down on about day 22 or 23.

Part (b)(i) is relatively straightforward and the information is given in 4.2, especially Fig. 4.11 and Fig. 4.12. You should be sure to include the influence of each hormone on the stimulation or inhibition of the production of other hormones. Part (b)(ii) refers to the use of sex hormones in contraceptive pills and fertility drugs.

Suggested answer

(a) (i) between days 8 and 15, most probably day 9 or 10
(ii) days 11 to 13
(iii) day 22 or 23
(iv) day 25

(b) (i) ❶ Follicle stimulating hormone (FSH), produced by the anterior lobe of the pituitary gland, causes Graafian follicles to develop in the ovary and stimulates the ovary to produce oestrogen.

❷ Oestrogen, produced by the ovary, causes repair of the uterus lining following menstruation and stimulates the pituitary to produce luteinising hormone (LH).

❸ Luteinising hormone, produced by the anterior lobe of the pituitary gland, causes ovulation to take place and stimulates the ovary to produce progesterone from the corpus luteum.

❹ Progesterone, produced by the corpus luteum, causes the uterus lining to be maintained in readiness for the blastocyst (young embryo) and inhibits the production of FSH.

The hormones are produced in the following sequence: FSH, oestrogen, LH, and lastly progesterone. Progesterone at the end of the sequence inhibits production of FSH. In turn, the production of the other hormones stops, including progesterone itself. The absence of progesterone means that the inhibition of FSH ceases and so progesterone production recommences. In turn, all the other hormones are produced. This alternate switching on and off of the hormones produces a cycle of events – the menstrual cycle.

(ii) Contraceptive pills and fertility drugs

2 The following chart provides information on seven plant species natural to Britain. It shows, for each, the period during the year when leaves are present and when flowers, if any, are produced.

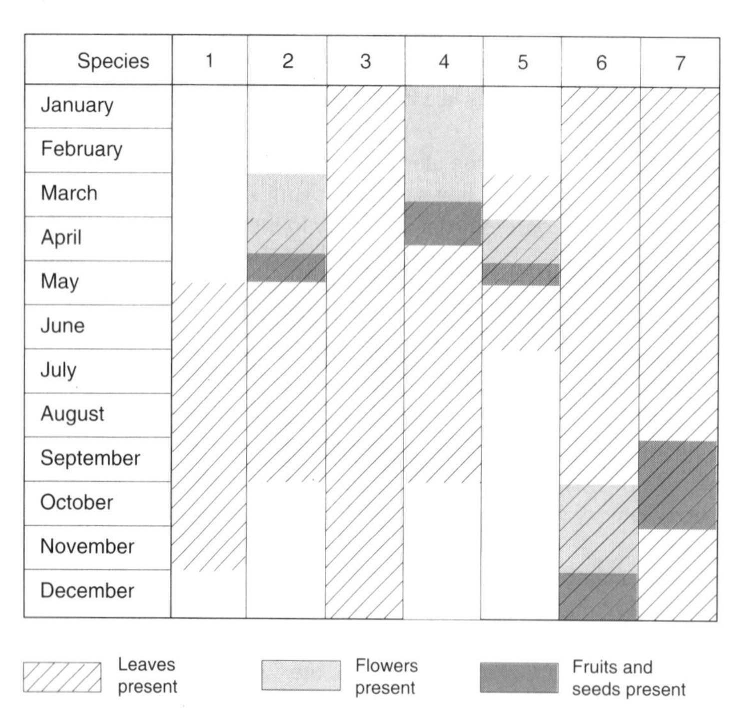

For each of the following plant types state with reasons which of the species in the chart are most likely to be:

(a) (i) a moss
(ii) deciduous trees (two species)
(iii) ivy
(iv) a fern
(v) a small woodland species, e.g. bluebell, lesser celandine

(b) Of the four flowering species, which two are most likely to be insect pollinated? Give a reason for your answer.

Time allowed 15 minutes

Tutorial note

In answering this question you will need to be aware of the life cycles of the plants listed, in particular their growing and dormant periods, as well as the type and timing of their reproduction. You should remember to give reasons in each case and, when reaching your answer, try to use more than one piece of information, e.g. not just if and when a species flowers but also how long it keeps its leaves and when its fruits and seeds appear.

Suggested answer

(a) (i) Species 3

Reason: mosses do not have flowers, fruits or seeds and it is therefore either species 1 or 3. As mosses do not lose all their leaves at one time but remain leafy throughout the year, this eliminates species 1.

(ii) Species 2 and 4

Reason: deciduous trees lose their leaves at some time during the year. Species 3, 6 and 7 cannot therefore be deciduous trees. Almost all British deciduous trees bear flowers (larch is one exception). This eliminates species 1 which has none. Of the remaining, species 5 is the least likely to be a deciduous tree as it loses its leaves in mid-June whereas most deciduous trees lose theirs during the autumn months.

(iii) Species 6

Reason: ivy is an evergreen plant and hence must be species 3, 6 or 7. Ivy is also an angiosperm and bears flowers. Of the three species mentioned only species 6 bears flowers.

(iv) Species 1

Reason: ferns are deciduous but do not have flowers. Only species 1 satisfies these conditions.

(v) Species 5

Reason: to obtain enough light to grow and produce flowers these plants emerge early in the year, before the canopy of the tree foliage has developed and shaded them. To achieve this early rapid growth they use carbohydrate stored in their bulbs (bluebells) or root tubers (lesser celandine) from the previous season's growth. As photosynthesis becomes more difficult due to lack of light when the taller plants and trees are in full foliage, they lose their leaves during mid-summer, soon after flowering. Species 5 is the only one that shows this type of life cycle.

(b) Species 2 and 5

Reason: being ectothermic most insects do not emerge from over-wintering until the temperature is more favourable, i.e. March and April. Fewer insects are active in January and February (the main flowering period of species 4) or October and November (the main flowering period of species 6). Species 2 and 5, which flower in March and April, are therefore more likely to be insect pollinated.

3 Give a description of the events which take place in reproduction beginning with the production of spermatogonia and oogonia and ending with the successful implantation of a mammalian blastocyst.

Time allowed 40 minutes

Tutorial note

This is a descriptive essay demanding a thorough knowledge of development in the mammal. You should be aware that both the spermatogonia and the oogonia are diploid and therefore meiosis has yet to occur before they form the sperm and ova, respectively. An examiners' report on a similar question noted that, while the process of gametogenesis was understood, there was nevertheless confusion about the timings of meiotic and mitotic divisions (see Fig. 4.3 for these details).

The next phase to describe is the release of the sperm, their journey to the uterus and their fusion with an ovum, including the processes of mating. It is probably worth a mention of courtship. Candidates too often ignore this process and accounts therefore infer that mating is a sudden event which takes place each time the male and female of a species happen to meet!

You should describe the developmental events which occur between the formation of the zygote and the implantation of the blastocyst. Examiners report that this part of the answer is frequently very sketchy. The nature of cleavage described in 4.4 and Fig. 4.22 should therefore be understood. Another weakness reported by examiners is a lack of understanding of implantation with little if any reference made to embryonic membranes. These details are given in 'Development of the zygote' in 4.2.

Finally, you should include a fairly detailed account of the hormonal control of all the events you have described.

Suggested answer

Gametogenesis Diploid spermatogonia grow rapidly to form primary spermatocytes which undergo the first meiotic division to form secondary spermatocytes (2n) and then the second meiotic division to form haploid spermatids, which differentiate into sperm. In the female the diploid oogonia grow into primary oocytes and reach the prophase of the first meiotic division *prior to birth*. The ovaries therefore contain primary oocytes which are haploid and these only undergo their second meiotic division after fertilisation by the sperm.

Mating Details of courtship. Process of mating including the route taken by the sperm: testes, epididymis, vas deferens (secretions of prostate and Cowper's glands), urethra. Importance of peristalsis. Sexual arousal, erection of the penis, introduction into vagina, ejaculation, swimming of sperm, contraction of uterus. Release of ova, movement along oviduct.

Fertilisation Penetration of vitelline membrane (acrosome of sperm), formation of fertilisation membrane and the combining of two haploid sets of chromosomes to give a diploid zygote.

Development of zygote Movement down oviduct (cilia), time period (one week). Holoblastic cleavage to give blastomeres and blastocyst.

Implantation into uterus lining. Development of the outer layer of blastocyst into trophoblastic villi, which project into lining. Formation of amnion, chorion and allantois – the last two combining to form the placenta.

Hormonal control The roles of the male hormones (ICSH and testosterone) and female hormones (FSH, LH, oestrogen and progesterone). See 4.2.

Question bank

1 When a yeast colony was grown in a nutrient medium and the number of cells were counted over a period of 20 hours, graph 1 was obtained. If the rate of cell division was measured over the same period, graph 2 was obtained.

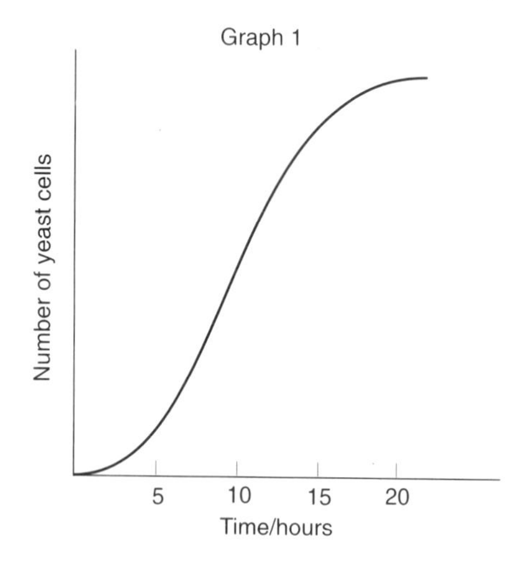

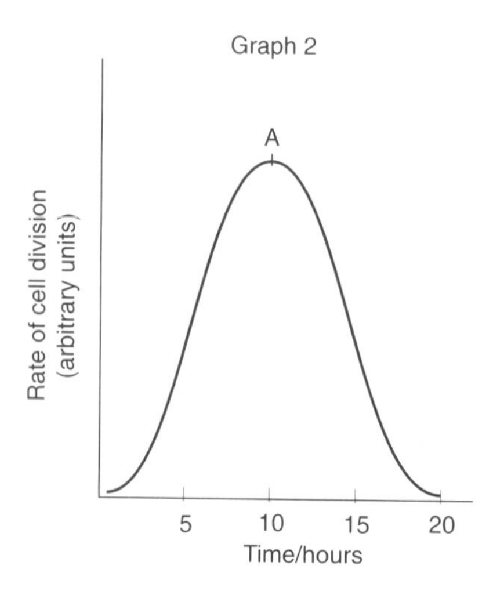

(a) Explain the shape of each graph.

(b) Suggest two possible reasons for the fall in growth rate after point A on graph 2.

Time allowed 10 minutes

Pitfalls

You will need to be careful about the term 'rate' and be aware that it is an increase or decrease over set time intervals (see the explanatory table in the points given below). Once you have understood this you will appreciate that the fall in graph 2 after point A does not mean that the number of yeast cells is diminishing, only that they are increasing less rapidly.

In suggesting reasons for the fall after point A, use the most likely events such as lack of food or build-up of waste rather than obscure factors (e.g. the introduction of a metabolic poison such as cyanide) or factors which you should assume remain constant unless you are told otherwise (e.g. temperature).

Points

(a) Graph 1 measures the actual number of yeast cells. As initial numbers are small, even when the cells are budding rapidly, growth is slow. However, as the number of budding individuals increases, numbers rise more rapidly and the graph increases exponentially. Some limiting factor causes the rate of increase to fall and so the graph begins to flatten until finally the number of new individuals exactly balances those dying, and so an equilibrium is established with no net increase in yeast cells. The graph flattens completely becoming parallel to the time axis.

In graph 2 the situation is more difficult and causes many students problems. It is essential to distinguish between 'growth' and 'rate of growth'. If, for instance, the actual number of yeast cells is measured at five-hourly intervals and the results obtained are 100, 350, 900, 1600, 2150, 2400 and 2400, then a graph of these plotted against time will have a shape similar to graph 1 – **a growth curve**. If, however, the increase for each five-hourly interval (i.e. the difference between adjacent numbers) is calculated, the following is obtained:

Hours	0		5		10		15		20		25		30
Number of cells	100		350		900		1600		2150		2400		2400
Increase in cells at five-hourly intervals		250		550		700		550		250		0	

This last set of figures when plotted against time produces a graph similar in shape to graph 2 – **a growth rate curve**.

To summarise:

growth curve = amount of growth time^{-1}

growth rate curve = amount of growth time^{-2}

The shape of graph 2 can now be explained. Initially when the number of yeast cells is small the rate of increase is low. As numbers increase so does the *rate* of increase until some limiting factor reduces the growth rate. Graph 2 therefore falls until the rate is equal to zero, i.e. actual numbers are not increasing. The growth rate curve is therefore a measure of the gradient of the growth curve.

(b) Choose from:

1. A build-up of waste products from the yeast cells, in this case ethanol, may begin to kill individuals.
2. The food supply (e.g. sucrose) of the yeast colony may be used up, so reproduction slows as less energy is available for budding.
3. An essential mineral such as nitrogen or phosphorus may be used up. Without these to produce the proteins for reproduction, the growth rate decreases.

2 A pollen tube of a flowering plant is

A the male gamete
B the embryo
C the male gametophyte
D the female gamete
E a germinating spore

Points

In multiple-choice questions it may be that more than one answer could be applied, but there is always one which fits the bill better than any other. It is sometimes best if you discount wrong options first and then seek to justify, with sound biological reasons, why the remaining answer is correct. This is the approach taken here.

Option A may be discounted because the two male nuclei of the pollen tube represent the gametes. The pollen tube is therefore more than just the gametes. Option B is incorrect because an embryo is the result of fertilisation. However, the pollen tube contains the gametes *for* fertilisation and must therefore precede fertilisation rather than follow it. Option D is clearly incorrect; the pollen tube is a male structure. Option E could easily mislead candidates. The pollen tube does arise from germination of the pollen grain which is the microspore and might therefore be considered a germinating spore. The question, however, makes no mention of the spore (pollen grain) only the tube derived from it. This tube is the gametophyte because it is derived from a spore and it is male because it is derived from the male spore (microspore). Option C is therefore the best answer.

3 Study the diagram below of a mammalian ovary.

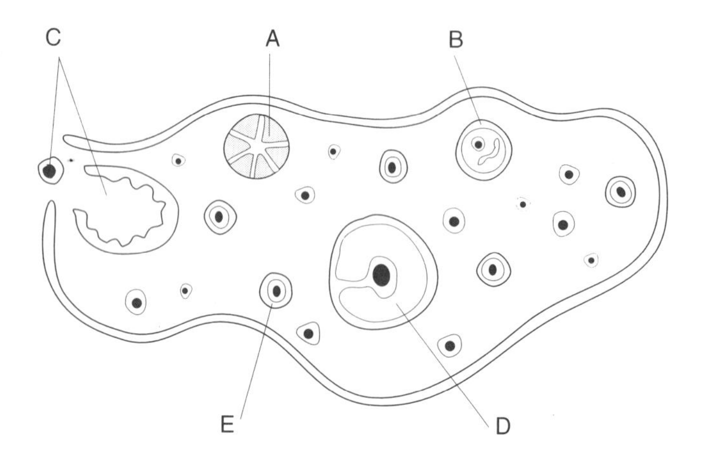

(a) Name the parts labelled A, B, C, D and E.

(b) State the correct developmental sequence of the structures labelled A, B, C, D and E.

(c) State whether each of the following is diploid or haploid:
 (i) germinal epithelium
 (ii) ovum
 (iii) secondary oocyte
 (iv) primary oocyte

Time allowed 5 minutes

Pitfalls

In labelling the parts A–E be prepared to give detail, e.g. rather than 'Graafian follicle', which could apply to two of the parts, you will need to give information about the stage of maturity of the follicle to distinguish between them.

Points

(a) A corpus luteum
B primary oocyte surrounded by follicle cells
C ovum being released from Graafian follicle at ovulation
D mature Graafian follicle before ovulation
E primary follicle

(b) The primary follicle (E) is present in the ovary at birth. After puberty it may mature into a primary oocyte (B) and later a Graafian follicle (D). The ovum from this Graafian follicle is released at ovulation (C) and the empty follicle develops into the corpus luteum (A).

The correct sequence is therefore E, B, D, C, A.

(c) (i) diploid
(ii) haploid
(iii) haploid
(iv) diploid

CHAPTER 5

ENERGETICS

Units in this chapter

Chapter objectives

You may have been asked, or asked yourself, what the difference is between living and nonliving. If you have, you may also have struggled to make the distinction. The difference is often described in terms of seven basic characteristics – reproduction, growth, respiration, nutrition, excretion, movement and sensitivity – which are only carried out by living organisms. There is, however, an alternative method which relates to the way each uses energy.

All matter tends to lose energy – hot objects cool, those thrown into the air fall to the ground. Put another way, material tries to attain its lowest energy state. Any system which is highly ordered has much energy. Under natural conditions it will tend to lose this energy and so become disordered. A house and a car both represent highly ordered systems with much energy. If left to themselves they will lose energy – slates fall from the roof of the house, the bodywork of the car rusts. They thus become disordered and random rather than ordered and organised. The state of random disorder is known as **entropy**. A disordered system has high entropy, an organised one has low entropy. Free energy and entropy are therefore inversely proportional; high free energy means low entropy and vice versa. The difference between living and nonliving is that living organisms, being highly ordered, have low entropy and much free energy. Nonliving things, however, have high entropy and little free energy. Living organisms only maintain their low entropy by the constant addition of energy to themselves, in the form of sunlight (photosynthesising plants) or food (animals). In this way they maintain an ordered system with much free energy despite the natural tendency to become disordered. The householder or car owner can do much the same; by using energy to replace the loose slates or to touch up rust spots on the bodywork of the car they can maintain an ordered system. The difference is that while living organisms can maintain low entropy for themselves, nonliving systems cannot.

In this chapter we will explore in much more detail how the basic laws concerning energy (laws of thermodynamics) apply to living organisms. We will look at how the energy that maintains low entropy is obtained (photosynthesis and heterotrophic nutrition) and how energy is released (respiration) to permit organisms to carry out those seven processes that distinguish living from nonliving.

5.1 ENZYMES AND ENERGY

THE LAWS OF THERMODYNAMICS

These apply to the conversion of one form of energy to another. Energy is defined as 'the capacity to do work' and takes a number of different forms, e.g. light, heat, chemical.

The first law of thermodynamics states that: energy cannot be created or destroyed but may be converted from one form into another. Much of the energy of the world is actually derived from sunlight. If you are reading this under artificial illumination, the light energy will almost certainly be from the conversion of electrical energy in the lamp. The electricity in turn will have been the result of the conversion of the kinetic (movement) energy of a turbine, which will have been derived from the heat energy used to expand water into steam. The heat may have been the result of the burning of coal (chemical energy), which is actually fossilised plant material that derived its energy from sunlight during photosynthesis. It is quite likely that the light energy you are reading this by was once sunlight captured millions of years ago. Only a little of the original sunlight will now reach you as light as much is lost as heat at each intermediate stage. However, energy is never destroyed. When biologists and others talk of energy being 'lost' they are not implying it has been destroyed, but simply that it has been converted to a form which cannot be usefully made available for a particular purpose.

The second law of thermodynamics states: all natural processes tend to proceed in a direction which increases the randomness or disorder of a system. In other words, all natural processes tend to high entropy. Organisms maintain low entropy only by using energy from their surroundings. These surroundings in turn have increased entropy. As the organisms and their surroundings represent one system the overall entropy is increased and the second law of thermodynamics is not violated.

ENERGY AND CHEMICAL REACTIONS

The chemical reactions which occur within living organisms are called **metabolism**. They are divided into two types. **Anabolism** is the build-up of simple chemicals into complex ones and **catabolism** is the breakdown of complex chemicals into simpler ones.

The products of a spontaneous chemical reaction have a higher entropy than the reactants, i.e. they have less free energy. Even so, such reactions often require an initial input of energy to enable them to occur. This is referred to as **activation energy** (Fig. 5.1).

All chemical reactions are reversible, the direction of the reaction being determined by the conditions. For example, a high pH may cause the reaction to proceed in one direction while a low pH reverses it.

Some reactions liberate energy and are termed **exogonic**; those which absorb free energy are **endogonic**.

ENZYMES

Enzymes are large globular molecules with catalytic properties and, until recently, all were thought to be proteins. However, discoveries have shown some (known as ribozymes) to be made of RNA, but the vast majority are protein in nature. Catalysts alter the rate of a chemical reaction without themselves undergoing a permanent change. In other words, they can be used over and over again.

Enzyme function

As we have seen, reactions need to exceed their activation energy if they are to take place. Enzymes reduce the need for activation energy and so allow reactions to take place more readily. In practice they allow reactions to take place at lower temperatures than would be the case without them.

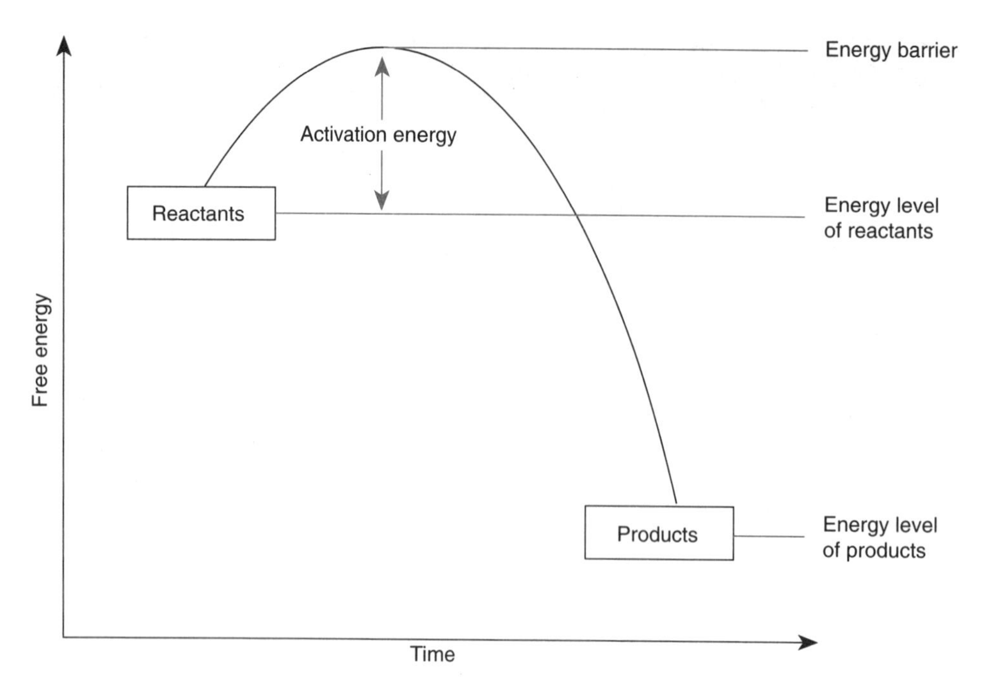

Fig. 5.1 Activation energy

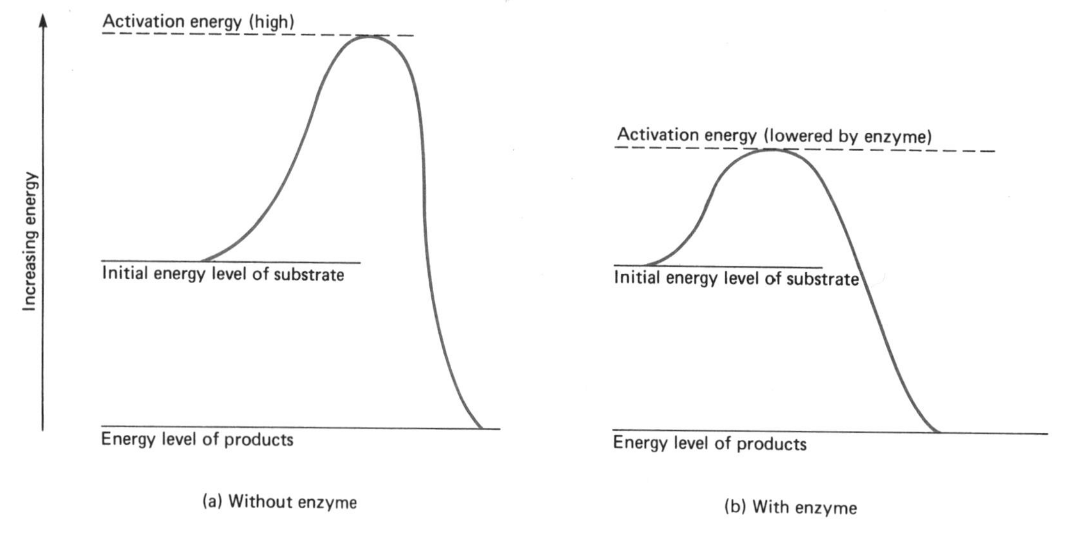

Fig. 5.2 How enzymes lower activation energy levels

At a molecular level it is thought that the substrate molecules fit precisely into the enzyme molecules in the same way that a key fits a lock. This theory is referred to as the **lock and key mechanism** (see Fig. 5.3). In practice it is probable that the enzyme moulds itself to some extent to the shape of the substrate. The part of the enzyme molecule into which the substrate fits is called the **active site**. The configuration of the enzyme molecule is due to ionic bonding, hydrogen bonding, disulphide bridges and hydrophobic interactions (see 1.1).

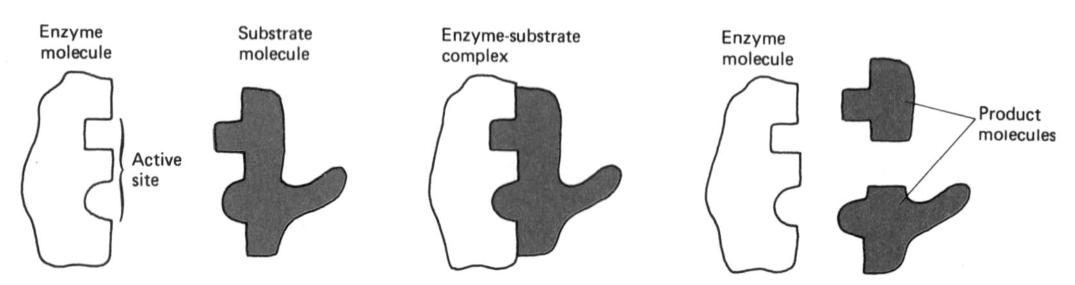

Fig. 5.3 Lock and key mechanism of enzyme action

Naming of enzymes

Enzymes are given two names. The **systematic name** can be long and complicated and is based on the classification of enzymes given in Table 5.1. The **trivial name** is shorter and easier to remember. Typically the enzyme's trivial name is based on:

1. the substrate
2. the type of reaction it catalyses
3. the ending '-ase'

e.g. DNA polymerase, citrate dehydrogenase.

Table 5.1 *Enzyme classification*

Enzyme group	Reaction catalysed	Example
Oxidoreductases	All reactions of the oxidation–reduction type	Dehydrogenases, oxidases
Transferases	The transfer of a group from one substrate to another	Transaminases, phosphorylases
Hydrolases	Hydrolytic reactions	Phosphatases, peptidases, amidases, lipases
Lyases	Additions to a double bond or removal of a group from a substrate without hydrolysis, often leaving a compound containing a double bond	Decarboxylases
Isomerases	Reactions where the net result is an intramolecular rearrangement	All-trans retinal →all-cis retinal
Ligases	The formation of bonds between two substrate molecules using energy derived from the cleavage of a pyrophosphate bond, such as ATP	Acetate + CoA–SH +ATP→ acetyl–CoA + AMP + P–P

Characteristics of enzymes

Specificity We have seen that the substrate molecule makes a precise fit into the active site of the enzyme molecule. While the enzyme molecule may be flexible up to a point, it should be apparent that the number of molecules which can fit precisely into the active site is very small – indeed it is frequently limited to just one type. Each enzyme is therefore specific to one type of reaction.

Reversibility Enzymes do not alter the equilibrium of a reaction, only the speed at which it is reached. They therefore catalyse the forward and reverse reactions equally. The enzyme carbonic anhydrase, for example, accelerates the conversion of carbon dioxide and water to carbonic acid in respiring tissues and the reverse reaction in the lungs of mammals. It is not the carbonic anhydrase but the pH which determines the direction of the reaction:

$$CO_2 + H_2O \xrightarrow{\text{lower pH (e.g. in tissues)}} H_2CO_3$$

$$CO_2 + H_2O \xleftarrow{\text{higher pH (e.g. in lungs)}} H_2CO_3$$

Temperature As temperature increases the molecules move faster (kinetic theory). In an enzyme-catalysed reaction this increases the rate at which enzyme and substrate molecules meet and hence the rate at which the product is formed. As the temperature continues to rise, however, the hydrogen and ionic bonds which hold the enzyme molecule in shape are broken. The molecular structure is disrupted and the enzyme ceases to function because the active site no longer accommodates the substrate. The enzyme is said to be **denatured**. The effect of these two influences is shown in Fig. 5.4.

pH We have seen that the efficient functioning of an enzyme depends upon its molecular shape and that this shape is determined, in part, by ionic and hydrogen bonding. Any change in pH affects this bonding and so alters the shape of the enzyme. Each enzyme therefore has an optimum pH at which its active site best fits the substrate. Variation either side of this pH leads to denaturation of the enzyme. Enzymes such as salivary amylase work best around pH 7, while pepsin functions well around pH 2. At the other extreme, arginase has an optimum of pH 10.

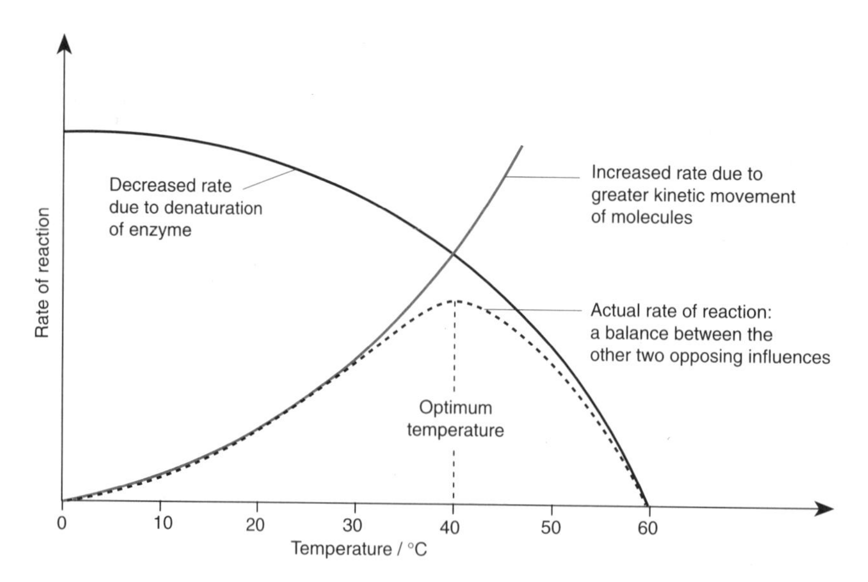

Fig. 5.4 The effect of temperature on an enzyme-catalysed reaction

Enzyme concentration Enzymes function efficiently in very low concentrations as the molecules can be used over and over again. Provided there is excess substrate, an increase in enzyme concentration will lead to a corresponding increase in the rate of reaction. Where the substrate is in short supply (i.e. it is limiting) an increase in enzyme concentration has no effect. See Fig. 5.5.

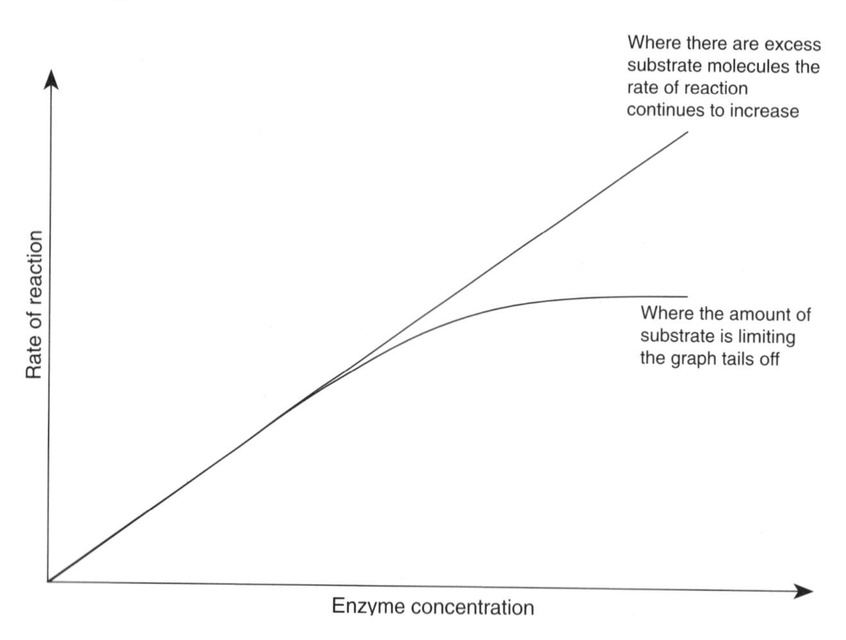

Fig. 5.5 The effect of enzyme concentration on the rate of reaction

Substrate concentration Provided there is an excess of enzyme molecules, an increase in the substrate concentration produces a corresponding increase in the rate of reaction. At the point where there are sufficient substrate molecules to occupy all active sites on the available enzyme molecules, the rate of reaction is unaffected by a further increase in enzyme concentration. See Fig. 5.6.

Inhibition Inhibitors that compete with the substrate for the active sites of the enzyme are known as **competitive** inhibitors. The greater the concentration of substrate, therefore, the more likely it is to occupy the active sites and the less the effect of the inhibitor. For example, malonic acid competes with succinate for the active sites of the respiratory enzyme succinic dehydrogenase.

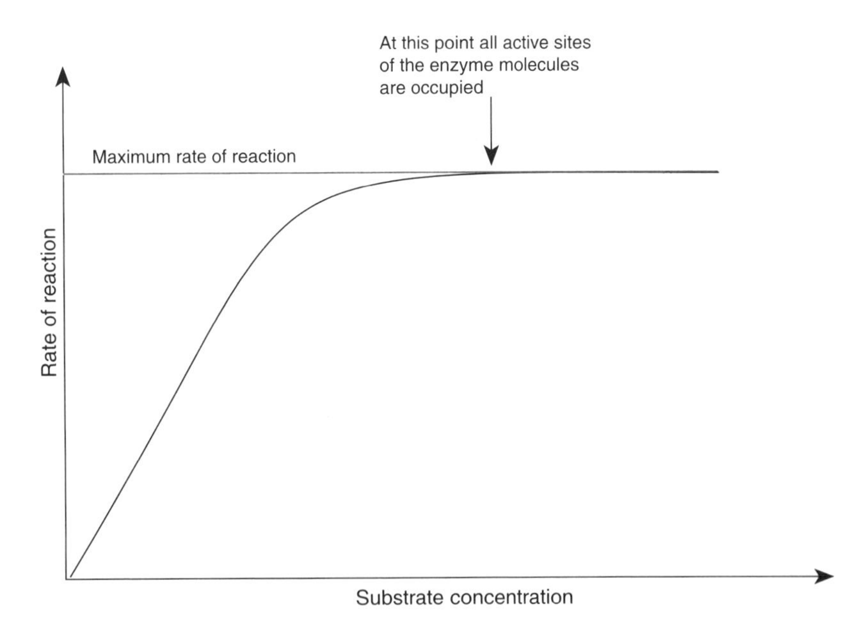

Fig. 5.6 The effect of substrate concentration on the rate of reaction

Noncompetitive inhibitors attach themselves to the enzyme at a site other than the active site. In doing so, however, they alter the shape of the active site in such a way that the substrate is unable to occupy it and the enzyme cannot function. As the substrate and inhibitor are not competing for the same site, an increase in substrate concentration does not diminish the effect of the inhibitor. An example of a noncompetitive inhibitor is cyanide which inhibits another respiratory enzyme, cytochrome oxidase.

Enzyme cofactors Cofactors are nonprotein substances which influence the functioning of enzymes. They include **activators** which are essential for the action of some enzymes, e.g. chloride ions are required for the activity of salivary amylase. **Coenzymes** are nonprotein organic cofactors which influence enzyme functioning but are not bound to the enzyme. Nicotinamide adenine dinucleotide (NAD) is a coenzyme for many dehydrogenase enzymes. **Prosthetic groups** are organic molecules which are bound to enzymes and are essential to their activity, e.g. the haem group is a prosthetic group for the enzyme catalase which converts hydrogen peroxide to water and oxygen.

METABOLIC PATHWAYS AND THEIR CONTROL

Within the tiny confines of a single cell, many hundreds of different reactions take place. The process is not haphazard, but is a highly structured system of metabolic pathways. The enzymes which control these processes are often attached to the inner membrane of cell organelles in a very precise sequence. This increases the chance of each enzyme coming into contact with its substrate and leads to greater efficiency. Within each organelle there may be the optimum conditions for the functioning of these enzymes, e.g. the pH may vary from organelle to organelle.

The level of metabolites is maintained at a constant level by a process called **homeostasis** (see 8.1). This is often achieved by the final product of a metabolic pathway inhibiting the enzyme at the start of the same pathway.

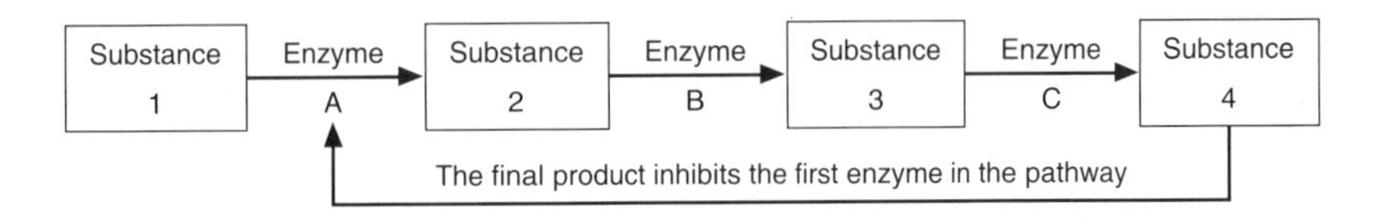

Fig. 5.7 Negative feedback control of a metabolic pathway

In the example shown substance 4, at the end of the pathway, inhibits enzyme A, the first in the pathway. If the concentration of substance 4 rises, inhibition of enzyme A occurs and so the amount of substance 4 is reduced. As the level of substance 4 falls, the inhibition diminishes and hence more substance 4 is produced. In this way the level of this metabolite is maintained between strict limits. It is an example of **negative feedback** (see 8.1).

BIOCHEMICAL TECHNIQUES

An individual cell may carry out up to 1000 chemical reactions, which must be carefully controlled. To assist control the conversion of one chemical to another takes place in a series of small steps known as a metabolic pathway. In addition the energy is released gradually rather than in a violent manner.

To determine the order of intermediates, a number of laboratory techniques are used:

- the addition of an intermediate which should lead to an increase in all intermediates between it and the product;
- the addition of an enzyme inhibitor which should cause an accumulation of the enzyme's substrate and the substances formed prior to it.

Centrifugation

Cells are broken down mechanically to form a homogenate suspension which is spun at a high speed causing the heavy structural components to separate out and form a sediment below a supernatant liquid. Particles of a different mass may be separated by centrifuging at different speeds (differential centrifugation).

Dialysis

Small molecules such as inorganic ions may be separated from larger ones such as protein by placing them within a membrane permeable only to the ions. The membrane is then placed in water into which the ions diffuse.

Chromatography

This is a means of separating out mixtures of chemicals by using their different solubilities in certain solvents. A concentrated spot of the solution to be separated is placed at one corner of a strip of paper. This is dipped in a solvent which carries the chemicals dissolved in it up the paper and deposits them – the least soluble first, the most soluble last. If necessary other chemicals may be used to make the spots visible, e.g. Ninhydrin colours amino acids. They are identified by their colour and R_f values:

$$R_f = \frac{\text{distance moved by spot}}{\text{distance moved by solvent}}$$

Spots may be further separated by running a different solvent at right angles to the first (two-way chromatography) or, instead of paper, a layer of suitable material (e.g. silica gel) may be used (thin layer chromatography).

Electrophoresis

This is similar to chromatography except that the spot is put in the centre of a strip of paper or a thin layer soaked in a suitable electrolyte. A potential difference is applied across it so that anions move towards the anode and cations towards the cathode, thus separating the substances, not on their solubilities, but on their relative electropositivity and negativity.

Isotopes

Since an isotope differs from the normal element its progress along a metabolic pathway can be traced in a number of ways. If radioactive, e.g. ^{14}C, its presence in intermediates can be determined using Geiger counters once all the intermediates have been separated by one of the above methods. If too little radiation is emitted, the chromatogram may be placed next to a photographic plate, which becomes exposed by those spots containing the radioactive material. The developed photograph is called an autoradiogram.

If the isotopes are not radioactive they are detected and measured using a mass spectrometer, which separates them according to their atomic mass.

X-ray diffraction

When a beam of X-rays is fired at a sample of the substance being studied, the atoms scatter the rays and the diffraction pattern so formed is recorded on a photographic plate. Rotation of the sample to allow patterns from different angles to be made gives a complete analysis. Computers are used to determine the precise positions of atoms from the patterns. This is used, for example, in determining the structure of DNA and protein.

Colorimetry

Many substances are detected by the production of a colour on the addition of an indicator or test reagent. The intensity of this colour is usually proportional to the amount of the substance present, e.g. the concentration of a starch solution may be determined by the intensity of the blue produced on the addition of iodine in potassium iodide solution. A colorimeter is an instrument which passes a beam of light through a sample of the solution under test (which is usually contained in a special flat-sided tube called a cuvette). The amount of light passing through alters the electrical resistance of a photocell, producing a deflection on the meter that measures the voltage across the cell. The scale shows either optical density and/or percentage transmission of light. The colorimeter has many applications, such as the measurement of the rate of starch hydrolysis by amylase over a period of time. It is necessary to draw a calibration graph of colorimeter readings against known concentrations of starch, from which the actual starch concentration for any given colorimeter reading can be found.

5.2 AUTOTROPHIC NUTRITION (PHOTOSYNTHESIS)

We saw in 5.1 that in order to maintain low entropy, all living organisms require energy. It is **autotrophs** which 'capture' this energy and use it to convert inorganic sources of carbon into complex organic molecules. A few organisms can use energy derived from the reactions of inorganic chemicals. These are the **chemoautotrophs**. Most, however, use sunlight as their energy source. These are the **photoautotrophs**.

LIGHT

Visible light represents that part of the electromagnetic radiation spectrum which lies between 380 nm (violet) and 750 nm (far red). Three properties of light that are of importance to organisms are:

- spectral quality
- intensity
- duration

The Particle Theory proposed by Einstein in 1905 states that light is composed of particles of energy called photons. The energy of a photon is not the same for all kinds of light but is inversely proportional to the wavelength – the longer the wavelength, the lower the energy. To be of use the light must be converted to chemical energy.

PIGMENTS

The main photosynthetic pigments are chlorophylls and carotenoids.

Chlorophyll This name covers a group of closely related substances, the most important of which are chlorophylls *a* ($C_{55}H_{72}O_5N_4Mg$) and *b* ($C_{55}H_{70}O_6N_4Mg$). They occur in plants in the approximate ratio of 2:1. Chlorophyll is usually contained within chloroplasts (see 1.2). The total weight of chlorophyll in green leaves varies between 0.55% and 0.20% of the fresh weight.

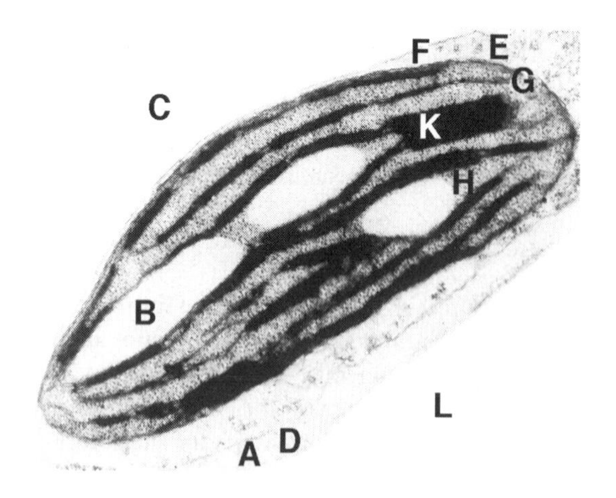

Key

Chloroplast

A Cell wall
B Starch grain
C Vacuole
D Plasmalemma
E Cytoplasmic ribosomes
F Tonoplast
G Chloroplast envelope
H Chloroplast stroma with ribosomes
K Granum
L Intercellular space

Fig. 5.8 Electron micrograph of a chloroplast in a leaf cell of Agrostis, *magnification 50 000 ×*

Carotenoids These are a large group of hydrocarbon pigments found throughout the plant. About 100 different carotenoids have been recognised. There are two major groups: carotenes and xanthophylls. They are important pigments in photosynthesis, especially in some groups of algae, e.g. fucoxanthin in the brown algae (Phaeophyta). They strongly absorb light in the violet end of the spectrum.

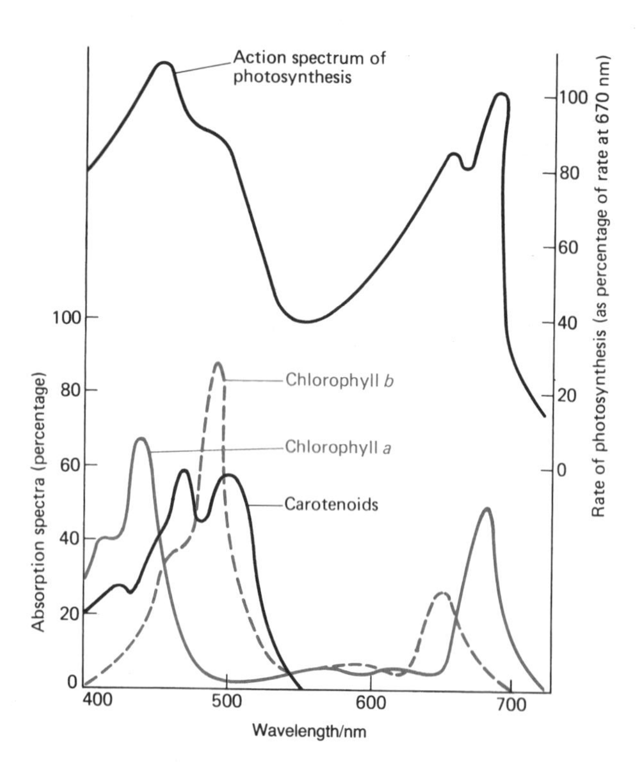

Fig. 5.9 Graph to show relationship between action spectrum for photosynthesis and absorption spectra of major photosynthetic pigments

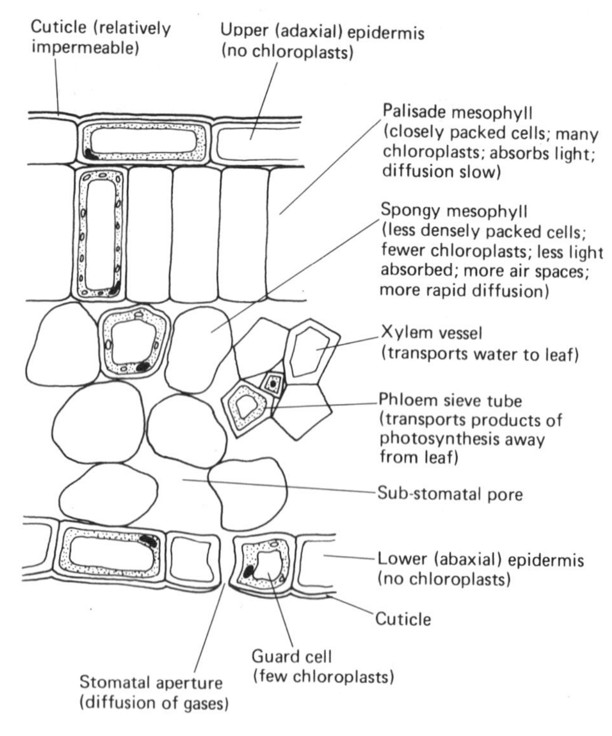

Fig. 5.10 VS part of leaf to show structure in relation to photosynthesis

MECHANISM OF PHOTOSYNTHESIS

The overall equation for photosynthesis given below is a simplified summary of a very complex series of reactions:

$$\underset{\text{carbon dioxide}}{6CO_2} + \underset{\text{water}}{6H_2O} \xrightarrow[\text{sunlight}]{\text{chlorophyll}} \underset{\text{glucose}}{C_6H_{12}O_6} + \underset{\text{oxygen}}{6O_2}$$

In 1932 Emerson and Arnold exposed plants to flashes of light lasting 10^{-4} seconds and varied the period of darkness between the flashes. They found that the yield of carbohydrate increased as the length of the dark periods increased up to a maximum dark period of 20 milliseconds (0.02 s) at 25 °C. If the temperature was reduced so was the yield of carbohydrate, although this could be compensated for by increasing the period of darkness. Increasing the duration of the flashes of light had no effect. These experiments suggest that photosynthesis has two stages:

1. a light-dependent stage (photochemical stage), which requires light, but is unaffected by temperature;
2. a light-independent or dark stage (chemical stage), which does not require light and is affected by temperature.

Before these stages can take place, the light essential to photosynthesis must be collected in a process known as **light harvesting**.

Light harvesting

Within the thylakoid membranes of the chloroplast (see 1.2 Figs. 1.16(a) and (b)) the chlorophyll molecules are grouped with the accessory pigments into units of several hundred molecules, each group being known as an **antennae complex**. Special proteins funnel the light entering the chloroplast into special molecules of chlorophyll a known as the **reaction centre chlorophyll molecule**. There are two types of reaction centre, which differ in their chlorophylls and their functions. These are known as **photosystem I (PSI)** and **photosystem II (PSII)**.

THE LIGHT-DEPENDENT STAGE

This takes place in the thylakoids of the chloroplast. When light is received by a chlorophyll molecule one of its electrons is lost, leaving the chlorophyll molecule positively charged and chemically less stable. This electron may return to the chlorophyll molecule, via the carrier system, thus stabilising it again. In doing so some of the energy it loses is used to combine adenosine diphosphate and inorganic phosphate into adenosine triphosphate (phosphorylation). As further light can again raise the electron's energy and allow the process to be repeated, it is termed **cyclic photophosphorylation**, and involves only photosystem I.

The electron may, however, not return directly to the chlorophyll molecule. It may combine with hydrogen ions (H^+) that result from the natural dissociation of water:

$$\underset{\text{water}}{H_2O} \rightleftharpoons \underset{\text{hydrogen ion}}{H^+} + \underset{\text{hydroxyl ion}}{OH^-}$$

In doing so a hydrogen atom is formed which is immediately taken up by a hydrogen acceptor such as **nicotinamide adenine dinucleotide phosphate** (NADP) which enters the dark reaction. The stability of the chlorophyll is restored in this case by the hydroxyl ion (OH^-) from the dissociation of water donating its extra electron to the chlorophyll molecule. The electron is transported via an electron carrier system and adenosine triphosphate (ATP) is yielded as before. The resulting OH is combined with others to form water and oxygen, the latter being evolved as oxygen gas:

$$4(OH) \rightleftharpoons 2H_2O + O_2\uparrow$$

Since a different electron is returned to the chlorophyll molecule the process is termed **noncyclic photophosphorylation**, or the **Z-scheme**, because of the zig-zag route taken by the electrons. Noncyclic photophosphorylation involves both photosystems I and II.

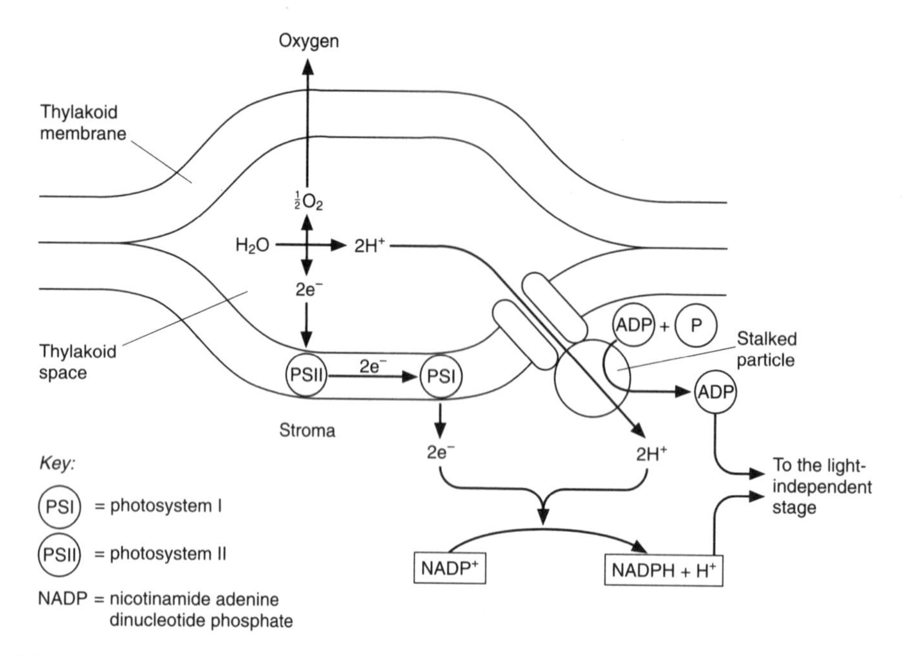

Fig. 5.11 Events of the light-dependent stage

THE LIGHT-INDEPENDENT (DARK) STAGE (CALVIN CYCLE)

This takes place in the stroma of the chloroplast. The reduced nicotinamide adenine dinucleotide phosphate (NADPH + H^+) from the light reaction is used to reduce carbon dioxide using the ATP formed in the light reaction as the source of energy. Carbon dioxide

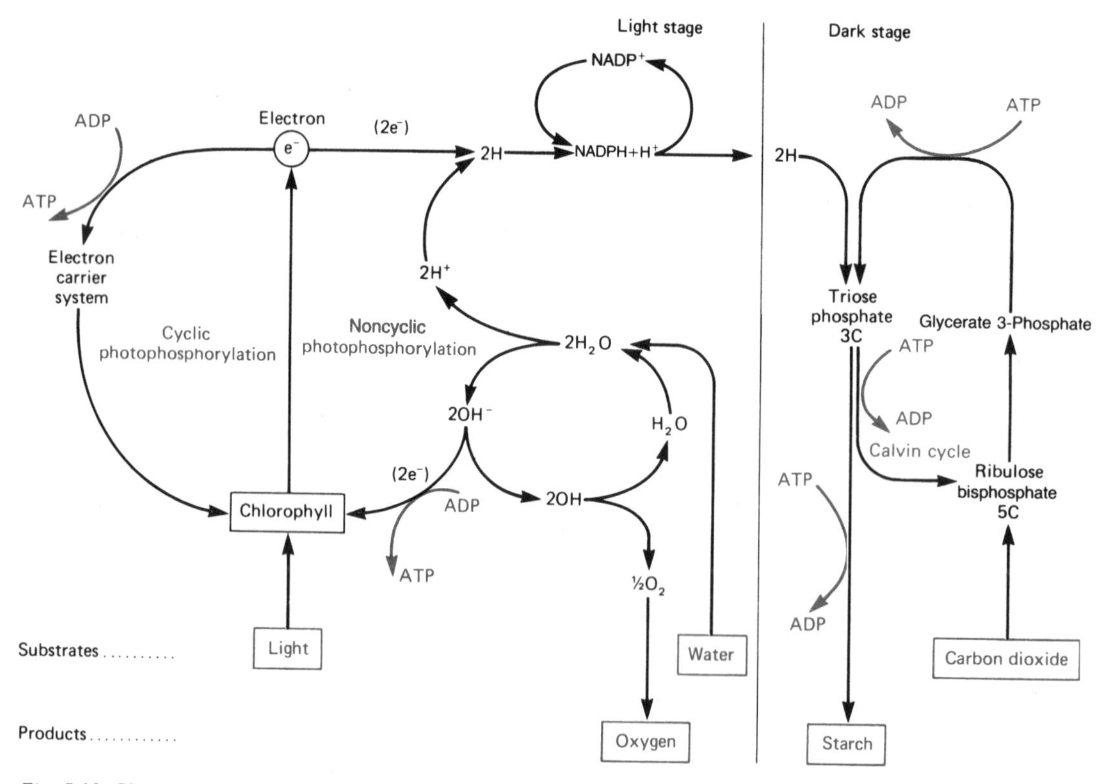

Fig. 5.12 Chemical reactions in photosynthesis

absorbed by the plant is combined with a five-carbon substance **ribulose bisphosphate** by means of an enzyme called **ribulose bisphosphate carboxylase oxidase** (RUBISCO). This results in an unstable intermediate, which immediately splits to give two molecules of the three-carbon substance **glycerate-3-phosphate** (GP). Most of this GP is used to reform ribulose bisphosphate but some is reduced by the NADPH + H^+ to give triose phosphate, which in turn is built into glucose phosphate which polymerises into starch by condensation.

IMPORTANCE OF PHOTOSYNTHESIS

1. Photosynthesis releases oxygen required by animals and plants for aerobic respiration.
2. It provides a store of useful chemical energy by converting inorganic substances to stable organic substances of high potential chemical energy. It has been estimated that plants annually fix 35×10^{15} kg of carbon.

Synthesis of nitrogenous compounds

The nitrogen of ammonia (NH_3) is brought into organic combination mainly by the formation of glutamate:

$$\alpha\text{-ketoglutarate} + NH_2 + \underset{\substack{\text{reduced}\\ \text{coenzyme I}}}{\text{CoI–}H_2} \xrightleftharpoons[]{\substack{\text{glutamate}\\ \text{dehydrogenase}}} \text{glutamate} + \text{CoI}$$

Once glutamate is available other amino acids may be formed by the transfer of its amino group to another carbon skeleton, a process known as **transamination**.

FACTORS AFFECTING PHOTOSYNTHESIS

Blackman's Law of Limiting Factors states that when the speed of a process is conditioned by a number of separate factors, the rate of the process is limited by the pace of the slowest factor. The main factors affecting the rate of photosynthesis are intensity and duration of light, carbon dioxide concentration and temperature. When temperature and the concentration of carbon dioxide are kept constant the rate of photosynthesis increases with an increase in light intensity and then levels off. When the carbon dioxide concentration is increased the rate of photosynthesis increases to a maximum before levelling off. If the temperature is then increased the rate of photosynthesis steadily rises instead of levelling off.

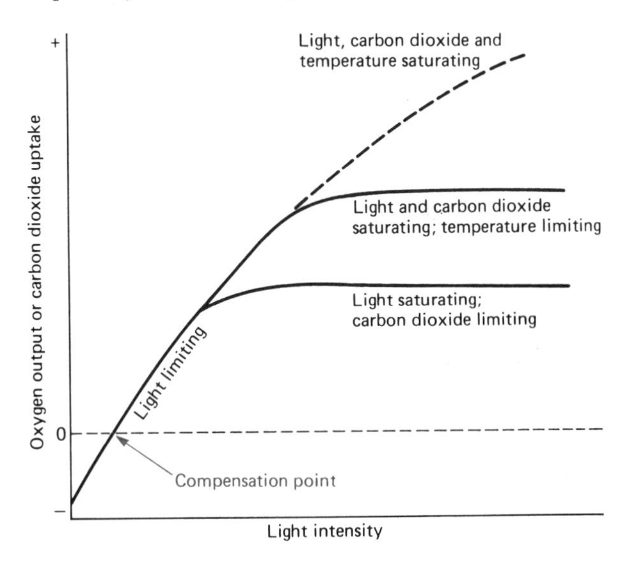

Fig. 5.13 Limiting factors in photosynthesis

Compensation point

This is the point at which the rate of carbon dioxide production from respiration is compensated exactly by the rate of carbon dioxide uptake in photosynthesis. Below the compensation point the rate of photosynthesis is less than the rate of respiration and so carbon dioxide is evolved. Above the compensation point the rate of photosynthesis exceeds the rate of respiration and so oxygen is evolved. At the compensation point there is no net exchange of gases.

LEAF ADAPTATIONS FOR PHOTOSYNTHESIS

The internal structure of the leaf is shown in Fig. 5.10. Overall, the leaf serves three main functions with respect to photosynthesis:

1. obtaining sunlight;
2. diffusion of oxygen and carbon dioxide into and out of it;
3. transport of water into it and sugar solution out of it.

Leaf adaptations to obtain sunlight

- The leaf arranges itself in a position which allows it to obtain maximum illumination.
- The leaf has a large surface area.
- The leaf is thin – thicker leaves would mean that the lower cellular layers received no light as it would be filtered out by the upper layers.
- The cuticle and epidermis on the leaf are transparent to allow light through to the chloroplasts.
- The palisade mesophyll layer of the leaf is densely packed with chloroplasts, which can also move within the cells to obtain maximum illumination.
- The chloroplasts are highly structured organelles which arrange chlorophyll in a manner that allows efficient photosynthesis.

Leaf adaptations for gaseous exchange

- Stomata are numerous on the underside of the leaf. These minute pores permit rapid exchange of oxygen and carbon dioxide across the leaf epidermis. They can be opened and closed (see 7.3) to help control water loss while ensuring efficient diffusion of the oxygen and carbon dioxide when needed.
- The spongy mesophyll of the leaf has many intercellular air spaces to ensure an uninterrupted gaseous layer between the outside and the photosynthesising palisade mesophyll cells. This speeds diffusion.

Leaf adaptations for the movement of liquids

- A network of veins containing xylem and phloem permeates the leaf and carries water to the photosynthesising cells and sugar product away from them.

USE OF PHOTOSYNTHETIC PRODUCTS

The carbohydrate products of photosynthesis can be converted into a number of other products:

- sucrose – a sugar which is readily transported around the plant and occasionally stored, e.g. in onion;
- starch – the main polysaccharide storage product of most plants;

- inulin – the polysaccharide storage product of a few species, e.g. *Dahlia*;
- cellulose – a polysaccharide which is the main structural component of cell walls;
- lipids – triose phosphate and glycerate-3-phosphate can be converted into glycerol and fatty acid, respectively. These can then be combined to form lipids which can be stored (e.g. in seeds), combined into cell membranes (phospholipids) or used to waterproof leaves (cutin);
- proteins – glycerate-3-phosphate can be converted into acetyl coenzyme A which is a starting point for making amino acids. These can then be polymerised into proteins. The many possible functions of these proteins are given in 1.1, Table 1.3.

5.3 HETEROTROPHIC NUTRITION (HOLOZOIC)

The evolution of autotrophic organisms inevitably led to the development of organisms that obtained their energy not by photosynthesis but by consuming the autotrophs. These were the **herbivores**. They developed various mechanisms for ingesting the food and digesting the cellulose walls that surrounded the plant cells. In turn there evolved organisms that obtained energy by consuming the herbivores. These were the **carnivores**. They developed a variety of mechanisms for capturing prey organisms as well as for ingesting them. Unlike autotrophs the energy source of heterotrophs is in complex form and contains some unwanted materials. The food must therefore be digested, the required materials absorbed and the unwanted ones eliminated. This digestion is carried out by enzymes each with an optimum pH. For enzymes to be efficient the food they act upon must have a large surface area. In addition to enzymes the digestive system therefore also produces substances to adjust the pH, water as a medium for the enzymes to act in and substances to increase the surface area of the food. To speed the absorption of the digested food there are methods to increase the surface area of the absorbing surface. To maintain a diffusion gradient this surface is well supplied with blood vessels and the food is kept moving. The surface is thin and moist.

There are five main types of heterotrophic nutrition.

- **Holozoic nutrition** This is the form of nutrition used by most animals and involves the ingestion of complex food which is broken down into simpler molecules before being absorbed.
- **Saprobiontic nutrition** Used by some bacteria and fungi, this method involves the external breakdown of complex food from dead organisms and its absorption by diffusion.
- **Parasitic nutrition** Used by a range of organisms, they derive their food from other living organisms usually in simple soluble form.
- **Mutualism** This is a close relationship between members of two species in which both derive some benefit.
- **Commensalism** This is a close relationship between members of two species, but where only one member benefits while the other neither benefits nor is harmed.

The term **symbiosis** is a general one that covers any two species living together in a close relationship. As such, it includes parasitism, mutualism and commensalism.

DIETARY NUTRIENTS

We spend much of our lives obtaining, preparing and eating food. That food is essential to our survival is obvious, but what exactly is it needed for?

1. It provides the energy needed to maintain all our body processes.
2. It provides essential nutrients which:
   (a) cannot be synthesised by our bodies;
   (b) replace materials that are lost through secretions or in urine;
   (c) replace material that has been incorporated into the body, e.g. calcium into bones.
3. It provides the basic materials for growth and repair.

There are six groups of nutrients needed by animals:

1. carbohydrates
2. fats
3. proteins
4. inorganic ions
5. vitamins
6. water

To these should be added **dietary fibre**, which, while not providing nutrients or energy, is necessary for the efficient functioning of the digestive system. The proportions of each nutrient differ from species to species, and with the activity, age, weight and sex of individuals of the same species.

Energy requirements

These are provided by carbohydrates and fats, details of which are given in 1.1. The quantity of energy each supplies is expressed in joules and a summary of some typical values for humans is given in Table 5.2.

Provision of essential nutrients

Proteins, inorganic ions, vitamins and water all provide material which is essential to the efficient functioning of the body. Table 5.3 lists essential inorganic ions and their functions in the body while Table 5.4 gives similar details for vitamins, as well as the symptoms which arise when they are deficient in the diet. Around 2.5 litres of water are needed daily in humans to replace that lost in urine, sweat, faeces and evaporation from the lungs.

Table 5.2 *Recommended daily intake of energy according to activity, age, weight and gender*

Age/years	Average body weight/kg	Degree of activity/ circumstances	Energy requirement/kJ Male	Female
1	7	Average	3200	3200
5	20	Average	7500	7500
10	30	Average	9500	9500
15	45	Average	11 500	11 500
		Sedentary	11 300	9000
25	65 (male)	Moderately active	12 500	9500
	55 (female)	Very active	15 000	10 500
		Sedentary	11 000	9000
50	65 (male)	Moderately active	12 000	9500
	55 (female)	Very active	15 000	10 500
75	63 (male) 53 (female)	Sedentary	9000	8000
Any	—	During pregnancy	—	10 000
Any	—	Breast feeding	—	11 500

Table 5.3 *Essential inorganic ions needed in the human diet*

Mineral	Major food source	Function
Macronutrients		
Calcium (Ca^{2+})	Dairy foods, eggs, green vegetables	Constituent of bones and teeth, needed in blood clotting and muscle contraction; enzyme activator
Chlorine (Cl^-)	Table salt	Maintenance of anion/cation balance; formation of hydrochloric acid
Magnesium (Mg^{2+})	Meat, green vegetables	Component of bones and teeth; enzyme activator
Phosphate (PO_4^{3-})	Dairy foods, eggs, meat, vegetables	Constituent of nucleic acids, ATP, phospholipids (in cell membranes), bones and teeth
Potassium (K^+)	Meat, fruit and vegetables	Needed for nerve and muscle action and in protein synthesis
Sodium (Na^+)	Table salt, dairy foods, meat, eggs, vegetables	Needed for nerve and muscle action; maintenance of anion/cation balance
Sulphate (SO_4^{2-})	Meat, eggs, dairy foods	Component of proteins and coenzymes
Micronutrients (trace elements)		
Cobalt (Co^{2+})	Meat	Component of vitamin B_{12} and needed for the formation of red blood cells
Copper (Cu^{2+})	Liver, meat, fish	Constituent of many enzymes; needed for bone and haemoglobin formation
Fluorine (F^-)	Many water supplies	Improves resistance to tooth decay
Iodine (I^-)	Fish, shellfish, iodised salt	Component of growth hormone thyroxine
Iron (Fe^{2+}) or (Fe^{3+})	Liver, meat, green vegetables	Constituent of many enzymes, electron carriers, haemoglobin and myoglobin
Manganese (Mn^{2+})	Liver, kidney, tea and coffee	Enzyme activator and growth factor in bone development
Molybdenum (Mo^{4+})	Liver, kidney, green vegetables	Required by some enzymes
Zinc (Zn^{2+})	Liver, fish, shellfish	Enzyme activator, involved in the physiology of insulin

Table 5.4 *Major vitamins needed in the human diet*

Vitamin/name	Major food sources	Function	Deficiency symptoms
A_1 Retinol	Liver, vegetables, fruits, dairy foods	Maintains normal epithelial structure; needed to form visual pigments	Dry skin, poor night vision
B_1 Thiamine	Liver, legumes, yeast, wheat and rice germ	Coenzyme in cellular respiration	Nervous disorder called beri-beri, neuritis and mental disturbances, heart failure
B_2 Riboflavin	Liver, yeast, dairy produce	Coenzymes (flavo-proteins) in cellular respiration	Soreness of the tongue and corners of the mouth
B_3 (pp factor) Niacin	Liver, yeast, wholemeal bread	Coenzyme (NAD, NADP) in cellular metabolism	Skin lesions known as pellagra, diarrhoea
B_5 Pantothenic acid	Liver, yeast, eggs	Forms part of acetyl coenzyme A in cellular respiration	Neuromotor disorders, fatigue and muscle cramps
B_6 Pyridoxine	Liver, kidney, fish	Coenzymes in amino acid metabolism	Dermatitis, nervous disorders
B_{12} Cyanocobalamine	Meat, eggs, dairy food	Nucleoprotein (RNA) synthesis; needed in red blood cell formation	Pernicious anaemia, malformation of red blood cells

Vitamin/name	Major food sources	Function	Deficiency symptoms
C Ascorbic acid	Citrus fruits, tomatoes, potatoes	Formation of connective tissues, especially collagen fibres	Nonformation of connective tissues, bleeding gums (scurvy)
D Calciferol	Liver, fish oils, dairy produce; action of sunlight on skin	Absorption and metabolism of calcium and phosphorus, therefore important in formation of teeth and bones	Defective bone formation (rickets)
E Tocopherol	Liver, green vegetables	Function unclear in humans; in rats it prevents haemolysis of red blood cells	Anaemia
K Phylloquinone	Green vegetables; synthesised by intestinal bacteria	Blood clotting	Failure of blood to clot

Materials for growth and repair

Although all nutrients are needed for proper growth and repair, proteins are especially important. There are nine amino acids which cannot be synthesised in humans and these so-called **essential amino acids** must therefore be provided in the diet. Details of the chemistry of proteins are given in 1.1 and their functions in Table 1.3.

Dietary fibre

Dietary fibre or **roughage** as it is often called comprises mostly the cellulose cell walls of plants. It is indigestible by human enzymes and so passes through the alimentary canal largely unchanged. It gives bulk to the food in the intestines and so stimulates peristalsis. In so doing it helps to prevent constipation and other intestinal disorders.

HOLOZOIC NUTRITION

The consumption and digestion of complex food material is carried out in seven stages:

1. obtaining food (may involve capturing food)
2. ingestion (feeding mechanisms)
3. physical digestion (by teeth, radula, etc.)
4. chemical digestion (mostly by enzymes)
5. absorption (of the required material)
6. assimilation (incorporating materials into cells)
7. elimination (removal of unwanted material)

As core syllabuses do not require knowledge of the first three stages we shall restrict discussion here to the final four only.

Chemical digestion of food

The following account refers to the digestion of food in a human, the digestive system of which is detailed in Fig. 5.14.

Chemical digestion involves the use of enzymes to break down complex molecules which are too large to pass through the alimentary canal wall into small ones which readily do so. Being highly specific, a different enzyme is needed for each main type of food in the diet. As each enzyme has an optimum pH at which it works, the alimentary canal is divided into regions which maintain acid, neutral or slightly alkaline conditions. A summary of the enzymes produced, where they are secreted and the role they perform is given in Table 5.5.

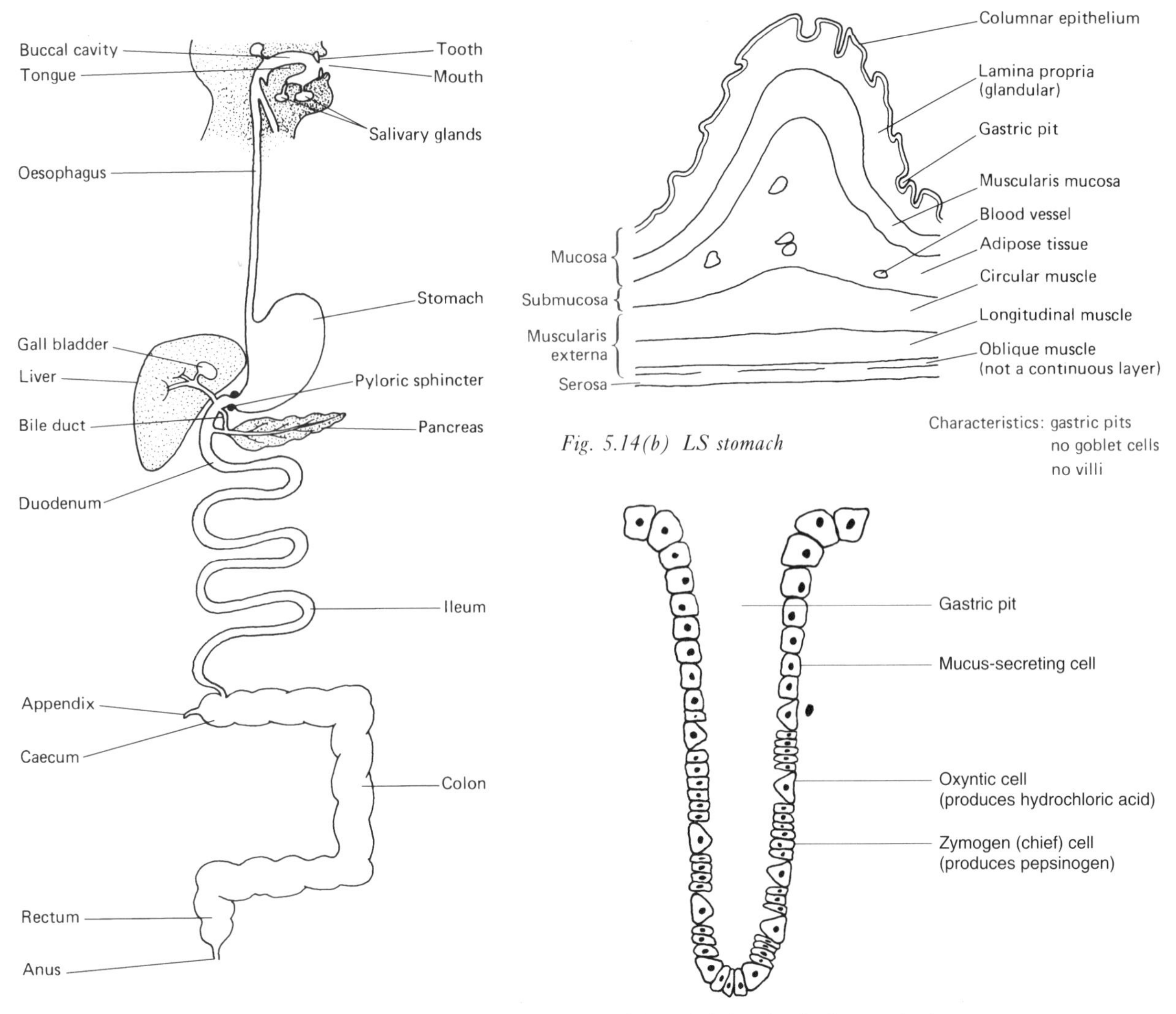

Fig. 5.14(a) Human digestive system

Fig. 5.14(b) LS stomach

Fig. 5.14(c) Cellular detail of a gastric pit

Absorption and assimilation

While the small, soluble products of digestion can be absorbed through the intestinal wall by diffusion, provided that a concentration gradient exists, this process can be wasteful. Most food is therefore absorbed by active transport. While some absorption can occur along most of the alimentary canal, it is in the small intestine, especially the ileum, where most occurs. To achieve this effectively the ileum:

1. is thin walled (usually a single-celled epithelial layer)
2. is moist
3. is well supplied with blood vessels (to maintain a diffusion gradient by removing absorbed material)
4. has a large surface area achieved by:
   (a) being almost 6 m long
   (b) being folded (folds of Kerkring)
   (c) having finger-like projections called **villi** over its surface
   (d) having minute projections called **microvilli** over the surface of the villi
5. is well supplied with mitochondria which provide energy for active uptake.

The structure of the intestinal wall is shown in Fig. 5.15 and details of the method of absorption and the fate of the products absorbed are given in Table 5.6.

Table 5.5 *Summary of digestion*

Organ	Secretion	pH	Site of action	Production induced by	Secretion (including enzyme)	Effect of secretion
Salivary glands	Saliva	Slightly alkaline	Mouth	Visual/olfactory expectation; reflex stimulation	Salivary amylase Mucin Salts	Starch → maltose via dextrin Sticks bolus Correct pH
Gastric glands of stomach	Gastric juice	Very acidic	Stomach	Gastrin (hormone); reflex stimulation	Hydrochloric acid Pepsin Rennin (chymase) Mucus	Provides correct pH Protein → polypeptides Caseinogen → casein Lubrication; prevents autolysis
Liver	Bile	Alkaline	Duodenum	Cholecystokinin induces release of bile; reflex action; secretin causes production of bile	Bile salts Bile pigments Sodium hydrogen carbonate Cholesterol	Emulsify fats Excretory products of haemoglobin breakdown Correct pH Excretory
Pancreas	Pancreatic juice	Alkaline	Duodenum	Secretin induces production of pancreatic juice; pancreozymin induces release of pancreatic enzymes	Mineral salts Trypsin Carboxypeptidase Chymotrypsin Amylase Lipase Nucleases	Neutralise acid chyme Protein → peptides + amino acids Peptides → amino acids Casein → peptides + amino acids Starch → maltose Fats → fatty acids + glycerol Nucleic acids → nucleotides
Small intestine wall	Intestinal juice	Alkaline	Small intestine	Mechanical stimulation of intestinal lining	Enterokinase Aminopeptidase Amylase Maltase Sucrase Lactase Nucleotidases	Trypsinogen → trypsin Peptides → amino acids Starch → maltose Maltose → glucose Sucrose → glucose + fructose Lactose → glucose + galactose Nucleotides → organic bases, pentose sugar and phosphoric acid

Table 5.6 *Absorption and fate of major digestive products*

End product of digestion	Form in which absorbed	Where absorbed	Mechanism of absorption	Fate of products	Storage	Use
Monosaccharides	Possibly as a complex, e.g. phosphate sugar	Capillaries of villi	Diffusion and active transport	Most remain in general circulation; excess converted to glycogen in the liver	Glycogen in liver	Respiratory substrate
Fatty acids and glycerol	Fatty acids and glycerol; some as droplets of neutral fat (chylomicrons)	Mainly lacteals in villi; some into capillaries	Diffusion and active transport	Fat enters blood stream from lymphatic system (in neck)	Fat under skin, in mesenteries, etc.	Respiratory substrate; insulation; protection; phospholipids (structural)
Amino acids	Amino acids	Capillaries of villi	Diffusion and active transport	Deamination in liver: nitrogenous residues converted to urea; carbon residues mainly in general circulation but some used for carbohydrate or fat synthesis	After deamination as fat or carbohydrate	Protein synthesis; products of deaminated proteins may become part of carbohydrates or fats

Most of the water we drink is absorbed by the stomach, but there is also a considerable amount of water which is secreted along with enzymes by the walls of the alimentary canal and the associated digestive organs, such as the liver and pancreas. The majority of this water is absorbed by the ileum, but the large intestine or **colon** plays a role in the final reabsorption which results in the production of semisolid faeces. The colon also houses bacteria such as *Escherichia coli*, which produce vitamin K that is absorbed by the walls of the colon.

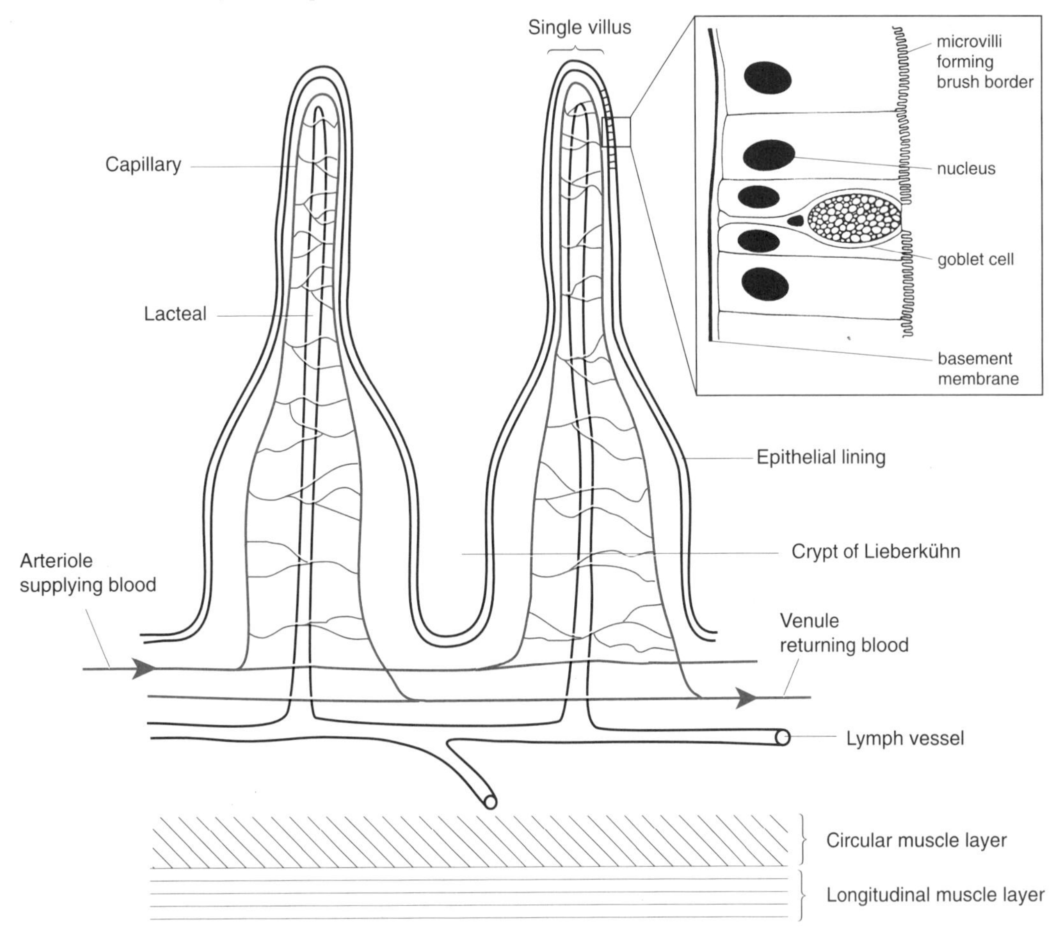

Fig. 5.15 Intestinal wall showing villi (LS)

Elimination (egestion)

Unwanted food is eliminated by periodic egestion from the organism, a process called **defaecation**. The majority of this material is not the result of metabolic activities within the organism and thus can be called egested material. The few materials, such as cholesterol and bile pigments, that are the product of internal metabolic processes are strictly speaking excretory products.

INTERCONVERSION OF FOOD PRODUCTS

Although each of the absorbed products has a particular role in the body, they can readily be interconverted to meet a temporary shortage of any one of them.

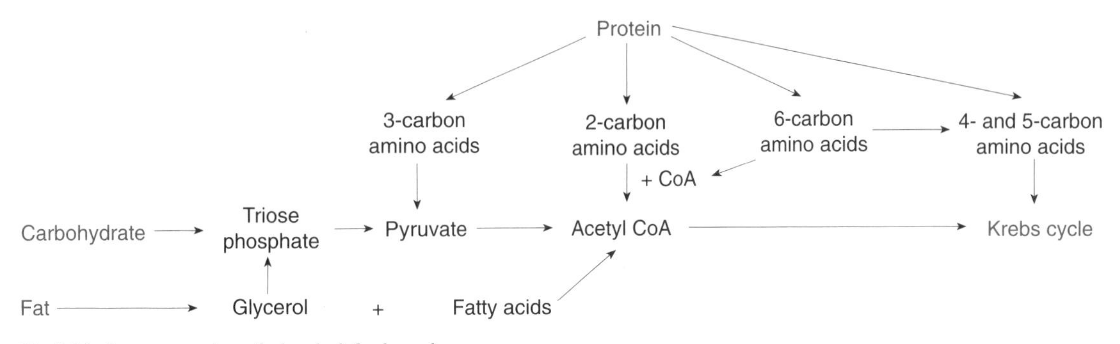

Fig 5.16 Interconversion of absorbed food products

5.4 HETEROTROPHIC NUTRITION (PARASITISM, SAPROBIONTISM AND MUTUALISM)

When autotrophic and holozoic organisms die, their remains comprise complex chemicals which may be broken down to simpler forms. The energy released in this process is utilised by groups of organisms called **saprobionts**, which speed up the breakdown of the dead and decaying remains. The energy released is used to build up their own complex structures. In addition saprobionts play a vital role in releasing valuable elements such as carbon, nitrogen and phosphorus in a form in which they can be absorbed by autotrophs and so be recycled. It is possible that such organisms arose when an autotrophic organism became unable to photosynthesise, e.g. lost the ability to make chlorophyll. For the organism to survive it would then need to feed heterotrophically. With no means of obtaining and ingesting complex food it would need to live on dead organisms and digest them externally, absorbing the soluble products.

In a similar way, any organism that lost, by mutation, the ability to obtain or utilise a particular substance would need to obtain it from some external source. One way would be to obtain it from another living organism. This is the basis of **parasitism**. In due course the parasite becomes increasingly dependent on its host for other nutrients. It is not in the parasite's interest to kill the host; indeed the less obvious it makes its presence the more likely it is to survive. The host and parasite may develop such a close relationship that they become mutually beneficial. This is **mutualism**.

SAPROBIONTIC NUTRITION

Saprobionts are organisms that obtain their energy from the dead remains of other organisms. The importance of saprobionts lies in their role in breaking down complex organic material and so releasing valuable elements such as nitrogen, carbon and sulphur, usually in the form of an oxide (e.g. nitrate, carbon dioxide, sulphur dioxide) or in a reduced state (e.g. ammonia, methane). Many bacteria and fungi are saprobionts, e.g. *Myxococcus, Mucor*.

Table 5.7 *Economic importance of some saprobionts*

Saprobionts	Major group	Carbon source	Effect
Cladosporium resinae	Fungi	Jet fuel	Blocks filters and damages tank linings
Lentinus and others	Fungi	Lignin and cellulose	Causes white rot of wood, e.g. in power station cooling towers
Bacillus subtilis	Bacteria	Proteins	Breaks down protein of wool to amino acids, reducing its commercial value
Serpula lacrymans	Fungi	Cellulose	Dry rot
Botrytis cinerea	Fungi	Hydrocarbons	Degrades and recycles small amount of hydrocarbons, including oil
Pseudomonas	Bacteria	Hydrocarbons	Degrades and recycles small amounts of hydrocarbons, including oil
Aerobacter and *Hydrogenomonas*	Bacteria	DDT	Their joint action breaks down pesticide residues in soil
Clostridium and others	Bacteria	Cellulose and similar polysaccharides	Important component of sewage microflora, breaking down cellulose
Lactobacillus bulgaricus	Bacteria	Lactose	Yoghurt production
Saccharomyces (yeast)	Fungi	Sucrose	Fermentation products include CO_2 and ethanol used for baking and brewing
Aspergillus niger	Fungi	Sucrose	Used in citric acid production
Candida utilis	Fungi	Molasses or potato starch	Used for large-scale production of vitamin B using cheap organic carbon sources

PARASITIC NUTRITION

Parasitism is an association between two organisms whereby one (the parasite) derives some benefit and the other (the host) does not, and may in fact be harmed. However, the most well-adapted parasites inflict little damage on their hosts.

Plants are parasitised mainly by fungi and bacteria; they are poor hosts for invertebrates because:

1. they remain in one position and so it is more difficult for the parasite to reach a new host;
2. their cellulose cell walls are difficult for animals to digest;
3. they have few internal cavities suitable for parasites.

The most successful invertebrate parasites of plants are nematodes. In animals the habitats most used are:

1. the body surface and its infoldings such as the buccal cavity, lungs, external gills and nostrils;
2. the alimentary canal and its associated organs such as the liver and bile duct;
3. internal tissues such as blood and muscle.

Table 5.8 *Methods used by parasites to escape from their hosts*

Habitat	Example	Method of escape
Lungs	*Bacillus tubercle*	Through the air
Intestines	*Ascaris*	In the faeces
Intestines (larvae encysted in muscle)	*Trichinella*	Disintegration of host tissues after death
Blood	*Schistosoma*	Urine (eggs in wall of bladder)
Blood	*Plasmodium*	Intermediate blood-sucking organism
Under skin	*Dracunculus*	Discharge larvae through skin of host

Parasites using the body surface as a habitat are termed **ectoparasites** (*ektos* = outside) and those using internal habitats are **endoparasites** (*endon* = within). Most parasites are endoparasites because the inside of the body provides an environment which is rich in nutrients and not subject to extremes of environmental conditions.

Transmission of parasites

The transmission of ectoparasites is relatively simple, e.g. by jumping in the flea *Pulex*. Since ectoparasites are subjected to varying conditions, existence away from the host can be tolerated temporarily. The transmission of endoparasites, however, presents three main problems:

1. leaving the host
2. living away from the host
3. entering a new host.

Parasites may live away from the host in:

(a) dormant stages, e.g. *Taenia*

(b) free-living stages, e.g. *Haemonchus*

(c) intermediate hosts, e.g. *Fasciola*.

To enter a new host the parasite may:

(a) be eaten, e.g. *Taenia*

(b) penetrate the skin or body surface of a new host after leaving an intermediate host, e.g. *Schistosoma*

(c) be injected by an intermediate host, e.g. transmission of *Plasmodium* by the mosquito.

The main features of parasites

1. A means of attachment is necessary to help the parasite maintain its position in/on the host, e.g. hooks, suckers.
2. Degeneration of certain unnecessary organ systems, e.g. the digestive system of gut parasites is reduced or absent because the parasite is surrounded by a large supply of predigested food. There are often special mechanisms developed to move a parasite from one host to another and so there is no need for a well-developed locomotory system.
3. Protective devices are often developed to protect the parasite from the host's defences, e.g. schistosomes absorb glycolipids from the host on to their surface to imitate the host tissue and thus forestall immunological attack.
4. Penetrative agents such as cellulases in fungal plant parasites.
5. Vector or intermediate host, e.g. *Fasciola* (liver fluke of sheep), requires an intermediate host, the snail, for the completion of its life cycle. The blood-sucking female mosquito *Anopheles* is the intermediate host for *Plasmodium* (malarial parasite) and is thus the vector of the disease malaria.
6. Production of many eggs because of the difficulty of finding a new host, e.g. *Diphyllobothrium* (fish tapeworm) produces 36 million eggs a day.
7. Dormant or resistant phases to overcome the period spent away from the host, e.g. eggs in *Ascaris*, encysted cercariae in *Fasciola*.

Table 5.9 *Some major diseases caused by parasites*

Parasite	Major group	Primary host	Disease caused	Secondary host, if any
Phytophthora infestans	Fungi	Potato	Potato blight	—
Puccinia graminis	Fungi	Wheat	Black stem rust	—
Eimeria	Apicomplexa	Poultry	Coccidiosis	—
Plasmodium	Apicomplexa	Man	Malaria	*Anopheles* mosquito
Schistosoma	Platyhelminthes	Man	Bilharzia (schistosomiasis)	Freshwater snail
Fasciola	Platyhelminthes	Sheep	Liver rot	Snail
Taenia solium	Platyhelminthes	Man	Occasionally cysticercosis	Pig
Wucheria bancrofti	Nematoda	Man	Elephantiasis	Mosquito
Onchocerca volvulus	Nematoda	Man	River blindness	Blackfly

MUTUALISTIC NUTRITION

This term is most commonly applied to a relationship in which both organisms benefit, e.g. *Zoochlorella* inhabiting the endodermal cells of *Hydra viridissima*. In this relationship *Zoochlorella* obtains protection, shelter, carbon dioxide (for photosynthesis) and nitrogen (from excretory wastes) from *Hydra*; *Hydra* obtains oxygen and carbohydrates from the photosynthesis of *Zoochlorella*. Lichens are a mutualistic relationship between an alga and a fungus in which the alga gains water, mineral salts and protection from desiccation and the fungus obtains oxygen and carbohydrates. The hermit crab *Pagurus* gains protection and camouflage from the sea anemone *Adamsia palliata* which lives on its shell. The sea anemone obtains food dropped by the hermit crab.

5.5 CELLULAR RESPIRATION

We explored at the beginning of this chapter the concept of entropy, and the fact that living organisms maintain low entropy by taking in chemical energy in the form of food. The chemical energy of the fats, carbohydrates and protein which comprise the food of many organisms must, however, be converted into forms such as ATP which can be used by cells. This involves the conversion of food into hexose carbohydrate such as glucose and its subsequent breakdown. This breakdown may involve the oxidation of glucose (**aerobic respiration**) or not (**anaerobic respiration**).

Respiration can be conveniently divided into two parts:

1. Cellular (internal or tissue) respiration – the biochemical processes which take place within living cells that release the energy from glucose.
2. Gaseous exchange (external respiration) – the processes involved in obtaining oxygen needed for respiration and the removal of gaseous waste such as carbon dioxide. This aspect is dealt with in 7.1.

Cellular respiration is a complex metabolic process of over 70 reactions which can be divided into three stages:

1. glycolysis
2. Krebs cycle (tricarboxylic acid cycle)
3. electron transfer system.

ELECTRON (HYDROGEN) CARRIERS

Before discussing these stages, it is necessary to mention a group of substances that are employed throughout cellular respiration. These are the electron (hydrogen) carrier or electron (hydrogen) acceptor molecules. Many reactions in cellular respiration involve the release of electrons at a high energy level. These electrons are collected by electron carrier molecules which pass them to electron carriers at lower energy levels. The energy released is used to form ATP from ADP. The electrons are initially released as part of a hydrogen atom which later splits into a proton and an electron. This is why the carriers have the alternative name hydrogen carriers. They act as coenzymes because they are essential to the functioning of the dehyrogenase enzymes that catalyse the removal of the hydrogen atoms. One of the most important is **nicotinamide adenine dinucleotide** (**NAD**) and others include **nicotinamide adenine dinucleotide phosphate** (**NADP**), **flavine adenine dinucleotide** (**FAD**) and **cytochromes**.

GLYCOLYSIS

This first stage in the breakdown of glucose is common to all organisms. In anaerobic organisms it is the only stage in respiration. It occurs in the cytoplasm of the cell. At this level it is not necessary to know all the intermediate compounds or enzymes but an overall appreciation of the major features is required.

The process of glycolysis produces two molecules of pyruvate for each molecule of glucose degraded. Effectively every substance after Stage 4 in the following diagram is doubled in quantity for each glucose molecule. The total energy yield is therefore two molecules of ATP directly and six molecules of ATP produced from the two reduced NAD molecules – a total of eight ATP molecules.

Stages of glycolysis

Stage 1 The glucose molecule has a phosphate group added to make it more reactive (= activation). The phosphate group is donated by ATP.

Stage 2 The glucose phosphate is reorganised into a fructose phosphate molecule.

Stage 3 The fructose phosphate is further activated by the donation of a second phosphate group by an ATP molecule.

Stage 4 The 6-carbon fructose bisphosphate is split into two 3-carbon triose phosphate molecules.

Stage 5 Hydrogen atoms are removed from the triose phosphate molecules and taken up by NAD. Inorganic phosphate is added to further activate the triose phosphate.

Stage 6 A phosphate molecule is lost and ATP is regenerated.

Stage 7 A phosphate molecule is lost and ATP is formed; a water molecule is also lost, for each triose phosphate.

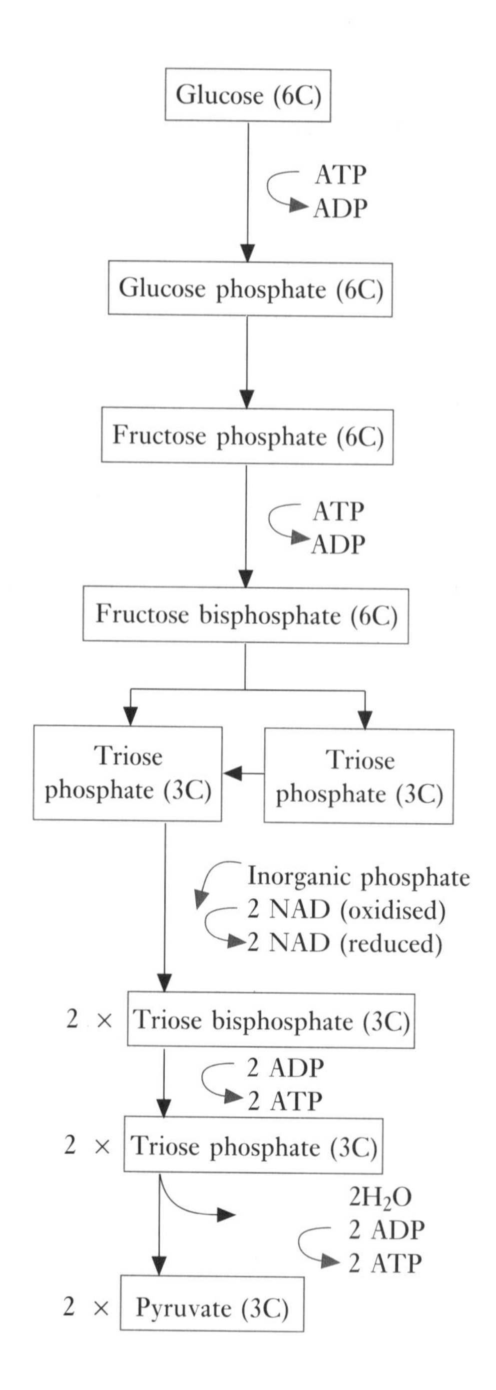

KREBS CYCLE

The pyruvate formed as a result of glycolysis may in the absence of oxygen be converted to a variety of substances to yield a little energy. These processes constitute the anaerobic pathways. In the presence of oxygen the pyruvate enters the Krebs cycle. Before entering the actual cycle one of the three carbon atoms of pyruvate is oxidised to carbon dioxide and a molecule of NAD is reduced by the addition of two hydrogen atoms (the NAD yields a further three ATPs). This leaves an acetyl group (CH_3CO) which is readily accepted by a coenzyme called coenzyme A. The resulting substance is acetyl coenzyme A. The 2-carbon acetyl group of this compound combines with a 4-carbon substance called oxaloacetate to give the 6-carbon molecule citrate. In a series of reactions, two carbon dioxide molecules are produced and the 4-carbon oxaloacetate molecule is regenerated in readiness to receive another 2-carbon acetyl group from acetyl coenzyme A. Other products include a total of eight hydrogen atoms which are used to reduce three molecules of NAD and one molecule of FAD. These reduced electron (hydrogen) carriers eventually pass on the hydrogen atoms to oxygen yielding 11 more ATP molecules for each pyruvate. In addition a further ATP molecule is yielded directly during the cycle to give a total of 12 ATPs per pyruvate molecule.

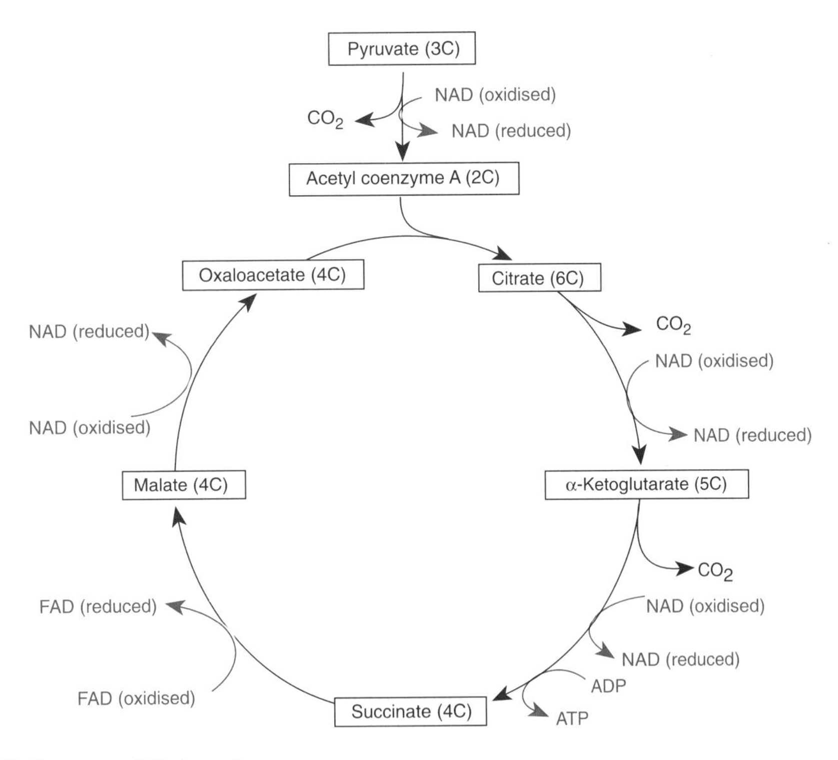

Fig. 5.17 Summary of Krebs cycle

Importance of Krebs cycle

1. It provides hydrogen atoms which ultimately yield the major part of the energy derived from the oxidation of a glucose molecule.
2. It is a valuable source of intermediates which are used to manufacture other substances, e.g. fatty acids, amino acids, carotenoids.

ELECTRON TRANSFER SYSTEM

Although the glucose molecule has been completely oxidised by the end of the Krebs cycle, much of the energy is in the form of hydrogen atoms which are attached to the hydrogen carriers NAD and FAD. These atoms are passed along a series of carriers at progressively lower energy levels. As they lose their energy it is harnessed to produce ATP molecules, three for each molecule of NAD and two for each one of FAD. The other carriers in the system are iron-containing proteins called cytochromes. The hydrogen atoms split into their protons and electrons during the pathway and the electrons are carried by the cytochromes. They recombine with their protons before the final stage where the newly reformed hydrogen atoms combine with oxygen to form water. Although oxygen, so closely connected with aerobic respiration, only plays a role at this final stage, it is nevertheless vital since it drives the whole process. If not for the ability of oxygen to act as the final hydrogen acceptor, the hydrogens would accumulate and the process of aerobic respiration would cease. The whole process whereby oxygen effectively allows the production of ATP from ADP is called **oxidative phosphorylation**. The electron transfer system occurs in the mitochondria. At each stage of the system the hydrogen atoms (or electrons) are passed from the reduced carrier to the oxidised carrier further along the pathway. Thus the reduced carrier becomes oxidised and able to accept more hydrogen atoms (or electrons) and the oxidised one becomes reduced.
e.g. for NAD and FAD

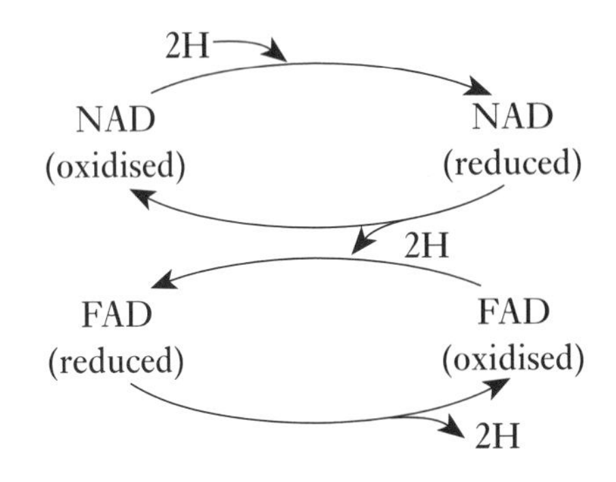

The full pathways can be summarised as follows.

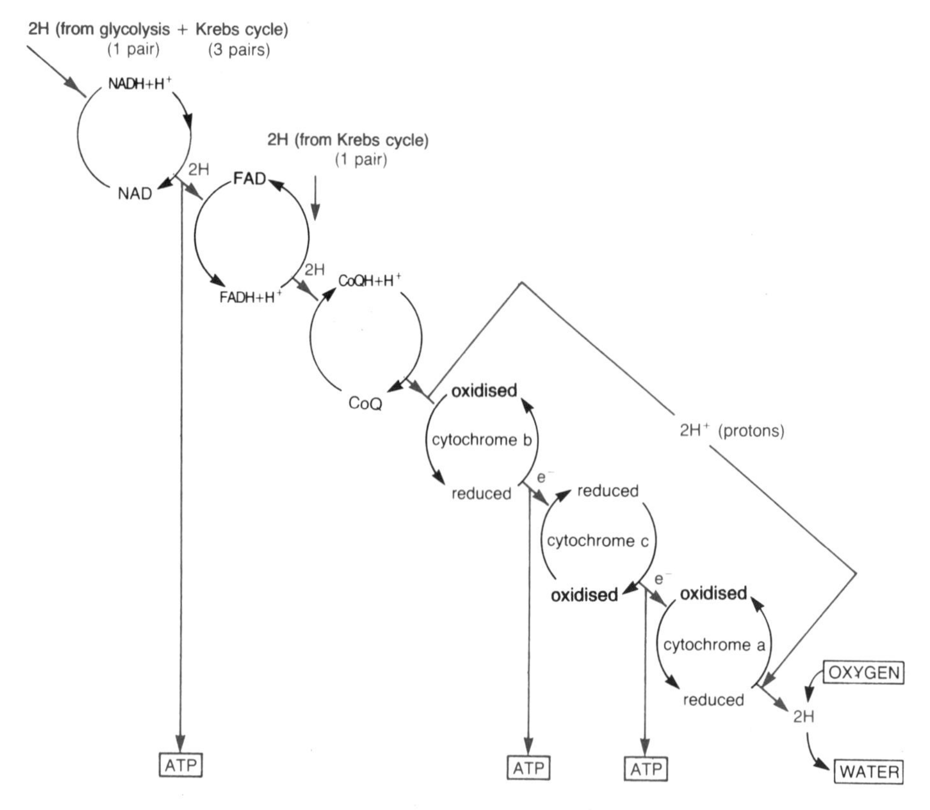

Fig. 5.18 The electron transfer system

ANAEROBIC PATHWAYS

If no oxygen is available the pyruvate molecules formed at the end of glycolysis do not enter the Krebs cycle but follow one of a number of anaerobic pathways. The anaerobic pathways are often referred to as fermentation. The pathways frequently use up 2H and so regenerate oxidised NAD from reduced NAD, allowing it to be used again in glycolysis.

Table 5.10 *Summary of three major anaerobic pathways*

Acetaldehyde fermentation	Alcoholic fermentation	Lactic acid fermentation
$CH_3COCOOH$ pyruvate	$CH_3COCOOH$ pyruvate	$CH_3COCOOH$ pyruvate
↓ → CO_2	2H → ↓ → CO_2	2H → ↓
CH_3CHO acetaldehyde, e.g. certain anaerobic bacteria	CH_3CH_2OH ethanol, e.g. yeasts and other plants. This process forms the basis of brewing and baking	$CH_3CHOHCOOH$ lactate, e.g. higher animals, especially in the muscles when oxygen usage exceeds supply, i.e. an oxygen debt is incurred

THE ROLE OF B VITAMINS

The group of vitamins collectively called vitamin B plays a major role in cellular respiration, particularly by acting as coenzymes. Some of the individual vitamins and their roles in respiration are given in Table 5.11.

Table 5.11 *The roles of some B vitamins*

Vitamin		Role in cellular respiration
B_1	Thiamine	Involved in formation of some Krebs cycle enzymes; forms part of acetyl coenzyme A
B_2	Riboflavin	Forms part of the hydrogen carrier flavine adenine dinucleotide (FAD)
B_3	Niacin (nicotinic acid)	Forms part of the coenzymes NAD and NADP; forms part of acetyl coenzyme A
B_5	Pantothenic acid	Forms part of acetyl coenzyme A

ATP AND ENERGY YIELDS

Adenosine triphosphate (ATP) is the form in which energy from the breakdown of glucose is temporarily stored. ATP consists of the organic base adenine, the 5-carbon sugar ribose and three phosphate groups.

The removal of the final phosphate to form adenosine diphosphate (ADP) releases 30.6 kJ mol^{-1} of energy:

$$ATP + H_2O \longrightarrow ADP + phosphate + 30.6\ kJ$$

and the removal of the next phosphate to form adenosine monophosphate yields a similar amount.

The amount of ATP produced during anaerobic respiration is small. For instance, in alcoholic fermentation glycolysis yields two molecules of ATP directly and two hydrogen atoms. However, the hydrogen atoms do not enter the electron transfer system as they are used to form the alcohol. The total yield is therefore only two ATPs or 61.2 kJ of energy from a potential of nearly 3000 kJ for the complete oxidation of a glucose molecule. The process is therefore only about 2% efficient.

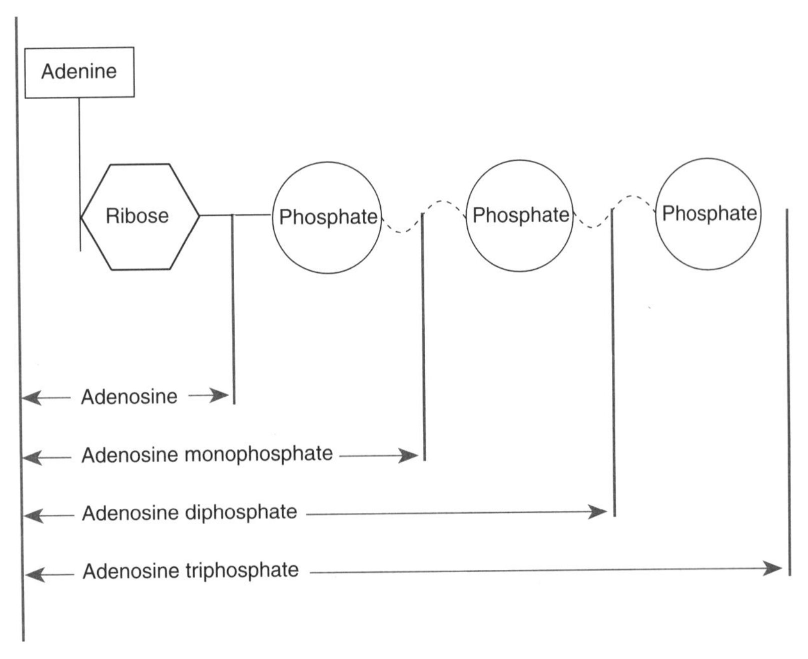

Fig. 5.19 Structure of adenosine triphosphate

In aerobic respiration the yield per glucose molecule is:

Glycolysis = 8 molecules of ATP (2 molecules of reduced NAD each yielding 3 ATP + 2 molecules of ATP formed directly)

Pyruvate $\longrightarrow$ Acetyl CoA = 6 molecules of ATP (2 molecules of reduced NAD each yielding 3 ATP)

Krebs cycle = 24 molecules of ATP (6 molecules of reduced NAD each yielding
3 ATP + 2 molecules of reduced FAD yielding
2 ATP + 2 molecules of ATP formed directly)

Total ATP yield = 38 molecules of ATP

This gives a total energy yield of 38 × 30.6 kJ = 1162.8 kJ, which is about 40.3% efficient.

One problem arising from the totalling of these ATPs that causes confusion is that the yields refer to the number of ATPs from a single glucose molecule. Early in glycolysis the 6-carbon sugar is split into two 3-carbon sugars. The yields from this point onwards are therefore doubled to take account of this.

Uses of ATP

In a metabolically active cell, up to 2 million molecules of ATP are required every second. This ATP is used for a variety of processes.

1. **Anabolic processes** ATP provides the energy necessary to build up macromolecules, e.g. proteins from amino acids.
2. **Active transport** Energy is needed to move materials against a concentration gradient, e.g. by the use of ion pumps.
3. **Movement** Muscle contraction, ciliary movement and the contraction of the spindle fibres during cell division all require energy from the hydrolysis of ATP.
4. **Activating reactants** Chemicals often require the addition of phosphate groups from ATP to make them more reactive, e.g. phosphorylation of glucose at the beginning of glycolysis.
5. **Secretion** ATP provides energy for the secretion of cell products.

RESPIRATORY QUOTIENTS

A respiratory quotient (RQ) is a measure of the ratio of carbon dioxide evolved to the oxygen consumed:

$$RQ = \frac{CO_2 \text{ evolved}}{O_2 \text{ consumed}}$$

For a hexose sugar such as glucose it can be seen from the equation

$$C_6H_{12}O_6 + 6O_2 \longrightarrow 6CO_2 + 6H_2O$$

that the ratio is $\frac{6CO_2}{6O_2} = 1.0$

For a fat such as stearic acid, however, the equation for its complete oxidation is

$$C_{18}H_{36}O_2 + 26O_2 \longrightarrow 18CO_2 + 18H_2O$$

giving $RQ = \frac{18CO_2}{26O_2} = 0.7$

Other RQs are:

malate 1.33
proteins 0.9 (though this varies slightly according to the particular protein)

In practice organisms rarely oxidise one compound alone and therefore RQs may lie between these values, e.g. for humans the RQ is typically 0.85 indicating the oxidation of a mixture of carbohydrate (1.0) and fat (0.7) rather than protein, which is normally only respired in extreme circumstances, e.g. starvation.

OTHER RESPIRATORY SUBSTRATES

Both fat and protein can be used as respiratory substrates without first being converted to carbohydrate. Fat must firstly be broken down into glycerol and fatty acids. The glycerol is then phosphorylated before entering glycolysis as triose phosphate while the fatty acids are broken down to 2-carbon fragments and enter as acetyl coenzyme A.

Proteins are used only as a last resort. They are firstly broken up into their component amino acids which then have their amino (NH_2) groups removed (**deamination**). The remaining portions then enter at different points depending on the number of carbon atoms they possess (see Fig. 5.20).

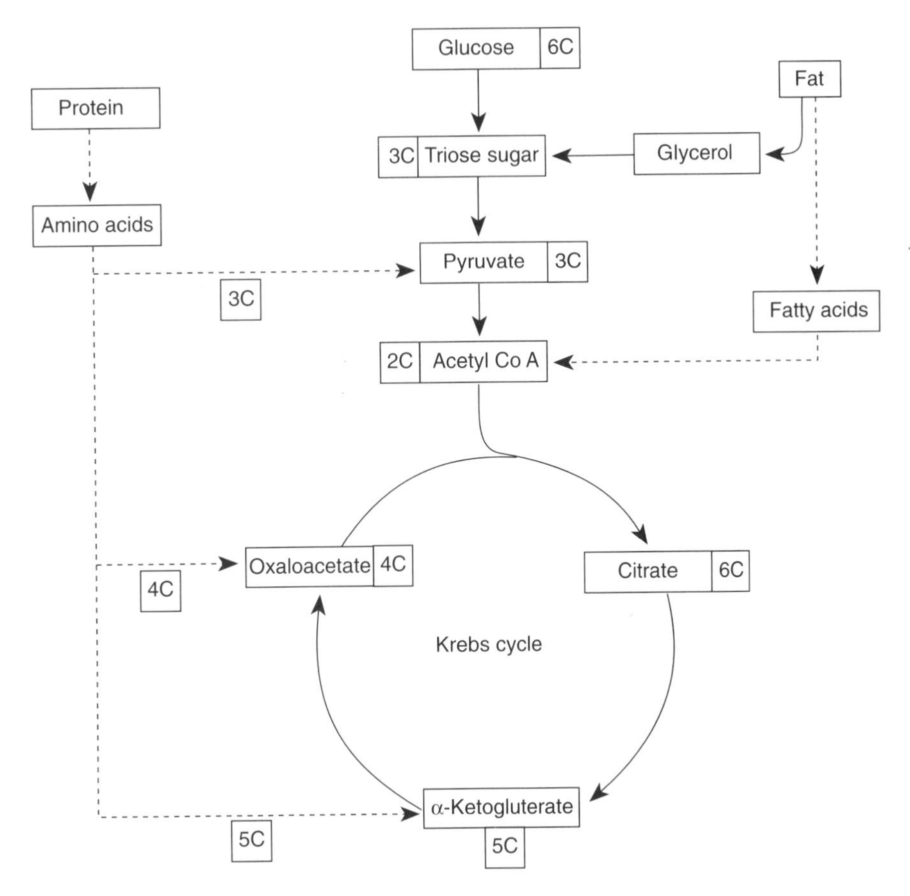

Fig. 5.20 Summary of respiratory breakdown of carbohydrates, fats and proteins

Chapter roundup

We have seen how many living organisms utilise energy in order to maintain low entropy and in doing so increase the entropy of their environment. Autotrophic organisms use energy sources such as sunlight to build up simple molecules such as water and carbon dioxide into complex ones such as carbohydrates, fats and proteins. In doing so they manufacture ATP in a process called **photosynthetic phosphorylation**. Heterotrophic organisms obtain their energy by consuming the complex molecules produced by autotrophs. Some heterotrophs eat plants and animals without living in close association with them; these are said to be **holozoic**. Others, known as **parasites**, live in close association with other living organisms, their hosts, deriving energy at their expense. **Saprobionts** live on the dead and decaying remains of organisms.

All living organisms, whether autotrophic or heterotrophic, break down complex molecules to simple ones in the process of respiration. In so doing they manufacture ATP in a process called **oxidative phosphorylation**. Where respiration requires a source of molecular oxygen it is termed **aerobic**; where it does not the name **anaerobic** is applied.

In the next chapter we will investigate the way in which energy flows through a biological system comprising many different organisms (an ecosystem) and how material within that system is recycled. In so doing we will explore the interrelations of organisms with each other and their environment – a study known as ecology.

Illustrative questions and worked answers

1 Five small discs cut from spinach leaves were floated on a small volume of buffered hydrogen carbonate solution in a flask attached to a respirometer. The discs were first exposed to bright light, then to dim light and finally left in the dark. Oxygen release was recorded as positive values and oxygen uptake as negative values.

The results obtained from the experiment are given below.

Light intensity	Time interval in minutes	Oxygen uptake or release for each 3 minute interval (mm^3)
Bright light	0–3	+57
	3–6	+64
	6–9	+58
	9–12	+60
Dim light	12–15	+16
	15–18	+3
Dark	18–21	−16
	21–24	−12
	24–27	−15
	27–30	−14

(a) Present the data in suitable graphical form. (5)

(b) Calculate the mean rate of oxygen release in bright light. (2)

(c) Explain the significance of the results obtained from this experiment. (8)

Time allowed 11 minutes

Tutorial note

A question of this type tests your ability to present data in a suitable graphical form, carry out a mathematical calculation and interpret the data. You can present the data either as a histogram or a point graph, but in both cases you should:

1. choose each axis carefully to make maximum use of the graph paper;
2. select a scale which enables you to plot points easily, e.g. choosing seven squares to represent one minute is very likely to lead to errors when plotting coordinates;
3. label each axis and state the units;
4. put a title on the graph.

In part (b) you need first of all to appreciate that the period of bright light is from 0–12 minutes. A possible error is that you might take only the figure given for the last period of bright light (9–12 minutes) because you assumed that this figure of 60 represented the cumulative total of oxygen production over the whole period. The fact that when this figure is divided by the time period of 12 minutes it gives a whole number of 5 $mm^3\ min^{-1}$, may reinforce this misguided view. You should appreciate that the figures given represent the oxygen produced in each 3 minute period. The total over 12 minutes is hence the sum of the first four figures, namely 57 + 64 + 58 + 60, giving a total of 239 mm^3.

In part (c) you will need to be aware that the rate of oxygen release is a measure of the rate of photosynthesis. You should appreciate also that the amount of oxygen released in bright light is not a true measure of the oxygen produced in photosynthesis as some has been used up in respiration and therefore not released from the leaves.

Suggested answer

(a) The finished graph should resemble one or other (not both) of the following.

Oxygen uptake (or release) of spinach leaf discs in different light intensities

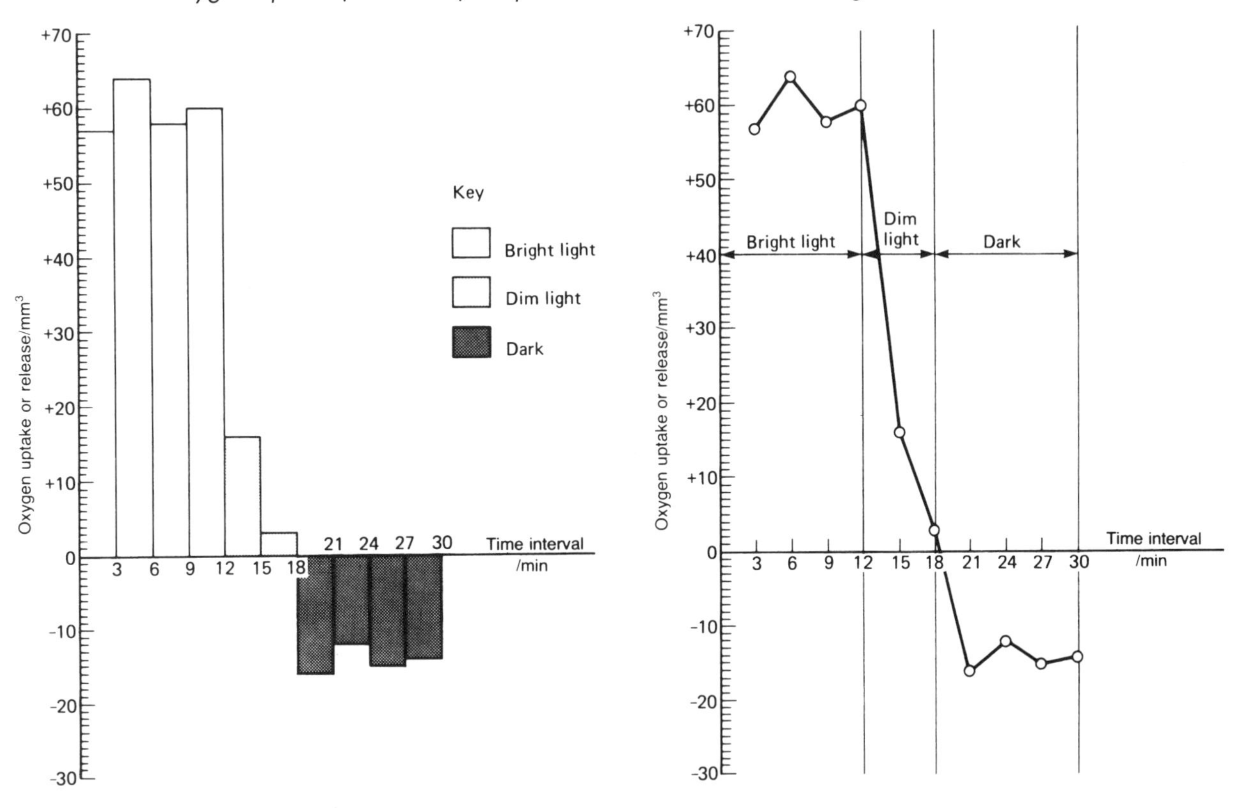

(b) Total oxygen released = 57 + 64 + 58 + 60 = 239 mm³.
Period over which oxygen is released = 12 minutes.
Rate = 239/12 = 19.9 mm³ min⁻¹ (don't forget the units).

(c) Oxygen release is a measure of photosynthesis. The rates of oxygen uptake in the dark and release in bright light are both fairly constant because:

1. in bright light some factor other than light intensity is limiting the reaction (e.g. CO_2 concentration) and therefore additional light has no further effect;
2. in the dark there is no photosynthesis and the rate of respiration (the reason for oxygen being taken up) is unaffected by light intensity.

In dim light the light intensity is the limiting factor and therefore a small change in light intensity produces a marked change in the rate of photosynthesis and hence oxygen release.

The point where the line crosses the *x*-axis is called the compensation point and represents the light intensity at which the oxygen released in photosynthesis is exactly counterbalanced by that taken up in respiration.

2 (a) Explain the term 'saprobiont'. (5)
(b) Compare and contrast saprobionts with parasites. (10)
(c) In what ways are saprobionts important to humans? (5)
Time allowed 35 minutes

Tutorial note

While you clearly require knowledge of saprobionts and parasites to produce a good answer, there are skills other than recall of information required here. Note the word 'explain' in part (a) – more than a short definition is required and time should be spent making clear the difficulties in fitting organisms into categories. The division between saprobionts and parasites is unclear at times; some organisms live parasitically until they have killed their host, at which point they live saprobiontically on the dead remains.

In part (b) the ability to make comparisons is essential. 'Compare and contrast' does not mean 'describe'. An account of parasites followed by one of saprobionts will yield low marks. It is for you, and not the examiner, to bring out similarities and differences. Avoid borderline points; rather keep to those which are clearly the same (e.g. both feed heterotrophically and absorb soluble food) or those which are obviously different (e.g. saprobionts derive energy from dead organisms, while parasites derive energy from living ones). Use specific examples to illustrate your points wherever possible, but take care they are appropriately chosen. *Mucor* (pin mould) is the example of a saprobiont most often studied, while *Taenia* (tapeworm) is a popular parasite example. Unfortunately both are atypical of their type. For example, most parasites are aerobic, while many saprobionts are anaerobic; however, *Taenia* and *Mucor* are exceptions to this.

A variety of examples is needed in part (c). It is insufficient to simply describe their importance in recycling minerals or in the decomposition of sewage. It is easy to forget that *Saccharomyces* (yeast) and mushrooms are both saprobiontic fungi.

Suggested answer

(a) Saprobionts are organisms which obtain energy from the dead remains of other organisms. There are problems distinguishing saprobionts from parasites, e.g. *Armillaria mellea* (honey fungus).

(b) (i) Similarities

Both:

1. are heterotrophic
2. absorb soluble food
3. have simple digestive systems where they are present
4. have sexual and asexual phases in reproduction, often involving resistant stages
5. produce large numbers of offspring

(ii) Differences

Parasites	Saprobionts
Energy derived from living organisms	Energy derived from dead organisms
Many stages in life cycle	Usually a single adult stage and spores
Very specific to their hosts	Use a variety of food sources
Nutritionally highly adapted	Simple methods of nutrition
Most plant and animal groups have representatives	Almost totally bacteria and fungi
Most are aerobic	Anaerobic and aerobic

(c) Importance of saprobionts

1. recycling of materials, e.g. carbon, nitrogen, phosphorus
2. brewing and baking, e.g. *Saccharomyces* (yeast)
3. making antibiotics, e.g. Penicillin
4. decomposition of wastes such as sewage
5. production of yoghurt and cheese
6. food source, e.g mushrooms
7. industrial processes, e.g. tanning of leather, production of vitamins.

Question bank

1 (a) What are the main features of
(i) *autotrophic* nutrition; (2)
(ii) *heterotrophic* nutrition? (1)

The diagram shows part of a transverse section through a mammalian ileum.

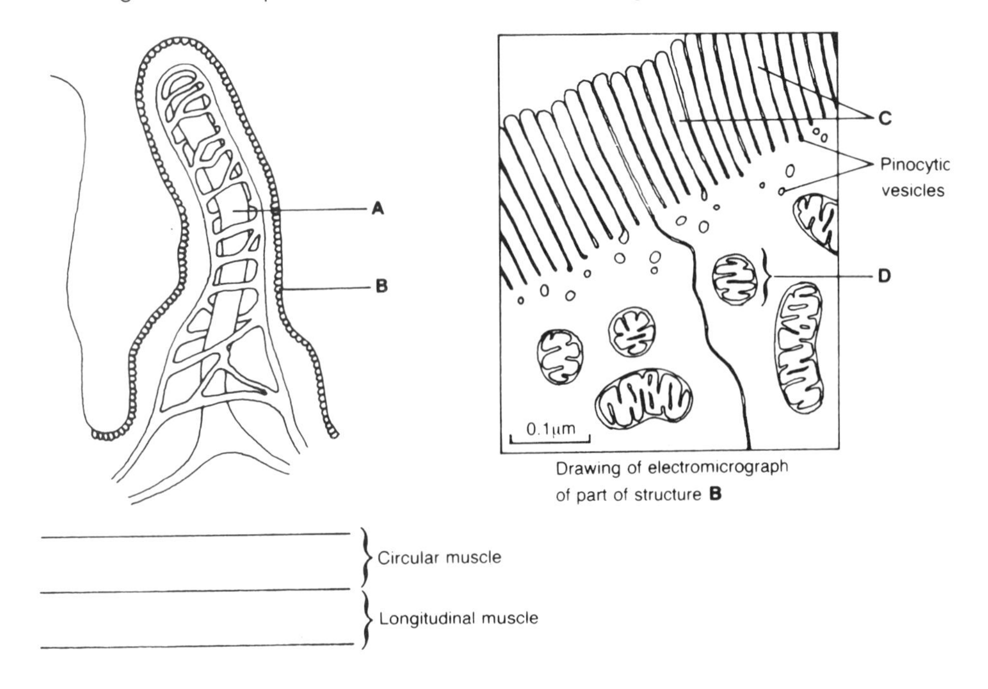

Drawing of electromicrograph of part of structure **B**

(b) Name the parts labelled A to D on the diagram.

(c) Briefly describe how *three* features, shown in the diagram, enable the ileum to carry out its function of absorption. (6)

(d) (i) Of what type of muscle do the layers of circular and longitudinal muscle consist? (1)
(ii) What is the function of this muscle in the ileum? How is this function achieved? (2)

(e) The duodenum and ileum receive secretions from the pancreas and liver.
(i) List the components of the secretion derived from the pancreas. (2)
(ii) List the components of the secretion derived from the liver. (2)
(iii) Describe how the flow of the pancreatic secretion is controlled. (4)

Time allowed 35 minutes

Pitfalls

In part (b) you should take care not to confuse the lacteal (structure A) with a blood vessel. Equally the microvillus (structure C) is very commonly labelled as a villus. You should remember that a villus is the much larger finger-like projection of the intestinal lining. The microvillus is a tiny projection of a cell membrane.

In part (c) you should pay attention to the phrase 'shown in the diagram'. Answers frequently contain features that are not illustrated. You should provide some detail in part (c). For example, when describing how the smooth muscle achieves peristalsis, rather than saying 'it contracts', you are likely to obtain more credit if you say 'by the alternate contraction and relaxation of the circular and longitudinal muscle'.

Your most likely error in part (e) is to only state the enzyme components of the secretions of the pancreas and liver and fail to include mineral salts, water, bile pigments, etc. Remember the liver has no enzymes in its secretions into the intestine. Also, candidates frequently believe that the hormones which stimulate the production of pancreatic juice pass from the duodenal wall to the pancreas along the intestine rather than in the blood stream.

Suggested answer

(a) (i) the building up of complex organic molecules from inorganic forms of carbon, such as carbon dioxide, using light or chemical energy.

(ii) the breakdown of complex organic molecules into simple ones, thus releasing energy.

(b) A lacteal
B epithelial cell of villus
C microvilli on epithelial cell of villus
D mitochondrion

(c) Any three from:

- Microvilli increase surface area for diffusion and active transport.
- Blood vessels within the villus remove absorbed nutrients and so maintain a concentration gradient, hence increasing the rate of inward diffusion.
- Mitochondria release the energy which is needed for active transport of nutrients into the epithelial cells.
- The single-celled epithelial layer is very thin thus minimising the distance over which diffusion and active transport occur.

(d) (i) smooth (involuntary) muscle

(ii) to maintain movement of material along the intestine by peristalsis which entails the alternate contraction and relaxation of the circular and longitudinal muscle.

(e) (i) mineral salts, trypsin, carboxypeptidase, chymotrypsin, amylase, lipase, nucleases

(ii) bile salts, bile pigments, sodium hydrogen carbonate, cholesterol

(iii) Secretin is produced by the duodenal wall when food enters the duodenum. It travels in the blood stream to the pancreas where it stimulates the secretion of mineral salts. Cholecystokinin-pancreozymin from the duodenal wall stimulates the pancreas to secrete its enzymes.

2 (a) There are three stages in the release of energy from a molecule of glucose: glycolysis, the Krebs cycle and the electron transfer system. What are the essential features of each of these processes? (10)

(b) In what circumstances would you expect anaerobic respiration of glucose to occur in:

(i) yeast? (2)
(ii) a flowering plant? (4)
(iii) a mammal? (4)

Time allowed 40 minutes

Pitfalls

You should note the words 'essential features' in part (a). To attempt a full explanation of all three processes would take such an inordinate amount of time that you would find yourself unable to complete the question paper. Keep to the major points. With 3–4 marks for each of the processes, 3 or 4 points on each would seem sensible.

Part (b) may appear deceptively easy, or impossible, depending on whether you can think of any circumstances. The problem lies in giving, as the mark allocation suggests, more than a single answer for each. In all cases think of the circumstances in which an organism, or part of it, may be deprived of oxygen. These might include external conditions, e.g. stagnant water, and internal ones, e.g. a restriction in blood supply. Equally, it may be that the demand for oxygen outstrips its supply on a temporary or permanent basis.

Suggested answer

(a) Glycolysis

❶ phosphorylation of glucose

❷ splitting of hexose to two triose molecules

❸ production of ATP directly
❹ production of ATP indirectly – via electron transfer system.

Krebs cycle

❶ loss of carbon dioxide to give a 2-carbon molecule
❷ combination of 2-carbon and 4-carbon molecules to give a 6-carbon molecule
❸ oxidation of two carbon atoms to carbon dioxide to generate a 4-carbon molecule
❹ production of ATP directly and indirectly (via electron transfer system).

Electron transfer system

❶ progressive transfer of electrons to carriers at lower energy levels
❷ the use of the energy associated with the electrons to generate ATP from ADP
❸ the reduction of oxygen to water.

(b) (i) Yeast

❶ in stagnant solutions
❷ centre of decomposing fruits and other organic matter.

(ii) Flowering plants

❶ young seeds, centre of fruits or large stems
❷ in roots in compacted or waterlogged soils
❸ in aquatic plants growing in stagnant ponds.

(iii) Mammals

❶ inefficiency of lungs, e.g. emphysema
❷ reduction in blood supply, e.g. haemorrhage, pressure on artery
❸ low oxygen-carrying capacity of blood, e.g. anaemia, bone marrow disease
❹ low cardiac output, e.g. slow heart rate, coronary thrombosis
❺ capillary network inadequate, e.g. angina
❻ high oxygen demands, e.g. strenuous exercise, pregnancy
❼ others, e.g. hibernation, sperm in oviduct, high altitude.

3 A plant with variegated leaves was supplied with radioactive carbon dioxide ($^{14}CO_2$) during an experiment. Leaf Y was kept in the dark and leaf X was illuminated. The radioactivity in the leaves was measured at the end of the experiment and found to be as shown.

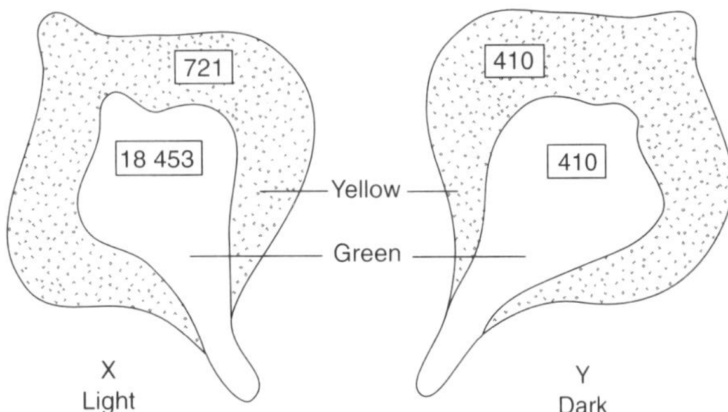

Radioactivity is shown in arbitrary units in boxes

The most likely explanation for the level of radioactivity found in the yellow zone of X is:

A Photosynthetic products diffuse into the yellow zone.
B Photosynthesis takes place in this zone but no storage of starch occurs.
C Some photosynthesis occurs in the yellow zone, but in the absence of chlorophyll the amount is small.

D Radioactive carbon dioxide diffuses into the leaf and accumulates in this zone.

Time allowed 1½ minutes

Pitfalls

Firstly you will need to appreciate that the level of radioactivity is a measure of the amount of ^{14}C present. Although supplied as $^{14}CO_2$ the ^{14}C can be incorporated into other organic molecules within the leaf. Secondly you should compare the figures for leaf X and leaf Y. When in the dark the levels of radioactivity are equal regardless of leaf colour. This level must be due to $^{14}CO_2$ which has diffused evenly throughout the air spaces of the leaf. This, in effect, is the control against which other levels of ^{14}C can be compared. In X (the leaf in the light) the green areas (i.e. those with chlorophyll) have 45 times as much radioactivity as those with no chlorophyll. This is due to ^{14}C being incorporated into organic molecules as a result of photosynthesis. The non-chlorophyll regions, however, show nearly double the radioactivity of the 'control' leaf Y.

Answer A seems probable, but you should never be content with what you think is the correct response until you have found sound biological grounds for rejecting the other options. Answers B and C both assume some photosynthesis occurs in the absence of chlorophyll and this is not the case. If D were correct, a similar uneven distribution of radioactivity would be expected in leaf Y. As it is not, D can be rejected.

Suggested answer

Option A – Photosynthetic products diffuse into the yellow zone.

CHAPTER 6

ECOLOGY

Units in this chapter

Chapter objectives

Having seen why energy is necessary to living organisms and how it is utilised by them, we will now turn our attention to the way in which this energy, along with essential chemicals, is passed amongst organisms. We will investigate the ways in which organisms interact with one another and to what extent they depend on each other for their survival. The interrelationships between the nonliving (abiotic) environment and living organisms will also be surveyed. As a major influence on our environment, we will look at our own role in determining the distribution of organisms and, in particular, the effects of the many pollutants produced by humans. The methods being utilised to conserve our environment will be discussed. We will begin, however, with a look at populations and the factors which control their size and distribution.

6.1 POPULATIONS

A **population** is a group of individuals of a single species which occupy a particular area at the same point in time. An example would be a population of woodlice on a rotten log or sticklebacks in a pond. A **community** is all the populations which occupy a particular area. In other words, all the animals and plants on the rotten log (woodlice, beetles, fungi, moss, etc.) or in the pond (sticklebacks, water boatmen, pondweed, algae, etc.) form a community.

If a population is small to begin with its growth will be slow (**lag phase**). As numbers increase so too does the rate of growth (**exponential phase**), but in time the numbers in the population will outstrip the supply of some factor needed to support it (e.g. food) and the death rate will increase. A point is reached where the death rate and birth rate are equal and the population size becomes stable (**stationary phase**). This point is referred to as the **carrying capacity** of the population. A change in some environmental factor (e.g. reduced food supply, diminished rainfall) can lead to an increase in the death rate and hence a fall in the population (**decline phase**).

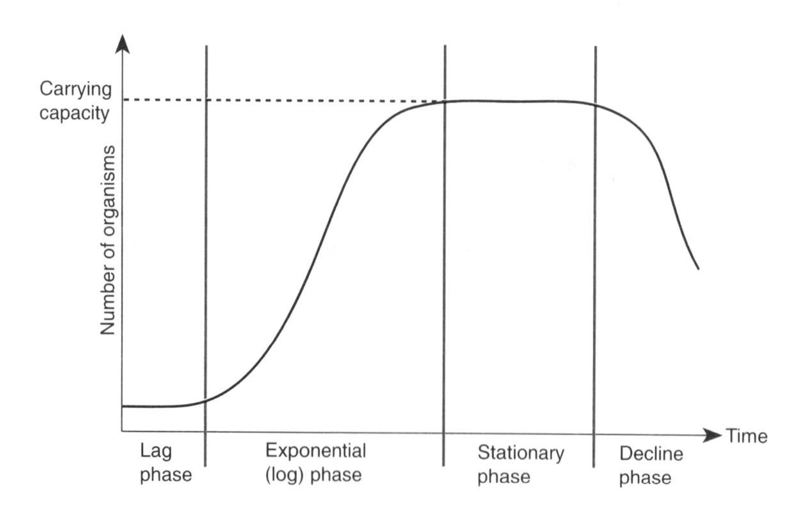

Fig. 6.1 Population growth and decline

FACTORS INFLUENCING POPULATION SIZE

Collectively the factors which limit growth are termed **environmental resistance**.They fall into two categories:

1. **Density-dependent growth** The size (density) of a population affects the growth rate. Consider, for example, that there is a fixed supply of food available; the larger the population is to begin with, the less food there is for each individual and the slower will be the growth rate. A small population can expand more rapidly.
2. **Density-independent growth** Imagine the introduction into a pond of a chemical that is toxic to water fleas. The size of the population will not influence its effect – if it is concentrated enough to kill the fleas it will do so whether there are ten or ten million. Typical density-independent factors include temperature changes and natural catastrophes like floods, storms and fires.

Regulation of population size

Whatever factors influence a population, it is the balance between the number of organisms being added to a population and those being lost from it which determines its size. Organisms added to a population depend on two factors:

1. **Fecundity** – the reproductive capacity of the population. In mammals this is referred to as the **birth rate (natality)**.
2. **Immigration** – the addition to a population of organisms from neighbouring populations.

Organisms are lost from a population in two ways:

1. **Mortality** – the death rate of organisms in the population, from whatever cause.
2. **Emigration** – the loss of individuals to other populations.

Edaphic (soil) factors

The soil clearly influences plant growth and also, directly or indirectly, animal population size. Edaphic factors include:

1. **Texture** This is determined by the proportion of sand, silt and clay particles. These in turn influence the water-holding capacity of soil. Clay soils with their smaller particles hold more water than sandy ones. Table 6.1 gives a comparison of clay and sandy soils.

Table 6.1 *A comparison of clay and sandy soils*

Clay soil	Sandy soil
Particle soil is less than 0.002 mm (2 μm)	Particle size from 0.02 mm to 2.00 mm
Small air spaces between particles	Large air spaces between particles
Poor drainage; soil easily compacted	Good drainage; soil not compacted
Good water retention leading to possible waterlogging	Poor water retention and no waterlogging
Being a wet soil, evaporation of water causes it to be cold	Less water evaporation and therefore warmer
Particles attract many mineral ions and so nutrient content is high	Minerals are easily leached and so mineral content is low
Particles aggregate together to form clods, making the soil heavy and difficult to work	Particles remain separate, making the soil light and easy to work

2. **Humus** This constitutes all dead and decaying remains of organisms. It acts like a sponge in retaining water and so improves sandy soils. Equally it lightens clay soils and helps them to break up. As it decomposes, the humus slowly releases minerals and so adds nutrients to the soil.
3. **Air** Roots derive their oxygen directly from the soil as do most soil animals. If the soil becomes waterlogged, the air is driven out and this can lead to anaerobic conditions and the subsequent death of the fauna and flora.
4. **Water** The water content of the soil is the difference between rainfall and evaporation or other losses such as uptake by plants. Water is an essential metabolite and a medium for the transfer of gametes.
5. **Minerals** Different species require different minerals and so the distribution and size of any population is dependent on the blend of available minerals. Table 1.1 in Chapter 1 lists some of the more important minerals and their functions.
6. **pH** The pH of the soil directly influences which plants grow: heathers prefer acidic and stonewort prefers alkaline soils. It also affects the physical properties of soils and the uptake of minerals.
7. **Temperature** This influences other soil factors such as water content as well as activity, germination and growth of the organisms in the soil.
8. **Biotic factors** Soils contain vast populations of organisms. These influence aeration and drainage (earthworms burrowing) and mineral content (bacteria breaking down humus) as well as having direct effects such as parasitising plants or feeding on their roots.

Climatic factors

1. **Light** This is essential for photosynthesis and affects flowering in plants as well as reproduction, hibernation and migration in animals. Three aspects of light are important: wavelength, intensity and duration.
2. **Water** The importance of water is dealt with in Chapter 7; at this stage it is sufficient to know that it is essential to all organisms, and the more plentiful it is, the larger and more varied the community is likely to be.
3. **Air and water movements** Both are important in dispersing seeds and spores and so determine the distribution of many species. The intensity of the movements may also influence the shape and survival of organisms, especially plants where transpiration rates are affected.
4. **Humidity** This affects the transpiration rate in plants and the rate of evaporation in animals.

5. **Temperature** Each species has an optimum temperature at which it lives, and it can only survive within a fixed temperature range. Temperature also indirectly influences distribution by affecting water availability and humidity.

Biotic factors

1. **Competition** Within a population individuals compete with each other for food, water, minerals, territory, shelter, nesting sites and mates. This is called **intraspecific competition**. They also compete with individuals from other species for these essentials. This is called **interspecific competition**.

2. **Predation** All organisms can provide food for others. The rate at which a population is consumed by organisms which use it as a food source is a major influence on population size. Cyclic changes occur in the prey and predator populations (see Fig. 6.2) because, as the prey population diminishes, the predators are increasingly deprived of their food source and their numbers likewise reduce.

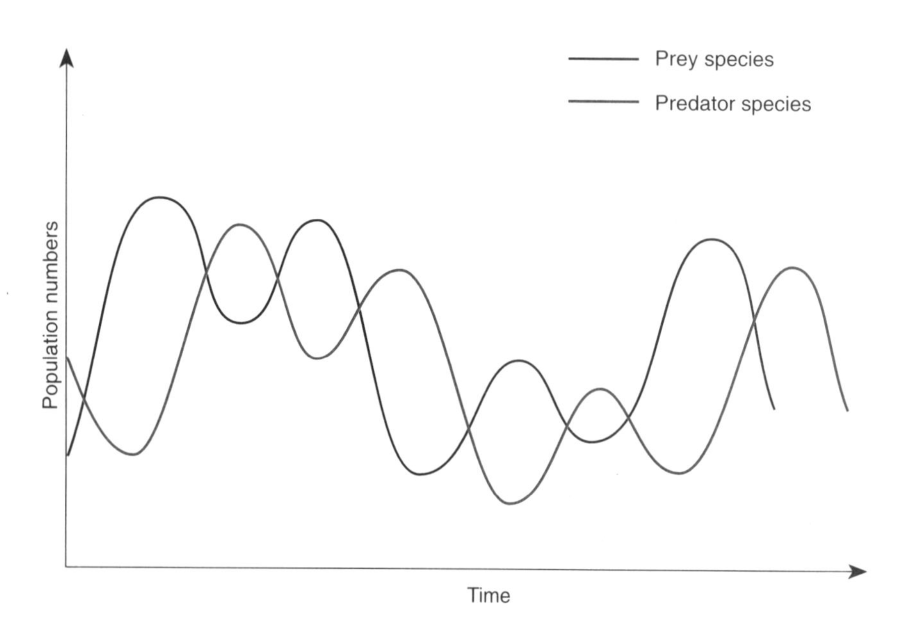

Fig. 6.2 Relationship between predator and prey populations

3. **Pollination and dispersal** Many plants use animals to aid their pollination and dispersal. The distribution and size of these two populations are clearly influenced by each other.

4. **Antibiosis** It is increasingly being discovered that organisms produce chemicals which repel others. These include antibiotics which inhibit bacterial growth, pheromones of ants which warn off others of their species and the use of territorial markers in mammals.

6.2 ENERGY AND THE ECOSYSTEM

It may surprise you to know that of all the sunlight which reaches the earth only 1% is actually captured by plants in photosynthesis. It is this 1% which maintains all life. The energy flows from one organism to another. Unlike inorganic ions, energy is not recycled but is lost as heat, and so must constantly be replaced by sunlight reaching the earth.

The thin film of land, water and air around the earth's surface which supports life is called the **biosphere**. This can be divided into areas called **biomes** whose nature is determined by the dominant plant types, e.g. tropical rainforest, grassland, desert. Biomes are subdivided into **zones** which in turn comprise small areas with specific characteristics called **habitats**. Examples of habitats include a pond, an oak wood and estuary, and a rocky shore. An individual member of a community is usually confined to a particular region of the habitat. This is called a **microhabitat**. The physical position of a species within a habitat, its behaviour and interrelationship with the biotic and abiotic environment is referred to as its **ecological niche**. The interrelationships of all elements, biotic and abiotic, in a biological system is called the **ecosystem**. While, in practice, an ecosystem might be considered as a small unit such as a pond, which is more or less self-contained, for the purposes of renewing energy flow and mineral cycling we will consider the biosphere as one single ecosystem.

ENERGY FLOW

The flow of energy through the ecosystem (ecological energetics) obviously begins with sunlight, which provides all the energy that maintains life on earth.

Trophic levels

The sun's energy is passed from one feeding level to another through the ecosystem. Each feeding level is termed a **trophic level**. Green plants represent one trophic level. As they manufacture complex molecules from simple ones they are called **primary producers**. All nonautotrophic organisms are dependent for their energy on feeding, directly or indirectly, off these producers. These are called **consumers**. **Herbivores**, which feed directly off plants, are known as **primary consumers**, whereas **carnivores**, which feed by eating the herbivores, are called **secondary consumers**. Those carnivores which consume other carnivores are termed **tertiary consumers**. Energy is thus passed along a hierarchy of trophic levels with primary producers at the bottom and consumers at the top. This is referred to as a **food chain**. When producers and consumers die, some energy remains locked up in the chemicals of which they are made. This can be utilised by a further group of organisms: the **decomposers** (saprobiontic fungi and bacteria) and **detritivores** (certain animals such as earthworms).

Food chains and webs

As energy is passed from one trophic level to the next in a food chain much is lost as heat during respiration, and so only a small proportion is left for the next stage (see Fig. 6.3). It is this which limits the length of any food chain because a situation is soon reached where there is an insufficient amount of energy left to sustain a viable population. Food chains are a useful idea for understanding how energy flows in an ecosystem but, in practice, organisms feed off more than one type of food. Feeding relationships are better depicted as a complex series of interconnections known as a **food web** (Fig. 6.4).

ECOLOGICAL PYRAMIDS

Since much energy is lost in moving from one trophic level to another, it follows that what remains can sustain far fewer organisms. The numbers of each organism therefore normally reduce as one moves up a food chain to give a **pyramid of numbers** (Fig. 6.5).

The concept of a pyramid of numbers has some drawbacks. For example, one large sycamore tree (producer) may support millions of aphids (primary consumer). It is therefore better to take into account the biomass of the organisms (**pyramid of biomass**). In the previous example the total mass of living tissue of the single sycamore tree exceeds the total mass of all the aphids living off it. Even pyramids of biomass have their problems because they can vary according to when during a year the biomass is measured. Therefore it is best of all to consider **pyramids of energy**. Here the total productivity of an organism for a given area over a given time (usually a year) is measured.

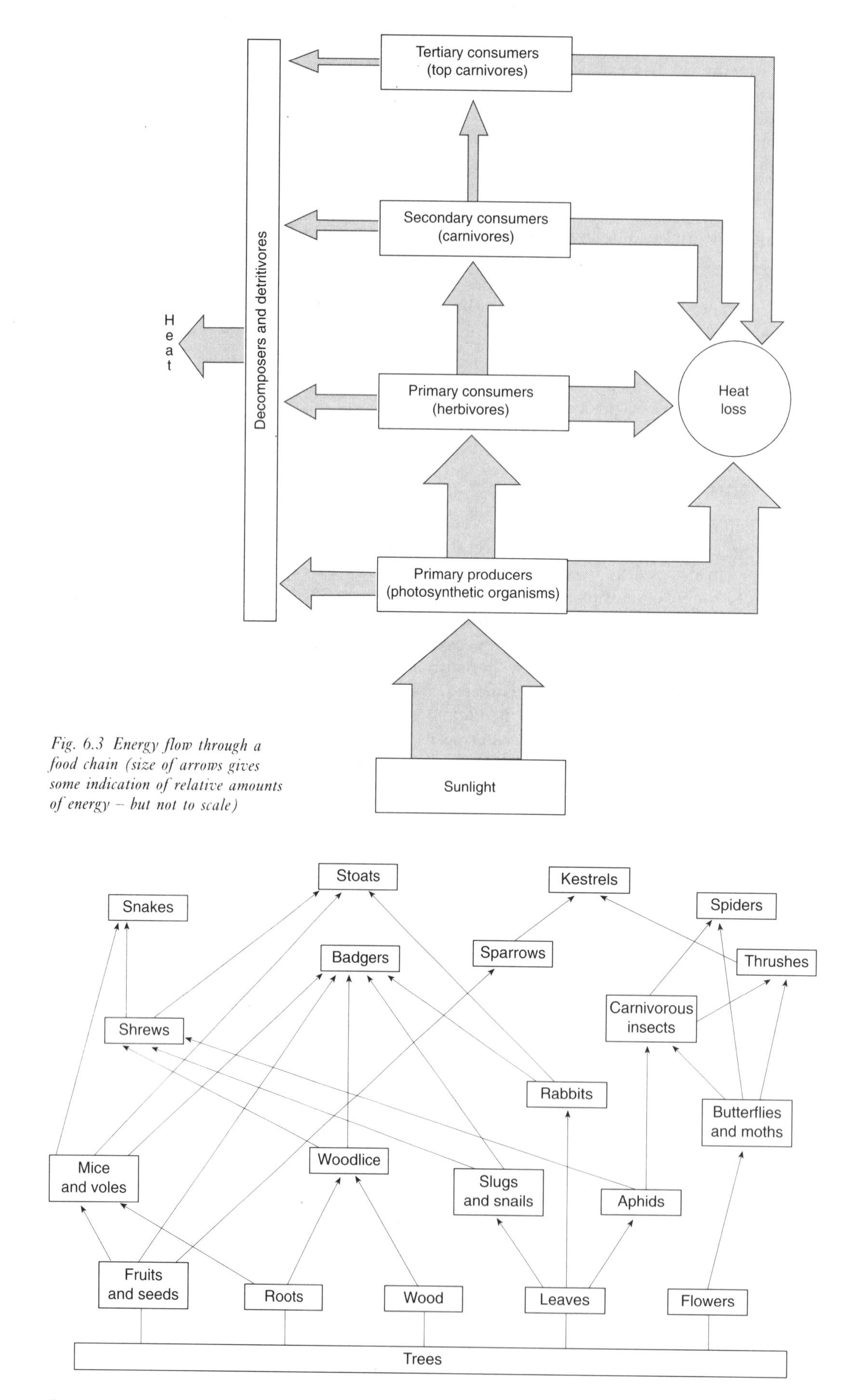

Fig. 6.3 Energy flow through a food chain (size of arrows gives some indication of relative amounts of energy – but not to scale)

Fig. 6.4 Typical food web found in a woodland habitat

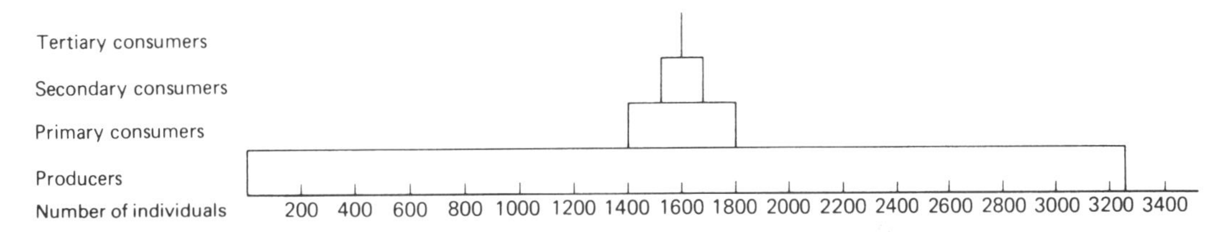

Fig. 6.5 Ecological pyramid of numbers

NUTRIENT CYCLES

In contrast to energy flow in ecosystems, which passes through in one direction, nutrients are constantly recycled. These nutrients normally exist in a geological form, which often acts as a reservoir of the nutrient, or in a biological form – namely the living organisms which incorporate and use them. We will look at two such cycles, for carbon and nitrogen.

The carbon cycle

Despite a mere 0.04% carbon dioxide, the atmosphere nevertheless forms a large and vital reservoir of carbon. Plants extract it during photosynthesis while heterotrophs consume the plants and respire the organic material to carbon dioxide again. The decomposers return to the atmosphere what carbon remains when organisms die. Fossil fuels form an important further reservoir of carbon.

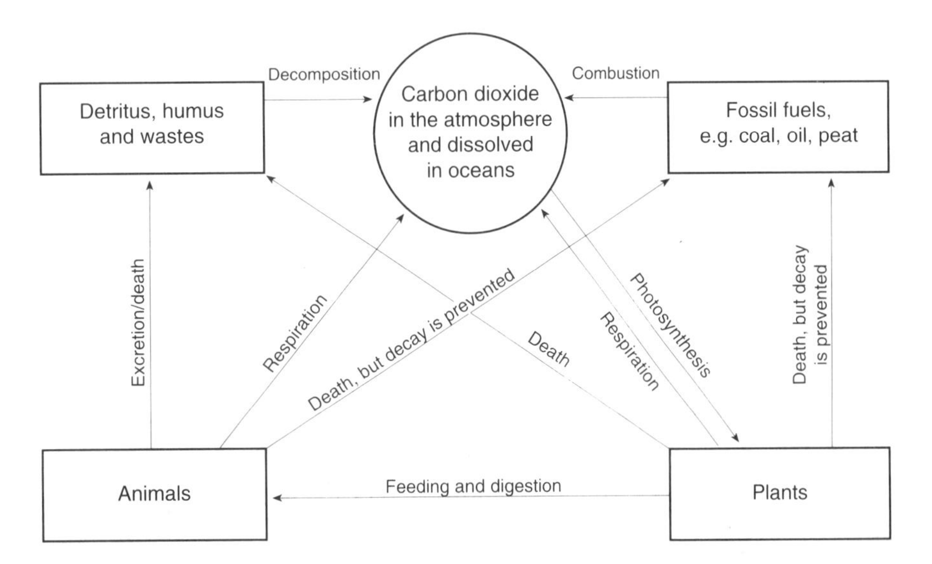

Fig. 6.6 The carbon cycle

The nitrogen cycle

Soil-dwelling bacteria are some of the few organisms which can utilise atmospheric nitrogen. They are able to convert it to nitrates which can be absorbed by plants and used during protein synthesis. Herbivores acquire their nitrogen from plants, whereas carnivores rely on animal protein for their nitrogen supply. The removal of nitrate from the soil is balanced by its return as a result of bacterial activity. The circulation of nitrogen in the community is similar to that of carbon. Nitrogen is passed from the plants to herbivores and along a food chain. Saprobionts eventually decompose the tissues of all the organisms concerned and so liberate the nitrogen once more. This is as ammonia, which reacts with carbon dioxide and water present in the soil spaces. At this point in the cycle, two species of bacteria present in soils play a vital part. One, called *Nitrosomonas*, is able to derive all the energy it needs from that released when ammonium carbonate is oxidised:

$$(NH_4)_2CO_3 + 3O_2 \rightarrow 2HNO_2 + CO_2 + 3H_2O + ATP$$

The nitrous acid reacts with salts in the soil to produce nitrites. These are acted upon by *Nitrobacter*, which derives its energy by oxidising nitrites to nitrates.

Another method of nitrate production is via a species of bacterium called *Rhizobium leguminosarum*, which lives in swellings on the roots of plants belonging to the family Leguminosae. They can incorporate atmospheric nitrogen into their protoplasm. The host plants absorb some nitrogenous molecules from the bacteria. Furthermore, the whole community eventually benefits from their activity because, when the host plant dies and is decomposed, a greater amount of nitrogen is released in the soil than would otherwise be the case.

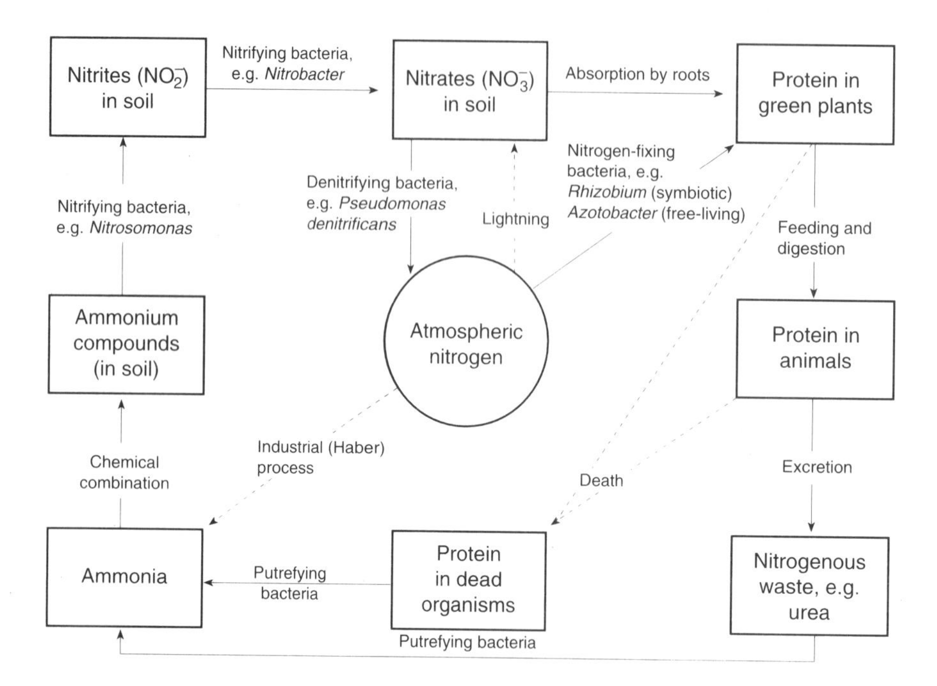

Fig. 6.7 The nitrogen cycle

6.3 MAN AND HIS ENVIRONMENT

Asked to describe the major human influences on the environment you would most likely refer to the greenhouse effect, damage to the ozone layer, acid rain and the many other recent examples of pollution. It is easy to think that it is only post-industrial man who has had a significant impact on the world. This is not so – preindustrial man also caused important changes to his environment.

Primitive man affected his environment by hunting, fishing and removing trees for fires and shelters. He was nomadic and therefore any effect he had on an area was temporary and the environment had adequate time to recover. About 11 000 years ago man began to cultivate his own crops and thus communities settled in one place in order to harvest and store the crops. The use of tools and later domestication of animals led to more efficient agriculture, and the ability to produce more food and hence support a larger population. Demands were created on the environment and trees were felled to provide shelters and more land for cultivation. The cultivation of crops was aided by domesticated animals (e.g. to haul ploughs) and therefore some of the crop had to be used to feed them, leaving less for human consumption.

With the advent of fossil fuels all this changed and little of the energy in the crops grown went back into growing more crops. Instead machines driven by fuels helped cultivate the crops, the majority of which were available to support a larger population which in turn required more efficient agriculture to support it. Fertilisers, pesticides and rapid transportation all developed and brought with them pollution problems in addition to those caused by burning the fossil fuels. Such fuels are a finite source of energy and hence new energy sources were needed. Nuclear energy was developed with its consequent environmental problems.

EXPLOITATION OF NATURAL RESOURCES

Agriculture

Probably the first attempts at cultivation began with the deliberate use of fire to clear large areas of land for planting; by doing so, man began to alter the structure of communities. Climax vegetation was destroyed and replaced by monoculture, e.g. a field of cereal crops replaced a forest of oak. It was soon discovered that if the same crop was grown for several successive seasons, the gross yield fell because crops make specific demands on the soil's resources. The problem could be avoided by crop rotation or by replacing lost materials with fertilisers. The latter solution raises another problem because excess use of fertilisers often causes pollution of rivers due to leaching and draining.

Ecologically, agriculture is contrary to the natural development of communities. Monoculture results in large areas planted with a single species of plant instead of the climax community of many species. As a result, weeds and pests compete with the single species for environmental resources. When man attacks the competitors with herbicides and pesticides, he creates additional problems of pollution of soil and water by toxic chemicals. Agriculture provides enormous food resources for those herbivores which share the same food as man. With many plants of one species grown close together, there is no problem of individuals finding new food plants at the dispersal stage. Thus, large numbers of herbivores can build up at the expense of crops. Whole ecosystems are therefore affected by the change in the original climax community, because land is artificially prevented from reaching its climax vegetation as a result of regular grazing, ploughing and the use of fertilisers and pesticides. Over the past 40 years in the UK we have doubled our food production as a result of greater use of fertilisers and pesticides, the development of new strains of plants and animals, increased mechanisation and changes in farm practices. So great has been our success in producing food that there are now large surpluses in Europe, and farmers are paid to **set aside** up to 20% of their land for purposes other than food production, e.g. tree planting.

Hedgerow removal is one example of a conflict of interest between agricultural production and conservation. Each year in the UK, around 8000 km of hedges are removed because, to the farmer, they:

1. restrict access by large agricultural machinery;
2. take up space which could be used to grow crops;
3. harbour diseases and pests of crops;
4. reduce crop yields by taking up water and nutrients.

From a conservation viewpoint however, hedges:

1. produce food for a wide variety of plant and animal species;
2. provide a habitat for many species;
3. act as wind breaks and therefore prevent erosion of the soil by wind;
4. provide a corridor along which organisms can move and disperse themselves;
5. add interest and variety to the landscape.

Renewable resources

Renewable resources are basically things that grow and are therefore replaced. This is not to imply that they are inexhaustible, for if they are utilised at too great a rate they will most certainly diminish and even disappear altogether. If, however, they are used at a rate equal to, or less than, the rate at which they grow they can be exploited indefinitely. This is referred to as a **sustainable yield**. A stock of fish, for example, will breed and increase in size at a certain rate. Provided the number caught does not exceed this increase, the fish stock should not get smaller. If, however, overfishing occurs and fish are removed faster than they are replaced, the stock may be reduced to a size which makes fishing uneconomic. Such a situation has been reached with North Sea herring.

Nonrenewable resources

Nonrenewable resources are, for all practical purposes, not replaceable. Oil, gas, coal and mineral ores are renewed at such a slow rate that we must consider them as finite resources. In the case of oil and gas the known reserves are unlikely to last longer than around 50 years.

Biological fuels

With nonrenewable fossil fuels unable to meet our long-term energy needs attention has to be turned to renewable biological fuels. Wood and straw can be burnt directly as fuels, but most plant species can be fermented by microorganisms into methane, methanol or ethanol, all of which make excellent fuels. In Brazil sugar cane wastes have been used to make **gasohol** and in Britain sewage can be converted to methane (**biogas**).

The population explosion

There has been a rapid increase in human population since the seventeenth century and this has inevitably had its effect on the environment. Ancient agricultural races have been replaced in certain parts of the world by populations founded on the industrial and technological revolutions. The results have been an increase in the use of nonrenewable resources (e.g. fossil fuels) and an increase in pollution of land, sea and air by products of fuel combustion. Solutions to these problems rely on cooperation between ecologists, politicians and economists. A cost-effective method of reducing pollution is usually not easy to find and conservationists must argue their case against industry and other cost-conscious sections of the community. There is no dispute about the scientific aspects of the ecological situations involved.

An increase in population imposes greater strains on food supply and its distribution. There are basically two contrasting points of view relating to the food supply problem.

1. Many agriculturalists believe that, with the use of fertilisers, herbicides and pesticides, the world can produce enough food for the predicted growth rate of population.
2. In contrast, many ecologists argue that at least half the world's population are already undernourished and they suggest that it will be impossible to feed adequately the probable population forecast for the year 2010.

There are three main methods of improving this situation:

1. stabilise the human population
2. increase the efficiency of food production and utilisation
3. develop new food sources.

At present we lose a great deal of food during its transportation from the source of production to the consumer. This is due to competition between man and other organisms, e.g. rats, mice, insects and moulds. Improvements in techniques for combating these is feasible.

Some food is lost for economic reasons because of high transport costs from regions of overproduction to regions of need. There are instances of countries burning surplus food or allowing it to rot while developing countries have populations suffering from malnutrition and starvation.

It is very easy to confuse biological and political problems in a discussion concerning human food supplies. The biological problems relate to the efficiency with which different types of food can be produced in different environments. Since energy is lost at each link of the food chain, the most efficient way for an omnivore like man to tap solar energy is to eat plants. The highest primary production comes from tropical crops, e.g. sugar cane which can grow all the year round but most of the plant's structure is not eaten by man. Usually only parts such as fruit, seeds, leaves or roots are used so that much primary production is wasted.

Plant proteins differ from animal proteins in the proportions of amino acids. Herbivores have to eat large quantities of food to obtain certain essential amino acids. Eating animals is economical in terms of the bulk of food consumed although it is wasteful of solar energy. Of the present supply of protein for human food 70% comes from vegetation and 30% from animals. Intensive livestock rearing is the most efficient way of producing those animal proteins which are traditional foods. The mass production of chicken and veal relies on processed diets which incorporate parts of organisms normally discarded as inedible. Occasionally antibiotics and hormones are added to the diets with subsequent dangerous side effects to the secondary consumers.

Ideally, protein should form about 10% of the total food intake in terms of energy; this constitutes the main difference between diets of well-fed and under-fed races of the world. An adequate diet should contain about 44 g protein per day. In the USA, the average consumption is about 64 g per day, whereas some developing countries have people surviving on 9 g per day.

Other possible sources of food could depend on advances in biotechnology. Certain bacteria can be used to convert petroleum into protein and this so-called single-cell protein production could be an answer to the world's food shortage. However, these sources can only be exploited through energy-consuming industrial processes and the amount of energy needed to produce edible protein may be so great that the source is totally uneconomic.

POLLUTION

Pollution is a difficult term to define. It is derived from the Latin word *polluere*, which means 'to contaminate any feature of the environment'. It may be broadly said to be 'adding to the environment a potentially hazardous substance or source of energy faster then the environment can accommodate it'. In this broad sense the definition includes not only man's activities but also certain natural processes.

Air pollution

This form of pollution is largely a result of burning fossil fuels such as coal, coke or fuel oil. **Smoke** contains carbon particles, carbon dioxide, sulphur dioxide and fluoride. Domestic combustion is of importance in those areas which have not introduced smokeless fuels. Smoke is a major health hazard as it affects the respiratory tract, increasing the incidence of bronchitis, and may even be a carcinogen. Normally, the cells lining the bronchial tubes produce mucus to moisten the surface and trap bacteria. When cells are diseased, they cannot function properly and are susceptible to bacteria. Whenever fog builds up in air heavily polluted with smoke, a condition called smog is produced. The dampness, combined with the effects of suspended carbon in smoke, increases the incidence of pulmonary disease.

The problem of reducing industrial smoke is complex; it has been tackled in a number of ways. In 1956, the British government introduced the Clean Air Act, which made it illegal to discharge smoke in many areas. This meant that households and factories had to use smokeless fuels, e.g. electricity, gas, oil or solid smokeless coals. The introduction of electric, diesel and gas turbine locomotives also reduced air pollution from the level it was when steam engines were in use. Modern technology allows fuel to be burnt more efficiently, and washing processes can be installed to remove soluble and heavy particles before the gases are sucked up the chimney where sieves or electrostatic precipitators are also installed.

To allow efficient combustion of petrol, **lead** 'anti-knock' agents are added and these lead compounds are released into the atmosphere along with carbon monoxide and various nitrogen oxides. The lead is easily absorbed through the lungs and may cause mental retardation in children. The trend towards the use of unleaded petrol has been accelerated by the government's decision to give it a tax advantage, thus making it cheaper than the leaded variety.

When fossil fuels are burned **sulphur dioxide** is discharged into the atmosphere. Coal and coke contain about 1–2% sulphur. Much of this can be removed but the cost is high. Control of this form of pollution depends on high chimneys to carry the fumes away. In some areas, problems have arisen when it was washed out from the waste gases by rain as sulphuric acid. This corrodes buildings. Plants are particularly vulnerable to sulphur dioxide pollution because the gas is absorbed through stomata and into the leaf cells in lethal doses.

Fluoride is a waste product of certain industrial processes involved in the manufacture of pottery, bricks, steel and aluminium. It is absorbed by plants and becomes concentrated in leaves. Cattle and sheep grazing on the plants become affected over a period of time. Bones become soft and joints stiffen. Methods used to get rid of fluoride depend on building higher chimneys. Other methods have proved very expensive.

The greenhouse effect

We saw in 6.2 that fossil fuels form an important reservoir of carbon. Since the industrial revolution the burning of these fuels has released much additional carbon dioxide into the atmosphere. This change to the composition of atmospheric gases has prevented some of the sun's radiation which would normally be reflected back into space from escaping. This has led to heat being trapped next to the earth in the same way that greenhouse glass traps heat. This is the so-called **global warming**. Its effects could be to expand the oceans which, along with some melting of the polar ice caps, could lead to a rise in sea level and the consequent flooding of many of the major cities of the world. Other pollutants, like **chlorofluorocarbons (CFCs)**, which are used in refrigerators, aerosol sprays and some plastic foams, can contribute to the greenhouse effect by depleting the protective ozone layer above the earth which reduces the intensity of the sun's radiation, in part by filtering out ultraviolet light.

Aquatic pollution

Freshwater pollution Man soon realised that rivers are the easiest and cheapest means of transport. Towns grew alongside them, and the simplest way to get rid of unwanted material, including sewage, was to put it into the river. Rivers became the first sewers and, before the population explosion of the industrialised communities, were efficient as such and still remained healthy. However, when the bulk of the waste became too much, bacteria thrived and began to deplete the oxygen from the rivers causing a **biochemical oxygen demand (BOD)**. In still waters the effects have been marked with many lakes in the world 'dying' as a result. Mass epidemics have spread via water systems because of pollution by pathogens. It was not until the 1800s that sewage works were constructed to treat the waste before discharge. Toxic wastes from industrial sources are numerous and varied. The main danger is from organic chemical and fertiliser manufacturers. Heavy metals act as noncompetitive inhibitors of enzymes and build up via food chains once they enter the environment. Pesticides, herbicides and fungicides drain away from fields, enter waterways and are concentrated through food chains so that top carnivores become poisoned. Excessive use of fertilisers can also give rise to a build-up of toxic by-products by leaching and draining. **Thermal pollution** becomes a problem where large quantities of heated water are discharged, e.g. from power stations. The consequent rise in temperature not only kills organisms directly, but also reduces the solubility of oxygen in water, thus killing aerobic organisms. Anaerobes can thrive in these conditions and compete successfully with other forms of life for environmental resources. Detergents are rich in phosphates and these provide nutrients for the vegetation in rivers and lakes. Toxic by-products from algal blooms result in the death of many organisms.

The prevention of these forms of pollution depends on treating waste products at the source so as to render them harmless. Cost seems to be the major problem because the expense of treatment must be met by increasing the cost of the end-product manufactured or by reducing profit margins.

Marine pollution

Oil pollution of the sea is one of the most emotive forms of pollution, probably because its effects are direct and easily visible. Illegal washing of oil tanks at sea and the accidental loss of oil by collision, wrecks and oil rigs provide the biosphere with one of the most unsightly forms of pollution. The oil directly affects vertebrates which come in contact with it, e.g. sea birds, seals and fish, with the physical effect of preventing gaseous exchange. Planktonic invertebrates are also poisoned and so whole food chains become affected.

Methods of treating oil pollution have had limited success. They include absorbent methods involving covering oil with an absorbent material such as straw and collecting the oil-soaked material. Another approach is to spray the oil with a solution of plastic. The plastic hardens to form a coating over the oil. Plastic foam and sawdust have also been used, but the problems with these methods include cost effectiveness and the difficulty of spreading the absorber evenly and then collecting it at sea. Alternatively, heavy materials may be added to the oil causing it to sink. Bacteria break down the oil quite rapidly when it is on the sea bed. Sand, treated chemically to make oil cling to the grains, has been used to sink oil and, because very small amounts of chemicals are needed to treat the sand, it is cheap and safe for marine life. Calcium sulphate and pulverised fuel ash, when treated with silicone, are other effective materials for sinking oil. Detergents have been used to reduce the surface tension of oil but unfortunately may damage wildlife when ingested.

Radioactive pollution

One of the unavoidable consequences of the use of nuclear energy is the production of radioactive waste. Highly reactive, long-lived wastes are usually concentrated, sealed into lead containers and dumped in deep-sea areas. Wastes with a lower level of activity are dispersed by the sea, which dilutes them to acceptably safe levels. The amounts which can be discharged are strictly controlled. One of the most dangerous effects of radiation is the damage it causes to human chromosomes. Mutations often result. These may be passed from one generation to the next before they exert their often damaging effects.

In considering the dangers posed by a radioactive substance a number of factors need to be taken into account. These include the half-life of the substance, its possible concentration in food chains and its role in an organism. For instance, ^{90}strontium is particularly hazardous to humans:

1. It is produced by nuclear explosions and is therefore constantly produced.
2. It has a half-life of 28 years and therefore persists.
3. It is readily absorbed by grasses and concentrated in cows' milk, which is consumed in large quantities by humans, especially babies, who are particularly vulnerable because their cells are rapidly dividing.
4. It becomes concentrated in human bones, which contain the bone marrow which is rapidly dividing to produce blood cells. This division may therefore become disturbed and leukaemia can result.

A radioactive substance released at a relatively harmless level may become accumulated to dangerously high levels along a food chain. The levels of the phosphorus isotope ^{32}P in a North American river were found to be:

1. in the water – 1 (arbitrary unit)
2. in phytoplankton – 1000
3. in aquatic insects – 500
4. in fish – 10
5. in ducks – 7500
6. in duck eggs – 200 000 (shells contain calcium phosphate)

Here again the main danger is that the highest levels are found next to the developing embryo, which comprises rapidly dividing cells.

Terrestrial pollution

There are two distinct types of terrestrial pollution: waste materials that are dumped and pesticides. Waste heaps, derelict mines and buildings present the following problems.

1. They are unsightly.
2. They waste possibly useful land.
3. They may create a danger to life, e.g. children trapped in mines, the Aberfan slag heap disaster of 1966.
4. They may become a health hazard by attracting vermin.
5. They cause urban decline, causing people to move away, which creates social decay.

Pesticides are numerous. It is essential that they are properly used. Often, however, they 'drift' from their targets or are leached into streams, rivers and lakes. A good pesticide should:

1. be specific only to the pest it is directed at;
2. be rapidly broken down and not persist in the natural environment;
3. not be accumulated through food chains.

Sound pollution

Sound is measured in decibels on a logarithmic scale which means that an increase of 10 dB represents a doubling of the intensity, i.e. a noise of 60 dB is twice as loud as one of 50 dB. Noise is annoying and stressful rather than dangerous, although at extremes (140 dB and higher) it can cause deafness even if the duration of the noise is short.

Table 6.2 *Summary of major pollutants*

Medium affected	Pollutant			
	Natural	Man-made		
		Domestic fires	*Internal combustion engine*	*Industrial wastes*
Atmospheric	Sulphur dioxide in volcanic smoke Radioactivity due to cosmic rays	Grit Smoke (carbon particles) Tars Carbon dioxide Heat (thermal)	Carbon monoxide Nitrogen oxides Ozone Lead compounds Various organic compounds Noise	Sulphur dioxide Hydrogen sulphide Ammonia Fluorine Sulphuric acid Heavy metals, e.g. zinc, copper Radioactive fallout from bomb tests and nuclear power stations Noise
Aquatic	Eutrophication by leaching of minerals Putrefaction in small ponds due to leaves Silting in lakes and rivers Leaching of toxic ions, e.g. copper, iron	Inert, e.g. washings from mines and quarries Putrescible, e.g. domestic sewage Toxic, e.g. copper, zinc, lead, mercury from industrial processes Tainting, e.g. dyes Eutrophication, e.g. nitrates from fertilisers, phosphates from detergents Thermal, e.g. cooling water from power stations Radioactive, e.g. nuclear power station wastes Oil, e.g. washings from oil tankers, accidents at sea		
Terrestrial	Local accumulation of toxic minerals, e.g. ferrous salts Radioactive minerals	Dumped wastes, e.g. slag heaps, spoil heaps, domestic rubbish Derelict buildings and land, e.g. disused houses, factories and mines, old railway lines and canals Pesticides (overuse, drift and persistence), e.g. inorganic, di-nitro, organophosphorus, mercuric and other compounds, chlorinated hydrocarbons, hormone weed killers		

CONSERVATION

You will probably have heard the word conservation used when some species, habitat or valuable resource is under threat. What then is conservation in a biological context? It is often wrongly thought to be maintaining a species, habitat or resource exactly as it is. This is preservation; conservation is more. It accepts that things, especially natural things, must adapt to meet changing conditions. Conservation is an attempt to maintain a balance, or status quo, so that one factor does not upset the delicate equilibrium that exists between the many living and nonliving components of the ecosystem. It often entails preservation, e.g. of an endangered species, but it is not exclusively this. Conservation involves careful and skilful management of a complex web of biotic and abiotic resources. It entails many strategies designed to combat the many pressures, often man-made, which endanger natural resources. These include:

1. Education – to permit people to understand the reasons for conserving nature.
2. Designation of national parks and nature reserves – to protect vulnerable species and habitats, e.g. Giant pandas and the bamboo forests of China.
3. Planned use of land – to restrict human activities which might threaten nature, e.g. designating land as an Environmentally Sensitive Area (ESA) or a Site of Special Scientific Interest (SSSI).
4. Legislation to protect wildlife – to prevent endangered species becoming extinct, e.g. Koala in Australia.
5. Controlling introduced species – to prevent them outcompeting indigenous species in the way that the grey squirrel has ousted the red squirrel in many areas of Britain.
6. Breeding in captivity – endangered species can be bred under controlled conditions before being returned to the wild, e.g. Przewalski's horse.
7. Ecological study of threatened habitats – to improve our understanding and aid in their proper management.
8. Control of pollution – measures such as reducing smoke emissions, control of pesticide and fertiliser use and prevention of oil spills help to avoid the destruction of species and their habitats.
9. Recycling – the more material that is recycled, the less needs to be obtained from natural sources, e.g. mining for metal ores often destroys sensitive habitats in mountainous regions.

Chapter roundup

In this chapter we have had a glimpse of the complex and intricate interrelationships of organisms. We have attempted to show how no organism can live in isolation, but that each is dependent on others for some aspect of its survival. Plants capture the sun's energy in photosynthesis and use it to manufacture the complex chemicals which form the food of herbivores. Herbivores are eaten by carnivores which in turn are prey to larger carnivores and parasites. Plants depend on saprobionts to decompose dead plants and animals, thus releasing essential chemicals such as carbon, nitrogen and phosphorus so that they can continue to photosynthesise and maintain the flow of energy into the ecosystem. Finally we reviewed the often adverse impact of man and his activities on the delicate balance of nature.

In the next chapter we will look at some of the uses to which organisms put the energy they acquire – especially in transporting material within themselves. In addition, we will see that some of this transport occurs without the need for an external energy supply.

Illustrative questions and worked answers

1 Discuss the effects on the environment of overfishing, the disposal of nuclear waste and the use of pesticides and fertilisers.

Time allowed 40 minutes

Tutorial note

In answering this type of wide-ranging environmental question you must take the greatest care to produce an objective and scientific answer. Examiners frequently report that candidates become carried away with emotional pleas for saving the planet when what they are actually looking for are clear and logical arguments supported by appropriate facts and examples. Avoid generalised terms like 'pollutants', 'substances' and 'chemicals' and try instead to give the precise names of materials. You should, in the absence of any indication of the marks carried for each part, distribute your time equally between the four areas listed in the question, i.e. spend around 10 minutes on each. The suggested answer below is somewhat generalised because actual examples will depend on those that are covered by your own syllabus. Those given represent the better-known examples.

Suggested answer

Overfishing reduces fish populations, e.g. North Sea herring stocks are almost exhausted, and affects the distribution of fish. It disrupts food webs because plankton numbers increase in the absence of fish to eat them, while the predators of the fish, e.g. seals, may decrease due to the reduction in their staple diet. Ultimately the population being fished may become extinct altogether.

The disposal of nuclear waste is difficult because most wastes have long half-lives and so remain potentially dangerous for many thousands of years, e.g. ^{239}Pu has a half-life of 240 000 years. No disposal method is absolutely safe, but burial in stable rocks minimises the risks. Dumping at sea more quickly leads to seepage. Direct contamination with a radioactive source may cause ionisation of chemicals and hence metabolic disruption of cells leading to death. The radioactivity may accumulate along food chains especially where the element is metabolically important, e.g. ^{32}P. It may also concentrate itself in particular organs causing localised damage, e.g. ^{90}Sr in bones.

Pesticides are used to remove unwanted pests (e.g. aphids, weeds, mildews) or vectors of human diseases (e.g. mosquitoes, tsetse flies). They are often not specific in their action and so kill beneficial species. Pesticides (e.g. DDT) also disrupt food webs as they can concentrate along food chains, killing organisms at the top. They may affect animal products, e.g. shells of birds eggs become thin and break easily. Many pesticides are slow to break down and so their effects can last for years. They may 'drift' to areas that were not meant to be affected. Overuse can lead to resistance in the pest.

Fertilisers are used to increase crop yield. They often replace organic fertilisers and so soil texture suffers. They may 'run off' into water courses, causing eutrophication and algal blooms. These algae die and decay causing oxygen deficiency and death of aerobic life in the water. Fertilisers may harm other plants and adversely affect the soil microflora and fauna.

2 (a) Describe what is meant by the following terms:

(i) community
(ii) ecosystem
(iii) food chain (9)

(b) Evaluate the use of studying food webs, rather than food chains, in ecology. (3)

(c) Consider the trophic levels of a pyramid of numbers and illustrate how energy is lost in passing through the levels. (5)

(d) Briefly describe three ways in which nitrogen is incorporated into a food chain. (3)

Time allowed 30 minutes

Tutorial note

You should guard against spending too much time on any one aspect in this type of question. There are so many parts that there is a danger of overelaborating on the early ones, leaving yourself rushed on the later ones. Be concise, but do not sacrifice detail and precision, especially in the descriptions in part (a).

In part (b) 'evaluate' is the key word. Both food webs and food chains have their advantages and disadvantages. 'Evaluate' means weigh up the pros and cons for each criterion. Do not assume webs are always better than chains or vice versa. Augment your answer with specific examples.

Part (c) must incorporate an explanation of heat loss during respiration, which accounts for most energy loss along with some lost in faeces and excretory products.

Suggested answer

(a) (i) A community is a localised group of several populations of different species interacting with one another and their abiotic environment.

(ii) An ecosystem is a localised group of communities along with their physical environment.

(iii) A food chain is a series of stages through which energy passes, always beginning with chemical energy incorporated into plant tissues.

(b) Food chains are easier to construct and they reduce the complexity of representing theoretical feeding relationships. They are not, however, realistic as most organisms have many predators and are prey to many others. Food webs are more representative and describe true feeding relationships.

(c) Little of the energy incorporated into plants passes into herbivores – as little as 0.1%. A similar loss occurs at subsequent stages in the chain. Most energy loss occurs as heat during respiration, with some lost in faeces and excretory products.

(d) (i) With the aid of *Nitrosomonas*, a genus of bacterium which lives in soils, ammonium compounds can be oxidised to nitrites which in turn are oxidised to nitrates by another bacterium called *Nitrobacter*. The nitrate ions are then absorbed by plants and their nitrogen is incorporated into plant protein which enters the food chain when the plants are eaten. The starting point for this sequence is ammonium compounds in the soil; these have been formed as a result of putrefaction of nitrogenous excretory products, such as urea, which have been produced by animals.

(ii) Electrical and photochemical fixation of atmospheric nitrogen results in the formation of nitrogen oxides, which then form nitrates and can be absorbed by plants.

(iii) Nitrogen-fixing bacteria, *Rhizobium*, in the root nodules of leguminous plants incorporate atmospheric nitrogen into their protoplasm. The host plants then absorb some nitrogenous molecules from the bacteria and incorporate them into their protein.

3 The number of lichen species growing along a 20 km transect from the centre of Belfast was recorded at 1 km intervals. The results are presented graphically below.

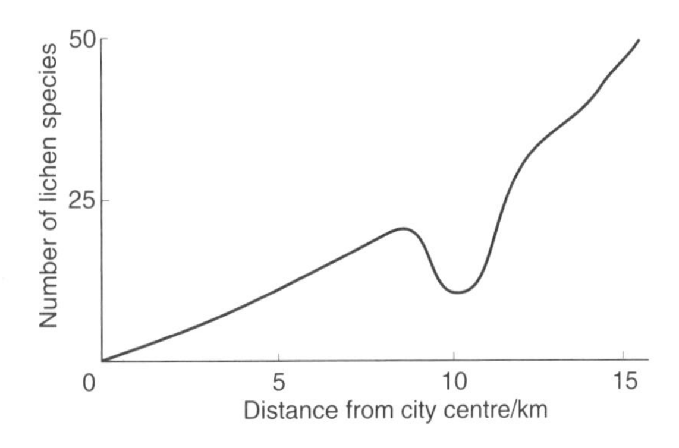

(a) What relationship is illustrated by the graph? (2)

(b) Explain, as far as possible, the relationship between the distance from the city centre and the number of lichen species. (3)

(c) Give one possible reason for the fall in the number of lichen species at a distance of 10 km from the city centre. (2)

(d) What is meant by the term indicator species? (3)

Time allowed 20 minutes

Tutorial note

You need to appreciate the general tendency for the number of lichen species to increase the further one gets from the city centre. This should be correlated with atmospheric pollution in general and sulphur dioxide concentrations in particular. The sources of this pollution (i.e. burning of fossil fuels by households and by industry) should be included. The reduced number of lichens at 10 km from the city centre must be due to a localised increase in sulphur dioxide levels – you should be able to deduce a likely source.

Do not be put off because you have never studied lichen distribution around Belfast. In this type of question you are not expected to know either the geography of the Belfast area, the precise nature of lichens or the air pollution. You are simply expected to make reasonable scientific suggestions based upon sound biological principles. Any such statements will attract credit, whether true or not in this specific case.

Suggested answer

(a) The number of lichen species increases more or less linearly with increasing distance from the city centre except at a distance of 10 km from the city centre where there is a fall in their number.

(b) Lichens are sensitive to air pollution, especially sulphur dioxide, the algal component of the lichen being killed at even relatively low concentrations of this gas. Different species show slightly different levels of tolerance. Industrial cities such as Belfast have high levels of sulphur dioxide as a consequence of domestic and industrial smoke produced from burning fossil fuels, especially oil. In older cities such as Belfast the industries and housing are situated in and around the city centre. Sulphur dioxide levels are therefore higher near the centre of the city and diminish slowly as one moves out from it. The pollution from sulphur dioxide is sufficiently large in the centre of the city to prevent any lichen species from growing. As one moves outwards the species more tolerant to sulphur dioxide appear first. As the sulphur dioxide levels fall even more, the less tolerant ones also survive. Finally, at some 15 km from the city centre around 40 lichen species exist.

(c) The levels of sulphur dioxide determine the number of lichen species and it is therefore most probable that the fall in lichen species 10 km from the city centre is due to an increase in sulphur dioxide levels at this point. Such an increase is likely to be due to industrial or domestic smoke from a large industry or small town at this distance from the city centre.

(d) This is a species whose distribution is largely determined by some particular factor. The presence or absence of an indicator species therefore gives some measure of the level of that factor. Using a number of indicator species it may be possible to determine the precise level of the factor at one point. One example involves determining the exposure scale, i.e. amount of wave action, on a sea shore. In the above case the level of sulphur dioxide pollution could be fairly accurately measured by looking at the number and types of lichen species present. For example, if species A is only found where the sulphur dioxide level is less than 2 ppm and species B where the level is less than 1 ppm, then an area where species A is present but species B is not probably has a sulphur dioxide level between 1 and 2 ppm. While other factors do affect the distribution of lichens, the use of indicator species nevertheless gives a remarkably accurate indication of the sulphur dioxide level in the atmosphere.

Question bank

1 Read the passage and answer the questions which are based on it.

'In 1949 synthetic organic pesticides were widely introduced into the Canete Valley of Peru in an attempt to control cotton pests in this important agricultural area. Many ecologists and evolutionary biologists argued against this approach, but their voices went unheeded, and massive applications of these pesticides began. Initially the program was very successful. Application of chlorinated hydrocarbons increased yields from 440 lb per acre in 1950 to 648 lb per acre in 1954. The entire valley was blanketed with these insecticides. They were applied from airplanes, and in many places trees were chopped down to facilitate the process. There were several fairly early repercussions from this campaign. When the trees were removed, many birds that nested in them disappeared, as did a number of other animals. Over the years the frequency of insecticide treatments had to be increased, and each year they had to be started earlier.

Soon the real troubles began. In 1952 the chlorinated hydrocarbon BHC failed to stop aphid infestation. In 1954 the chlorinated hydrocarbon toxaphene failed against the tobacco leafworm. In 1955 there was a major infestation of boll weevils. New previously unencountered pests began to appear in the valley, and old pests began to show very high levels of DDT resistance.

In 1955 synthetic organophosphates were used in place of the chlorinated hydrocarbons, and the frequency of treatments was increased from two or three times monthly to once every three days. In that one year, despite this intensive effort, yields dropped over 300 lb per acre. In five years of intensive warfare against insect pests, all of man's initial successes had been wiped out, and he was worse off than when he began.'

(Reproduced from 'The New Higher Paper II Interpretation Question' Arnold, B., Mills, P. R., Aberdeen College of Education Newsletter 34 November 1979 by permission of the editor.)

(a) What is a pesticide? (1)

(b) What evidence is there that the use of pesticides was initially successful? (1)

(c) How might the removal of trees lead indirectly to an increase in the number of pests? (2)

(d) Why did the whole valley have to be blanketed with insecticide and not just the cotton fields? (2)

(e) Why weren't all the aphids killed in 1949? (1)

(f) In what way did the changes occurring as a result of the addition of the pesticides to these populations of insects illustrate the principle of natural selection? (3)

(g) Why was it that new, previously unencountered pest species were able to invade the Canete Valley after 1955? (2)

(h) Describe the levels of pesticide you would expect to find in a top carnivore in this ecosystem. Give a reason for your answer. (2)

(i) In what ways were the farmers worse off after five years of using pesticides? (3)

Time allowed 20 minutes

Pitfalls

You should initially read and comprehend the passage as the answers to some questions, e.g. (b), are found within it. Most, however, go beyond the content of the text itself, although some useful hints and points lie within the passage. This style of question is becoming increasingly popular on examination papers as the need to understand scientific articles is a major requirement of today's scientists. Read the whole article through to obtain the general ideas and principles involved before reading it again in greater detail, pausing and re-reading sections to be sure you have fully absorbed the facts. Do this before attempting any questions. A common mistake is to start answering the questions before the passage is fully understood, or even before it has been read.

As the marks for each part are few, you should keep answers short and precise while including relevant detail, sometimes from the passage.

Points

An outline of the main points to be made in each part is given below.

(a) A chemical used to kill pests.

(b) Increased yields.

(c) Fewer predators. Thus, less predation, or numbers of pests not controlled, or equivalent.

(d) The pests may occur outside the cotton fields. The pests may not have their full life cycle in the cotton fields.

(e) Some were resistant.

(f) An idea of how this particular example demonstrates variation, change in the environment, survival of those best suited to the new conditions, spread of this character through the whole population (any three points).

(g) Ecological niches available, lack of competition, immigration (any two points).

(h) Levels would be high. Idea of accumulation as pesticide passes through the food chain.

(i) Overall decrease in yields, increase in number of pests, increase in different kinds of pests, pesticides are now ineffective (any three points).

2 The diagram below shows pyramids of biomass and numbers in the same ecosystem. (Neither is drawn to scale).

(a) What are the advantages and disadvantages of the two methods of measurement?

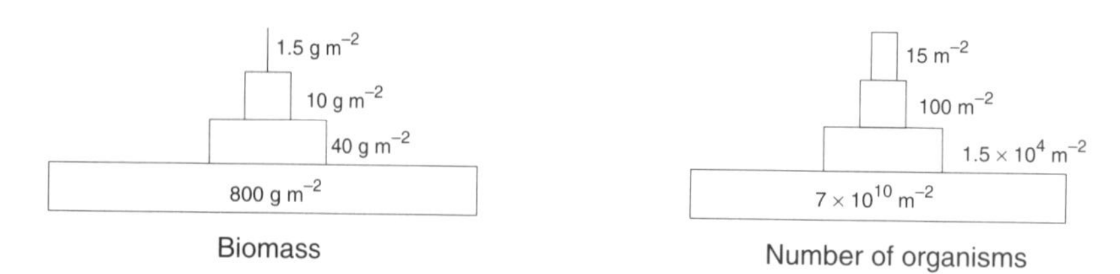

(b) What other kind of pyramid could be used to give further information about the four trophic levels?

(c) Why are there seldom more than four trophic levels in each pyramid?

Time allowed 6 minutes

Pitfalls

You will need to be brief in your responses with only six minutes in which to answer. There could be a tendency to overelaborate your answers to parts (a) and (c). In (a) you should remember to present both sides of the story in each case and in (c) you must stress the large loss of energy in moving from one trophic level to the next.

Points

(a) Number pyramids are easier to measure but take no account of size, no account of juveniles or immature forms whose diet may differ from the adults and they may be impossible to draw to scale (e.g. millions of aphids on a single tree).

Biomass pyramids give a more reliable indication of the amount of energy at each trophic level, but this is difficult to measure accurately and may vary from one time of year to the next.

(b) An energy pyramid, giving the number of kilojoules at each trophic level.

(c) Make reference to the Law of Conservation of Energy (First Law of Thermodynamics). There is much energy loss at each trophic level with only around 10% continuing on to the next stage. After four levels there is rarely enough remaining to sustain a viable population at a fifth level.

3 The table shows the amount of DDT measured in parts per million (ppm) found in a variety of organisms associated with a large freshwater lake.

Where the DDT level was measured	DDT/ppm
Water	0.0003
Phytoplankton	0.006
Zooplankton	0.04
Herbivorous fish	0.39
Carnivorous fish	1.8
Fish-eating birds	14.3

(a) Calculate the concentration factor from water to herbivorous fish. (2)

(b) What principle is illustrated by the data? (1)

(c) Briefly explain the reasons for the change in DDT levels in the different organisms. (2)

Time allowed 8 minutes

Pitfalls

In making the calculation in part (a) you should take care to use the correct figures – because herbivorous fish are not at the top of the food chain shown, it is possible that fish-eating birds, which are, might be chosen in error. Watch the decimal points!

You will need to provide some detail in your explanation in part (c). It will not be sufficient to say 'because one animal eats the other'. Two points are important:

❶ Organisms consume large quantities of material at the previous trophic level.

❷ DDT accumulates in tissues rather than being excreted.

Points

(a) The level in the water = 0.0003 ppm
The level in herbivorous fish = 0.39 ppm
Concentration factor = 0.39/0.0003 = 1300 times

(b) The data illustrate the concentration of DDT along a food chain.

(c) The DDT level in the water is relatively low. Some diffuses into the phytoplankton where it remains and accumulates, especially in any fatty material. Zooplankton feed on phytoplankton. During their lifetime they consume many times more than their own body weight of phytoplankton, i.e. they consume an increasing amount of DDT. This DDT accumulates within their tissues and is not excreted. As one looks down the table each animal consumes a larger weight of the animal above it and hence an increasingly larger amount of DDT. The levels of DDT therefore increase rapidly as one moves down the table (i.e. as one moves along the food chain).

CHAPTER 7

TRANSPORT AND EXCHANGE

Units in this chapter

Chapter objectives

We have seen in Chapters 5 and 6 how organisms obtain the energy so essential to their existence. We must now turn our attention to the ways in which the raw materials needed to release this energy, and the other essential substances, are exchanged with the environment and transported within the organism. In particular, we will investigate how animals obtain the oxygen needed to release energy in mitochondria, and how it and other materials are transported by the blood around the body using the pumping of the heart. We will then look at the rather different mechanisms plants use to move materials up and down themselves. Finally consideration will be given to how a favourable water balance is maintained in organisms and the methods by which unwanted wastes are removed.

7.1 GASEOUS EXCHANGE

We saw in 5.5 that much of the energy utilised by organisms for osmotic, mechanical, electrical and chemical work is derived from ATP formed in the mitochondria during oxidative phosphorylation. This process requires oxygen which the organism must obtain from its environment of air or water. Regardless of source, the oxygen enters the organism by diffusion. In order to obtain it in adequate quantities and remove the carbon dioxide produced, a relatively large, moist surface must be exposed to a source of oxygen. The distance between this oxygen source and the cells requiring or transporting it must be small and the diffusion gradient must be maintained if the process is to satisfy the needs of the organism. In small organisms such as protozoa and unicellular algae the surface area is sufficiently large compared with volume so that diffusion over the whole body surface satisfies their needs.

As organisms became multicellular and grew in size they were only able to utilise the whole body surface for obtaining oxygen if their energy demands were small and their shape became flattened (e.g. platyhelminthes) or the centre contained no respiring material (e.g. cnidarians). Any further increase in size or metabolic rate required some specialised gaseous exchange system to compensate for a lower surface area/volume ratio and greater oxygen demand. Such systems included tubular ingrowths (e.g. in insects) to carry air directly to the cells or organs of gaseous exchange situated in one part of the organism and supplied with a means of transporting gases from it, to and from the other cells. In water this second system takes the form of gills and on land lungs are used. Whichever system is used the respiratory surface must have a constant supply of the oxygen-carrying medium flowing over it, to maintain the diffusion gradient.

Air and water as respiratory media

Since the volume of oxygen in a given volume of air is much greater than in the same volume of water, an aquatic animal must pass a greater volume of the medium over its respiratory surface in order to obtain the oxygen necessary for cellular respiration. Water is also far denser and has a greater viscosity than air at the same temperature and so it is more difficult for the organism to extract the necessary oxygen from it. It would, therefore, seem easier for animals to obtain oxygen from air then from water, but this is not the case since terrestrial animals must avoid desiccation and therefore have relatively impermeable skins. Terrestrial animals and the more active aquatic ones do not rely on the general body surface for respiratory exchange but have developed specialised respiratory surfaces. The simplest of these surfaces are gills, which may be external (e.g. annelida, amphibian tadpoles) or internal (e.g. fish). Insects have a tracheal system. Terrestrial vertebrates and some invertebrates (e.g. pulmonate snails) have developed lungs. For maximum efficiency internal respiratory surfaces such as internal gills and lungs need to be ventilated.

Table 7.1 *Comparison of water and air as respiratory media*

Property	Water	Air
Oxygen content	0.04–9.0 cm^3 $litre^{-1}$	105–130 cm^3 $litre^{-1}$
Oxygen diffusion rate	Low	High
Density	Relative density of water about 1000 × air	
Viscosity	Water about 100 × air	

General properties of respiratory surfaces

1. **Large surface area/volume ratio**: the body surface is adequate in very small organisms; infoldings of a restricted part of the body provide a large respiratory surface, e.g. lungs and gills, in larger organisms.
2. **Moist**: surfaces are moist so that diffusion may occur in solution.
3. **Thin**: diffusion is only efficient over short distances because the rate of diffusion is inversely proportional to the distance between the concentrations on the two sides of the respiratory surface.
4. **Transport**: this must occur in order to maintain the diffusion gradient. In large metazoans an efficient vascular system is required.

MECHANISMS OF GASEOUS EXCHANGE

Small organisms

With their large surface area/volume ratio single-celled organisms can successfully obtain oxygen and remove carbon dioxide by diffusion across their whole surface. Even certain multicellular organisms use the same method successfully, because their energy requirements are relatively small and no cell is very far from the surface. To achieve the latter requirement some cnidarians are hollow, while platyhelminths (flatworms) have a flattened body. The body surface must remain moist to permit diffusion to occur and so these organisms are restricted to aquatic habitats or very moist terrestrial ones.

Terrestrial flowering plants

By virtue of their autotrophic mode of nutrition the supply of gaseous oxygen to the tissues of flowering plants is facilitated for three reasons:

1. When phosynthesising oxygen is produced as a by-product and may be used directly for respiration.
2. Photosynthesis requires specialised structures – the stomata – in the leaves to allow diffusion of carbon dioxide and these can be utilised to obtain oxygen from the atmosphere when no photosynthesis is taking place.
3. The roots have a thin surface and large surface area to facilitate the collection of water by osmosis for photosynthesis. This feature allows oxygen to diffuse into the root easily from the soil spaces.

The stems of plants obtain their oxygen either through **stomata** (herbaceous flowering plants) or **lenticels** (woody flowering plants). The central tissue of woody stems and roots is composed of dead xylem cells. Therefore no living cell of a plant is ever far from a source of oxygen and no specialised respiratory surface is required as diffusion over the whole surface will suffice.

Insects – tracheal system

The respiratory system of an insect comprises a series of tubes, **tracheae**, with paired openings to the exterior called **spiracles**. The tracheae branch to form terminal **tracheoles**, often less than 1 μm in diameter, which reach more or less every part of the insect's body. In insects blood does not carry oxygen and carbon dioxide. Except at their very ends the tracheae and tracheoles are filled with air; oxygen and carbon dioxide are transported primarily by diffusion. Ventilation of the larger tracheae occurs through movements of the muscles or exoskeleton. The spiracles are surrounded by hairs to prevent the entry of dust and parasites and they may be opened and closed by valves. The larger tracheae are lined by cuticle which must be shed at each moult and this, together with the rate of diffusion in narrow tubules, is the main factor limiting the size of insects. See Fig. 7.1.

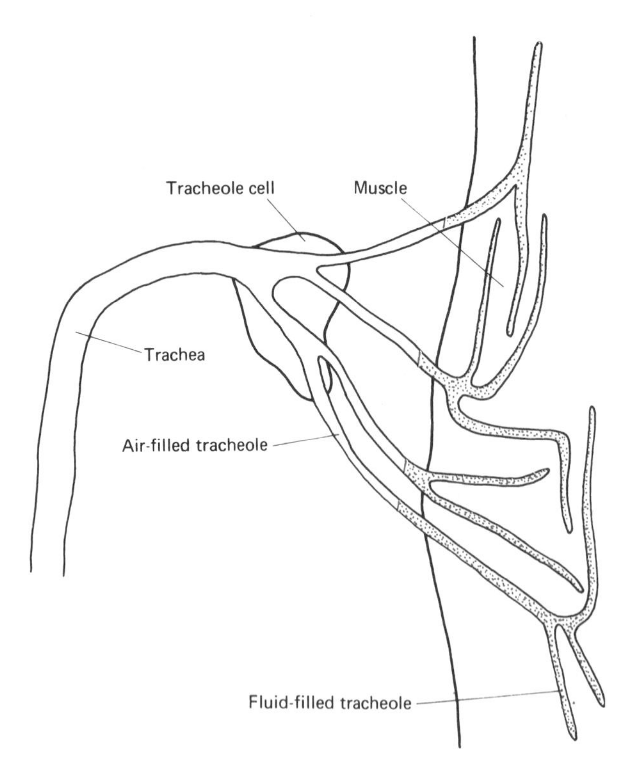

Fig. 7.1 Tracheoles in an insect

During exercise lactate accumulates in the muscle, raising its osmotic pressure. Water passes from the tracheole into the muscle by osmosis, thus drawing air further down the tracheole.

Fish – gills

Bony fish possess four pairs of gill arches which support **gill lamellae**. These lamellae form a double row arranged in a V-shape (see Fig. 7.2). The gills are covered by a bony flap called an **operculum**. The movement of water and blood across the gills is in the opposite direction. This so-called **counter-current principle** maintains a constant diffusion gradient between the blood and the water. It ensures that blood already containing much oxygen meets water rich in oxygen, while blood with little or no oxygen meets water with only slighly more oxygen present. The differences between the counter-current system and the parallel-flow system (where blood and water move in the same direction) are illustrated in Fig. 7.3.

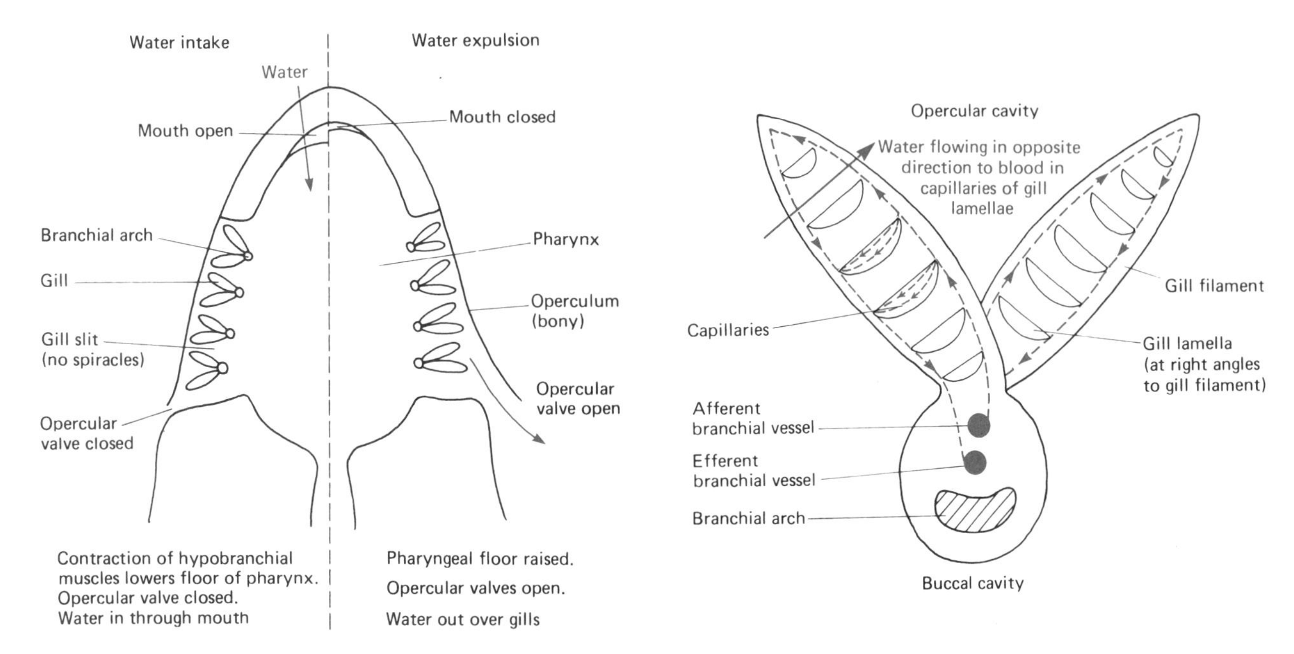

Fig. 7.2(a) Water flow across the gills

Fig. 7.2(b) Water flow across gill filaments

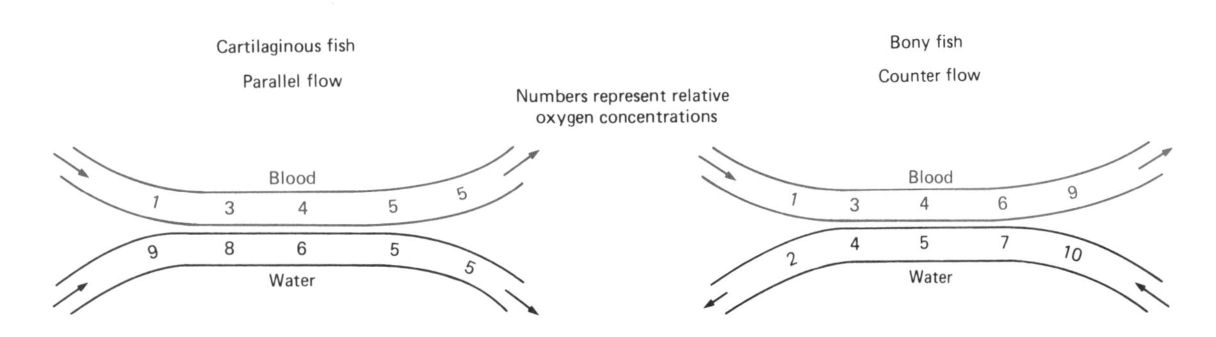

Water and blood flow side by side in the same direction. Oxygen is exchanged by diffusion. The oxygen concentration in the blood leaving the gills can never exceed that of the water leaving the gills. Equilibrium is soon reached.

Blood flows in the opposite direction to water. At each point of contact, the water has a higher oxygen concentration than the blood. Therefore diffusion can occur over the whole region and much higher blood oxygen concentrations can be reached than with parallel flow.

Fig. 7.3 Comparison of parallel- and counter-flow systems

Mammals (e.g. human) – lungs

Ventilation, which is clearly essential for a respiratory surface so deep inside the body, is brought about by the diaphragm and the ribs. The diaphragm is a fibrous sheet of tissue whose muscular edges are attached to the thoracic wall. Contraction of these muscles flattens the diaphragm and increases the volume of the thoracic cavity. Between the ribs are the **intercostal muscles** whose contractions cause the ribs to move. The external intercostal muscles contract to move the ribs upwards and outwards, at the same time raising the sternum. This also causes the volume of the thoracic cavity to increase. As the volume increases the pressure decreases to below that of the atmosphere and air rushes in. This is **inspiration** (breathing in). **Expiration** is brought about when the diaphragm muscle relaxes and the internal intercostal muscles contract decreasing the volume of the thoracic cavity and forcing air out.

Gaseous exchange at the alveoli The wall of each alveolus comprises an epithelial layer which is less than 0.5 μm thick. As each human lung contains 350 million alveoli, you can see that the total surface area of the lung is enormous – around 90 m^2. Around each alveolus is a network of blood capillaries (Fig. 7.4(c)). These capillaries are extremely narrow – so narrow that red blood cells can only pass through singly and even then are squeezed. This has two advantages:

1. It slows down the blood cells allowing more time for them to absorb oxygen.
2. As ᵗhe red blood cells are flattened against the capillary endothelium a large surface area is available for absorbing oxygen.

Oxygen in inspired air dissolves in moisture inside the alveolus, and then diffuses across the epithelium of the alveolus and the capillary endothelium into the red blood cells. Here the oxygen combines with **haemoglobin** to form **oxyhaemoglobin**. Carbon dioxide diffuses in the opposite direction.

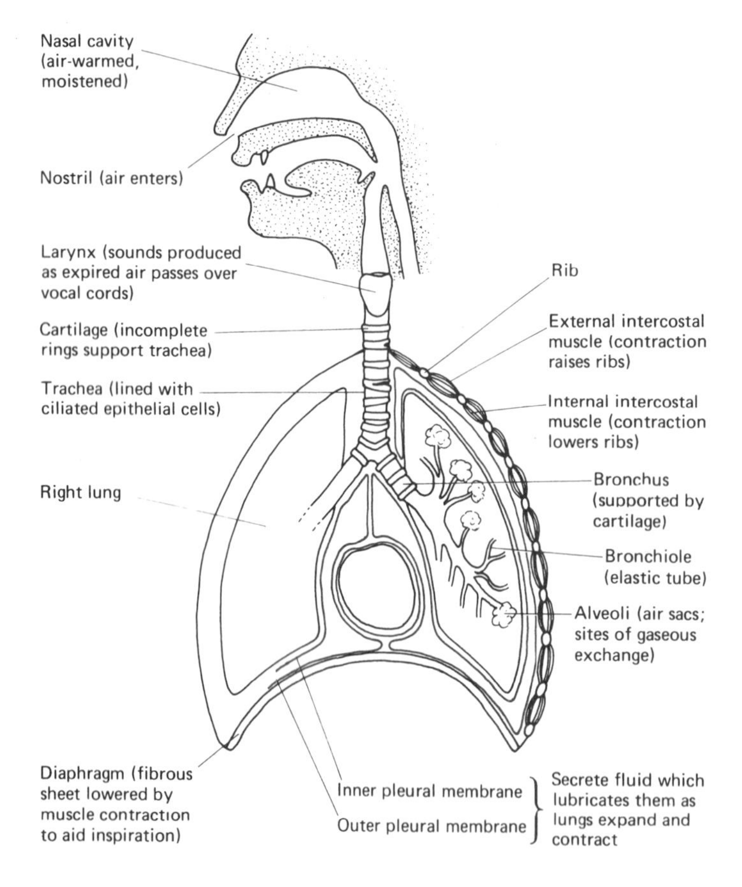

Fig. 7.4(a) Human respiratory system

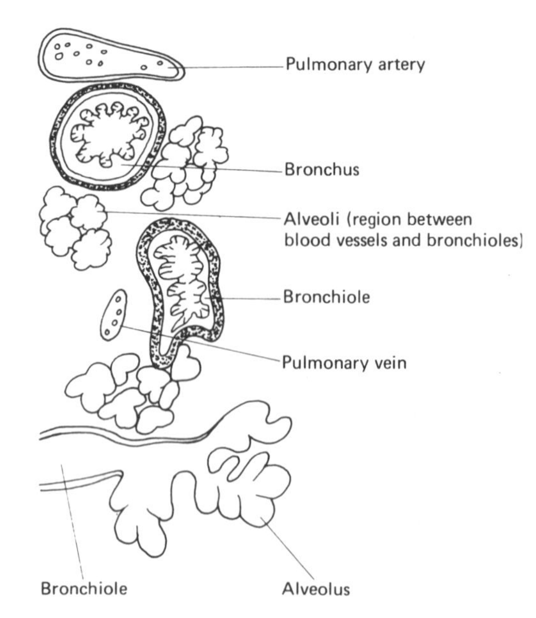

Fig. 7.4(b) LS part of mammalian lung

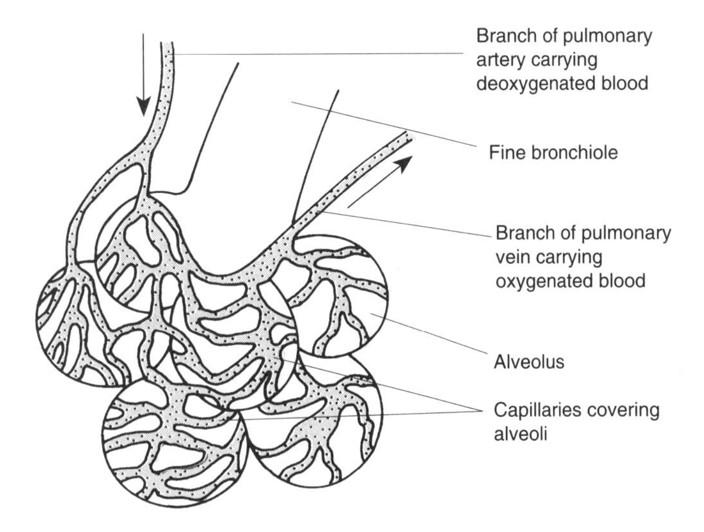

Fig. 7.4(c) External appearance of a group of alveoli

Measurement of lung capacity This can be made using an instrument called a **spirometer**. During normal breathing you will use only a small proportion of the lung's capacity – around 0.5 dm^3. This is known as the **tidal volume**. If, however, you take in as much air as possible you will probably inhale a further 1.5 dm^3. This is called the **inspiratory reserve volume**. In the same way, if you continue to expel air until no more can be removed another 1.5 dm^3 can be exhaled – the **expiratory reserve volume**. The volume of air you can exchange between the deepest breath in and the deepest breath out is therefore around 3.5 dm^3 and is known as the **vital capacity**. At no time are the lungs completely empty – there is always around 1.5 dm^3 present and this is referred to as the **residual volume**. The relationships between these various capacities are shown in Fig. 7.5.

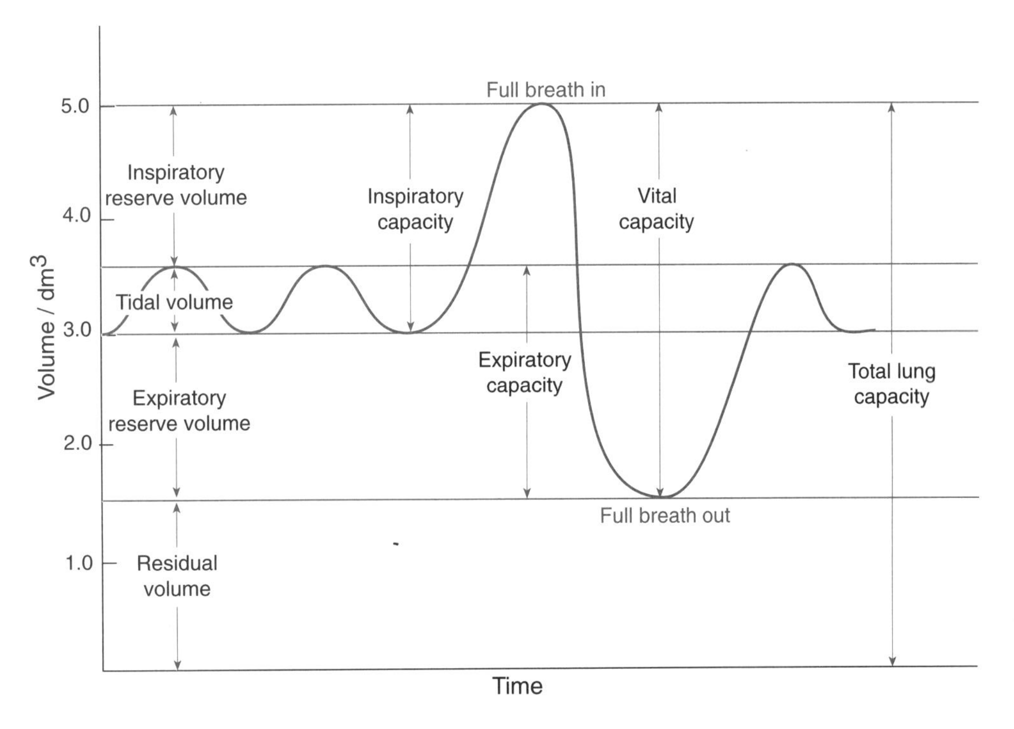

Fig. 7.5 The relationships of various lung capacities

Control of respiration The respiratory centre, which is in the medulla oblongata at the base of the brain, is particularly sensitive to changes in the concentration of carbon dioxide in the blood; a slight increase in the level of carbon dioxide causes deeper, faster breathing until the carbon dioxide concentration returns to normal. Changes in breathing are brought about by signals from the respiratory centre to the spinal nerves controlling the intercostal muscles and the muscles of the diaphragm. The respiratory centre is also sensitive to signals from other parts of the body, e.g. chemoreceptor cells in the aorta and carotid artery which detect when the oxygen concentration in the blood decreases. It is possible to bring breathing under voluntary control within certain limits but it is normally under involuntary control.

7.2 TRANSPORT IN ANIMALS

As animals grew in size and complexity, tissues and organs with specific functions developed, each organ dependent on the others for some essential process or chemical. The need arose for a system to transport materials, especially food, oxygen, carbon dioxide and wastes, between the various organs. This system must service every cell either by flowing freely over them (open system) or having capillaries within diffusing distance of them (closed system). The fluid must be circulated by cilia, body movements or a specialised pump – the heart. The heart takes many shapes and forms depending on the organism and its environment. It may be tubular (insects), two chambered (fish), three chambered (amphibians) or four chambered (birds and mammals). Both the beating of the heart and the distribution of the blood must be carefully controlled in order to meet the varying demands placed upon it at different times. The blood itself must be capable of carrying large volumes of many different substances, particularly oxygen. Pigments for carrying oxygen have developed and are either suspended in the plasma or, if there is a possibility of removal during excretion, in special cells. In addition the blood contains the body's defence and immune system because it is ideally situated to fight infection in all parts of the body. Finally, the liquid nature of the blood, necessary for transportation, creates problems if the body is damaged and leakage occurs. The blood has the capacity to clot in such circumstances.

MAMMALIAN BLOOD

Mammalian blood comprises:

1. red blood cells (erythrocytes)
2. white blood cells (leucocytes)
3. platelets
4. plasma

The basic structure of blood cells was considered in 1.3 (Fig. 1.20) and so in this section we will concentrate on the functions of blood. The main blood function of transport is summarised in Table 7.2.

Table 7.2 *Transport functions of blood*

Materials transported	Examples	Transported from	Transported to	Transported in
Respiratory gases	Oxygen	Lungs	Respiring tissues	Haemoglobin in red blood cells
	Carbon dioxide	Respiring tissues	Lungs	Haemoglobin in red blood cells, hydrogencarbonate ions in plasma
Organic digestive products	Glucose	Intestines	Respiring tissues/liver	Plasma
	Amino acids	Intestines	Liver/body tissues	Plasma
	Vitamins	Intestines	Liver/body tissues	Plasma
Mineral salts	Calcium	Intestines	Bones/teeth	Plasma
	Iodine	Intestines	Thyroid gland	Plasma
	Iron	Intestines/liver	Bone marrow	Plasma

Materials transported	Examples	Transported from	Transported to	Transported in
Excretory products	Urea	Liver	Kidney	Plasma
Hormones	Insulin Antidiuretic hormone	Pancreas Pituitary gland	Liver Kidney	Plasma Plasma
Heat	Metabolic heat	Liver and muscle	All parts of the body	All parts of the blood

Oxygen transport

As the solubility of oxygen in water is low – just 0.58 cm^3 of oxygen is dissolved in 100 cm^3 of water at 25 °C – active organisms have developed special chemicals to carry oxygen. These chemicals, which have a high affinity for oxygen, are called **respiratory pigments**. In humans the respiratory pigment is **haemoglobin**, a protein with a relative molecular mass of 68 000. If this haemoglobin molecule were in the plasma it would be possible for it to be excreted during ultrafiltration in the kidneys (see 7.4). To avoid this unwanted occurrence, haemoglobin molecules are contained within red blood cells. These cells lack a nucleus when mature, thus leaving a bigger proportion of the cell available for haemoglobin. The biconcave shape of red blood cells increases the surface area available for gas exchange. Lacking a nucleus shortens their life span, and red blood cells therefore only live for 120 days before being broken down by the liver and spleen at the rate of between two and ten million every second.

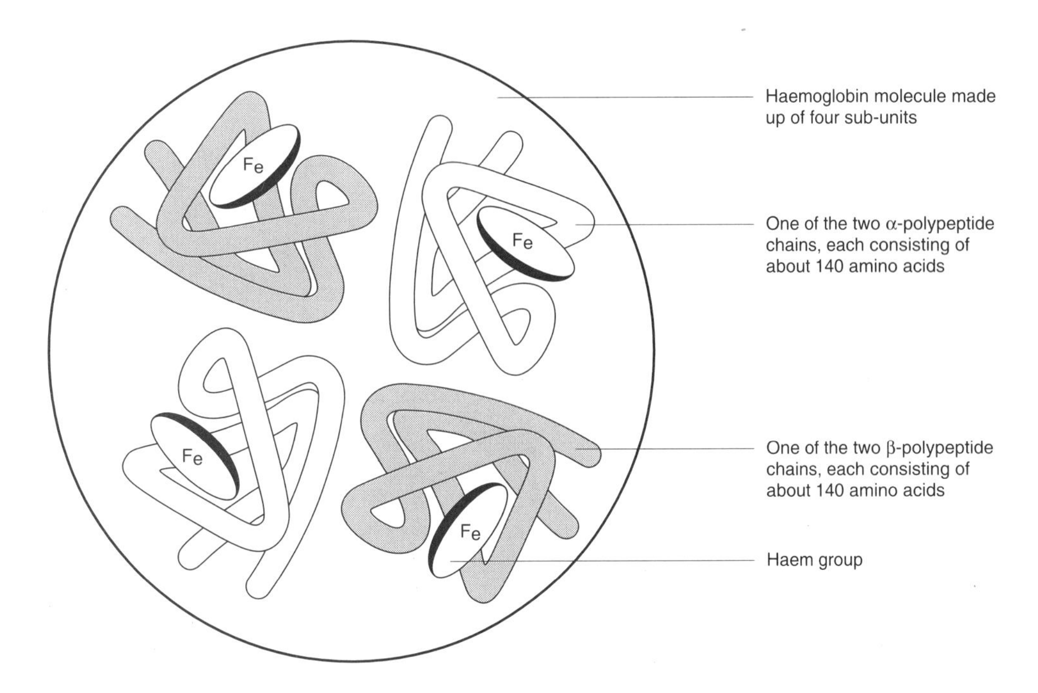

Fig. 7.6 Structure of haemoglobin molecule

When a respiratory pigment like haemoglobin is provided with increasing amounts of oxygen it absorbs it rapidly at first and then rather more slowly. This relationship between the oxygen tension and the rate at which haemoglobin takes up oxygen is shown in Fig. 7.7, which is known as an **oxygen dissociation curve**.

Oxygen dissociation curves vary in different species and according to different environmental conditions. The following should be helpful to you when interpreting oxygen dissociation curves.

1. The more that a curve is displaced to the right, the less easily the haemoglobin picks up oxygen, but the more readily it releases it.
2. The more that a curve is displaced to the left, the more easily the haemoglobin picks up oxygen, but the less readily it releases it.

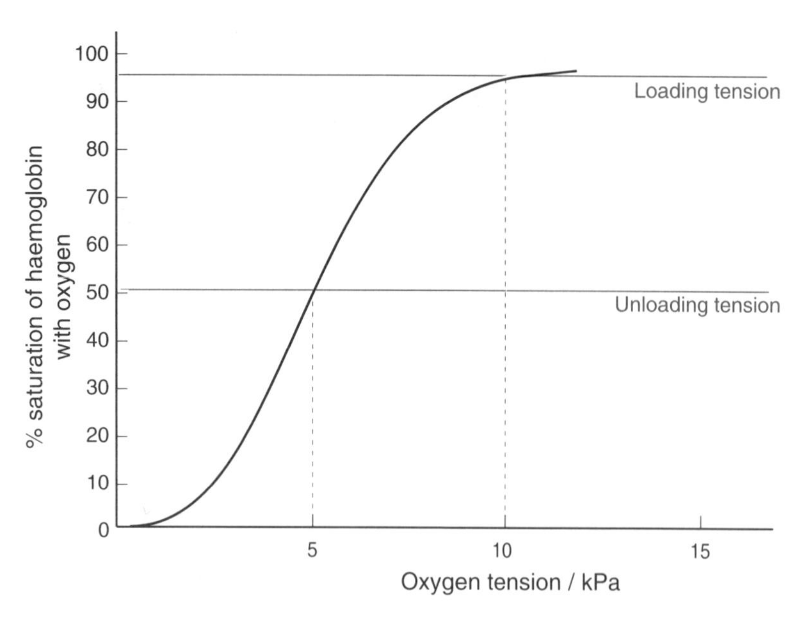

1 When a respiratory pigment is exposed to a gradual increase in oxygen tension it absorbs oxygen rapidly at first and then progressively more slowly.

2 Because 100% saturation of haemoglobin does not occur at normal environmental oxygen tensions, the **loading tension** is usually taken at 95% saturation.

3 The **unloading tension** is where the pigment is 50% saturated.

4 The partial pressure of oxygen in the lungs is usually less than in the atmosphere because air in the lungs includes the residual air from which some oxygen has already been removed.

Fig. 7.7 Oxygen dissociation curve for the haemoglobin of an adult human

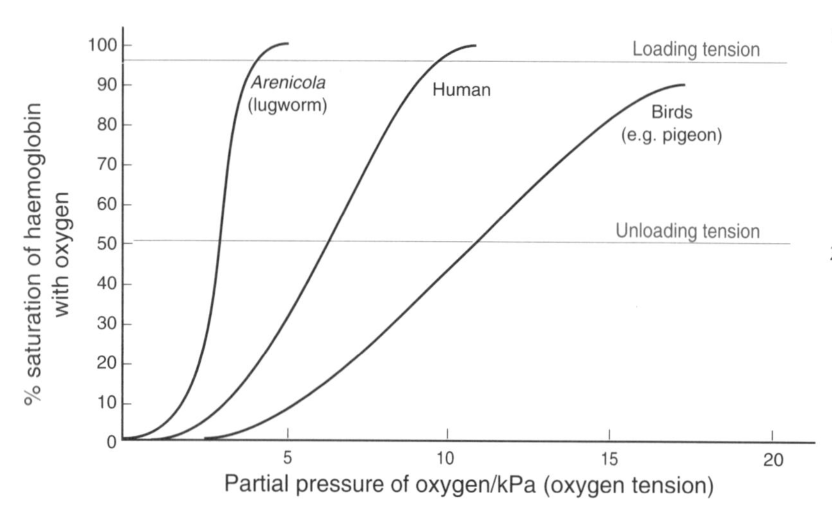

1 Because there is less oxygen in water than in air the dissociation curves for aquatic organisms are usually to the left of those for terrestrial ones. This is accentuated in *Arenicola* which lives in muddy waterlogged burrows where circulation of the water is limited.

2 The dissociation curve for birds is to the right of that for humans because the oxygen is given up to the respiring tissues by the pigment more readily, i.e. at a high partial pressure of oxygen. This is necessary if birds are to obtain oxygen fast enough to allow the high metabolic rate required by flight.

Fig. 7.8 Oxygen dissociation curves for haemoglobin in three organisms in different environments

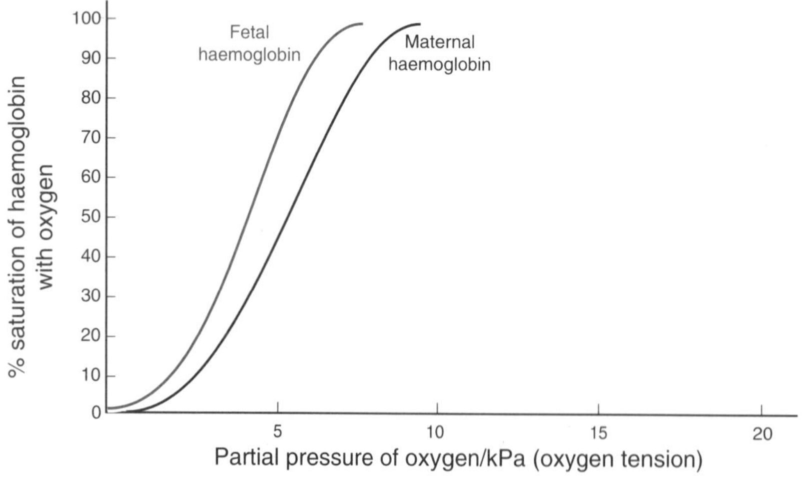

The curve for fetal haemoglobin is to the left of the mother's because, if the fetus is to obtain oxygen through the placenta, its haemo-globin must have a greater affinity for oxygen than the mother's haemoglobin.

Fig. 7.9 Comparison of oxygen dissociation curves for fetal and maternal haemoglobin of humans

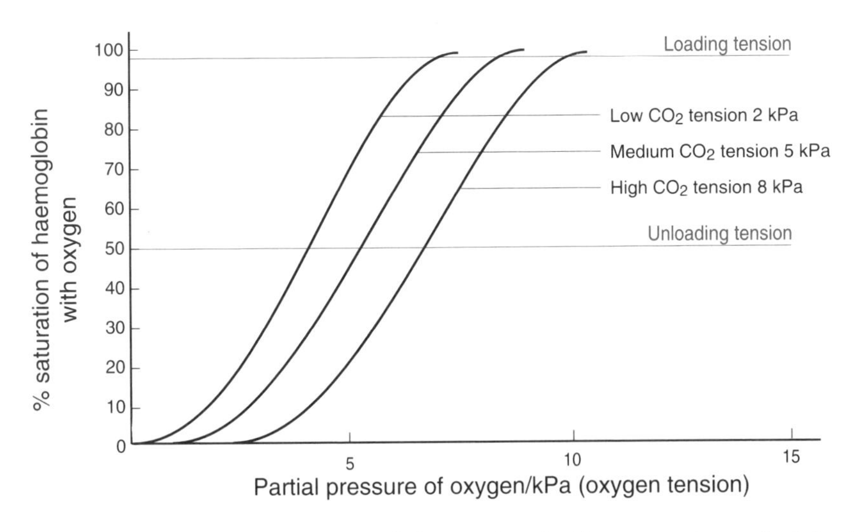

1. The partial pressure of carbon dioxide in the blood alters the oxygen dissociation curve – the Bohr effect.
2. Where there is a higher concentration of carbon dioxide, i.e. in respiring tissues, the haemoglobin will release its oxygen more readily. Therefore a higher partial pressure of carbon dioxide in the blood shifts the oxygen dissociation curve to the right.
3. This is the main factor responsible for oxygen being taken up in the lungs and released to respiring tissues.
4. Animals living in regions of low oxygen tensions are less sensitive to carbon dioxide effects as any shift to the right would mean they could not absorb oxygen from the low partial pressure of oxygen in their environment.

Fig. 7.10 Oxygen dissociation curves at different carbon dioxide concentrations – the Bohr effect

Carbon dioxide transport

Carbon dioxide is carried from the tissues to the respiratory surface in three ways:

1. In solution in the blood plasma – about 5% of carbon dioxide is carried in this way.
2. In combination with haemoglobin – carbon dioxide may combine with amino groups in the protein part of the haemoglobin molecule:

$$\text{Hb}-\text{N}\begin{matrix}\diagup \text{H} \\ \diagdown \text{H}\end{matrix} + CO_2 \rightleftharpoons \text{Hb}-\text{N}\begin{matrix}\diagup \text{H} \\ \diagdown \text{COO}^-\end{matrix} + H^+$$

haemoglobin + carbon dioxide ⇌ carbamino-haemoglobin + hydrogen ions

About 10% of the total carbon dioxide is carried in this way.

3. As hydrogencarbonate ions – some 85% of the total carbon dioxide is transported in this form. The carbon dioxide combines with water to form carbonic acid which dissociates into hydrogen and carbonate ions. The reaction is catalysed by the enzyme carbonic anhydrase:

$$H_2O + CO_2 \xrightarrow{\text{carbonic anhydrase}} H_2CO_3 \longrightarrow H^+ + HCO_3^-$$

water + carbon dioxide → carbonic acid → hydrogen ion + hydrogen carbonate ion

The hydrogen ions produced combine with haemoglobin, which thereby loses its oxygen. The hydrogencarbonate ions diffuse out of the red blood cell into the plasma where they combine with sodium ions from the dissociation of sodium chloride. It is largely in the form of the sodium hydrogencarbonate so produced that the carbon dioxide is transported in the blood. The negatively charged chloride ions from the dissociation of the sodium chloride diffuse into the red blood cell, replacing the hydrogencarbonate ions lost and thereby maintaining the ionic balance.

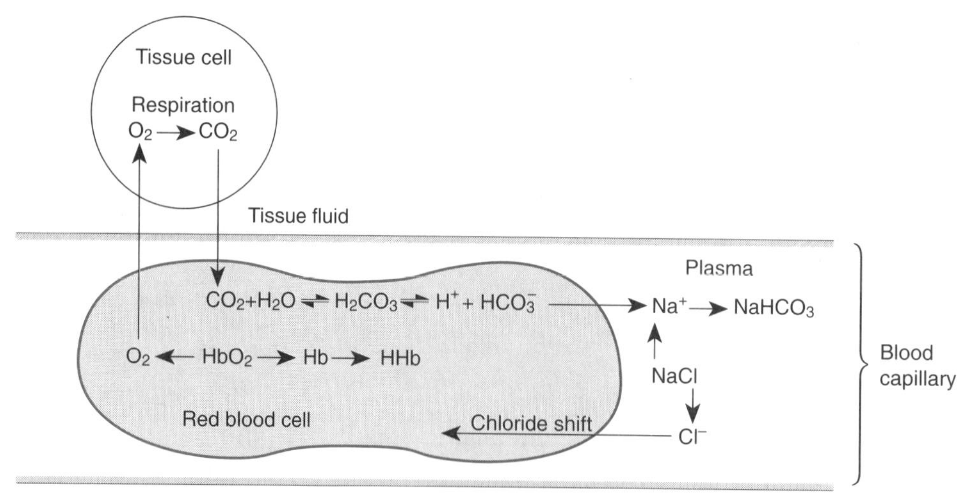

Fig. 7.11 The chloride shift

DEFENCE FUNCTIONS OF THE BLOOD

Phagocytosis

Phagocytosis is the take-up of large particles by cells via vesicles formed from the plasma membrane. In the blood the process is carried out by two types of white blood cell – **neutrophils** and **monocytes**. They do this for two reasons:

1. to destroy foreign organisms such as bacteria, which may pose a threat;
2. to remove the organism's own dead or damaged cells and other debris.

The process of phagocytosis by a neutrophil is summarised in Fig. 7.12.

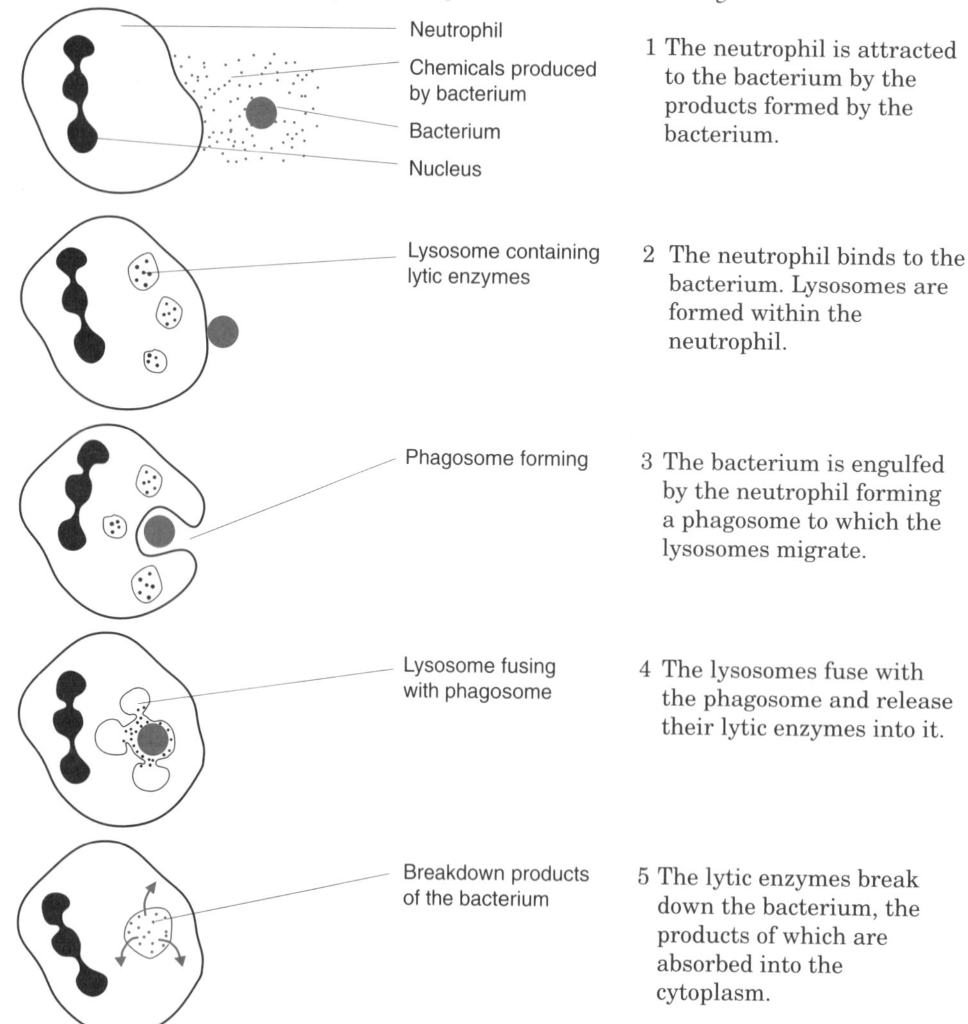

Fig. 7.12 Phagocytosis of a bacterium by a neutrophil

Production of antibodies – the immune response

Antibodies are chemicals which help combat disease. They are produced by a type of white blood cell called **lymphocytes**. They help confer on an organism the ability to resist disease (**immunity**). Foreign material, such as a bacterium, entering the body has a chemical coat which acts as an **antigen**. This causes the lymphocytes to produce an antibody which is specific to that antigen. There are a number of different types of antibodies, each with different functions:

1. **Agglutinins** cause foreign matter to clump together.
2. **Antitoxins** neutralise dangerous toxins produced by foreign material.
3. **Lysins** break down foreign matter.
4. **Opsonins** stimulate other white cells to engulf foreign material.

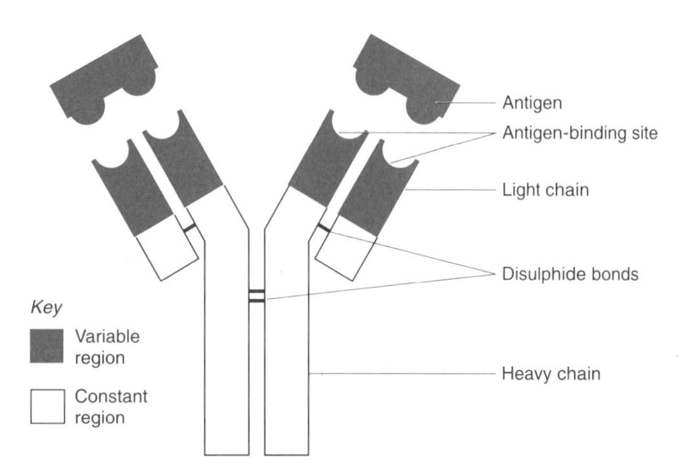

Fig. 7.13 Structure of an antibody

Immunity can be induced artificially. You are likely to have been immunised (vaccinated) against a number of diseases, in which case one of two basic methods will have been used.

Passive immunity The necessary antibodies are passed into a person in some way. As the recipient does not produce the antibodies for himself, the immunity only lasts as long as the antibodies persist in the body – rarely more than a few weeks.

Active immunity The individual is induced in some way (e.g. by injecting dead disease-causing organisms) to produce its own antibodies. This type of immunity lasts for longer – often for an organism's lifetime. Active immunity can be acquired naturally, e.g. recovery from diseases such as measles usually confers a lifelong immunity to the disease.

Acquired immune deficiency syndrome (AIDS)

AIDS has received much attention in the news in recent years. It is a disease caused by the **human immunodeficiency virus (HIV)**, which attacks certain types of lymphocytes and so prevents them from carrying out their immune functions effectively. Individuals who are infected with HIV are said to be HIV positive, but as the virus remains dormant, on average for eight years, it is often a long time before they develop full-blown AIDS. Such people may quickly succumb to any infection, which they contract because they lack the mechanism to resist it. The virus can be found in almost all bodily fluids of an HIV-positive person, but only in blood, and seminal and vaginal fluids is its concentration high enough to transmit the disease to others.

Blood clotting

The problem with a liquid transport medium is preventing it being lost when a vessel is broken in some way. **Platelets**, or **thrombocytes** as they are often called, are cellular fragments in blood which bring about clotting (**thrombosis**) when a leakage occurs. They initiate a complex chain of events which results in a blood clot:

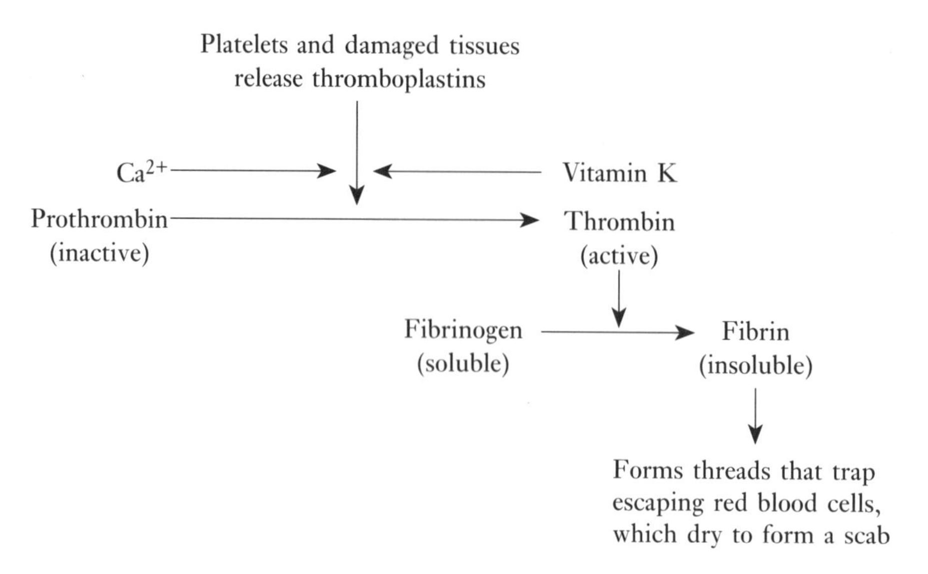

BLOOD GROUPS

The membrane of a red blood cell contains polysaccharides which act as antigens. There are many different systems by which blood is grouped but the ABO system is the best known. Here there are two antigens A and B, each of which has a corresponding antibody. The composition of the four possible groups is given in Table 7.3.

Table 7.3 *Blood group antigens and antibodies*

Blood group	Antigen (on red blood cell)	Antibody (in plasma)
A	A	b
B	B	a
AB	A and B	none
O	none	a and b

You will notice from the table that an antigen is never found with its corresponding antibody. This is because when both are present together the red blood cells clump (**agglutination**) and break down (**haemolysis**). It is therefore necessary during transfusions to ensure antigens and their corresponding antibodies do not mix. In practice it is possible to add relatively small quantities of antibody to the corresponding antigen without causing harm, but never the reverse. Table 7.4 shows the results of adding small quantities of one blood type to another.

You will see from this table that blood group O can safely donate blood to anyone in small quantities. People with group O are therefore known as **universal donors**. You will also notice that group AB can safely receive blood from anyone. Individuals with group AB are referred to as **universal recipients**.

Table 7.4 *Compatibility of blood groups in the ABO system*

Donor's blood group	Recipient's blood group			
	A	B	AB	O
A	safe	harmful	safe	harmful
B	harmful	safe	safe	harmful
AB	harmful	harmful	safe	harmful
O	safe	safe	safe	safe

Rhesus system

In addition to A and B another antigen is important. First discovered in the rhesus monkey it is called the **rhesus factor** or **antigen D**. An individual with antigen D on their red blood cells is said to be **rhesus positive**, one without is **rhesus negative**. It is important during transfusion to match the rhesus antigen as well as the A and B ones. There is no naturally occurring antibody to antigen D, but one can be artificially stimulated by the addition of antigen D blood to a rhesus-negative person.

LYMPH

Owing to the relatively high blood pressure in the capillaries a watery solution of low protein content, some salts, nutritive materials and phagocytic cells leaves the capillaries and bathes the tissues. This is called **tissue fluid**. Having had the nutritive materials and oxygen removed and wastes added by the cells, most of the tissue fluid re-enters the capillaries by osmosis. About 1–2% is returned in separate vessels called **lymphatics**. This is the **lymph**. Lymph moves along these vessels as a result of hydrostatic pressure and respiratory and muscular movements which squeeze the lymph vessels. The direction of flow is controlled by valves in the lymph nodes, especially in the armpits and groin. These nodes are also the site of lymphocyte production. The lymphatic vessels finally drain into the vena cava near its entrance to the heart via two major vessels: the right lymphatic duct, which drains the upper right side of the body, and the thoracic duct, which drains the remainder of the body.

CIRCULATORY SYSTEMS

Open blood system

In an open blood system, which occurs in arthropods and most molluscs, there are no capillaries connecting the arteries to the veins. Arterial blood passes into sinuses so that major tissues are bathed in circulating fluid. These fluids slowly work their way back to the open ends of veins or to the ostia of the heart. The organs lie directly in the blood-filled **haemocoel**.

Closed blood system

In a closed blood system, as found in vertebrates and annelids, a continuous network of minute capillaries unites the smaller arteries with the veins. Nutrients are transferred to cells by fluids filtering through the walls of the capillaries into the tissue spaces. There are two types of closed blood system: single and double circulatory.

Single circulatory system (fish) Blood passes through the heart only once per circuit of the body. Being forced through the gill capillaries lowers the blood pressure considerably and it still has to pass across another capillary system before returning to the heart.

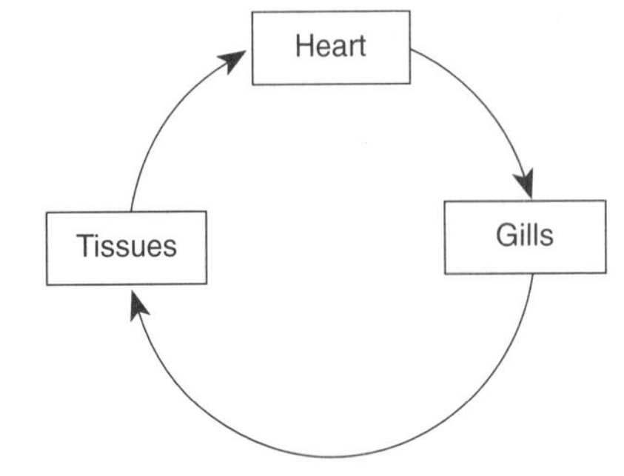

Fig. 7.14 Single circulation of fish

Double circulatory system Blood passes through the heart twice per circuit of the body. This enables blood pressure to be maintained.

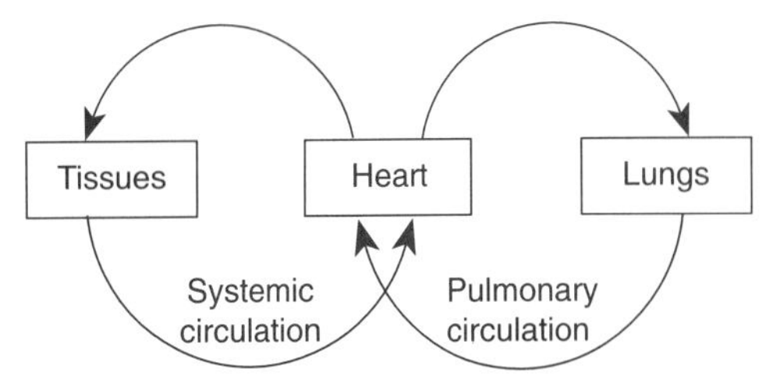

Fig. 7.15 Double circulation

THE HEART

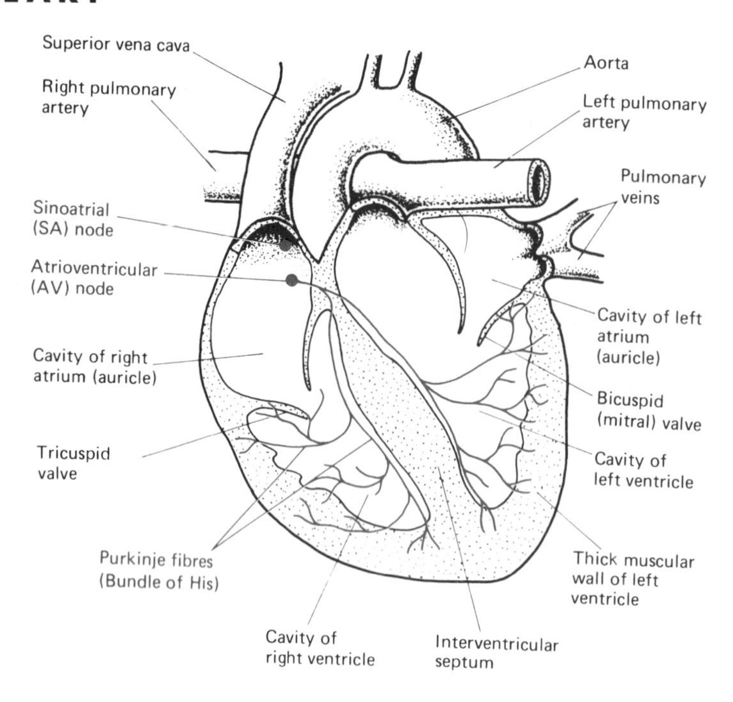

Fig. 7.16 Diagram of a vertical section through the human heart (ventral view)

The simplest form of heart is a muscular contracting region of a blood vessel, e.g. in an earthworm. In fish this contractile region is S shaped and divided into a receiving region, the atrium, and a pumping region, the ventricle. In amphibians there are two atria and a single ventricle, whereas in birds and mammals there are two atria and two ventricles arranged so that the left atrium and ventricle carry oxygenated blood and are completely separated from the right atrium and ventricle, which carry deoxygenated blood. The left atrium and ventricle are separated by the bicuspid or mitral valves and the right atrium and ventricle by the tricuspid valve. The aorta and pulmonary arteries have semilunar valves to prevent backflow of blood into the ventricles. The left ventricle pumps oxygenated blood to the whole body except the lungs and has a thicker muscular wall than the right ventricle, which pumps deoxygenated blood to the lungs only. The heartbeat consists of a contraction phase (**systole**) and a relaxation phase (**diastole**). In a resting adult human the heart pumps about 70 times per minute, pumping about 60 cm^3 of blood per beat from each ventricle. The total volume of blood pumped each minute is called the **cardiac output** and varies according to the amount of physical exercise undertaken and the emotional state of the individual.

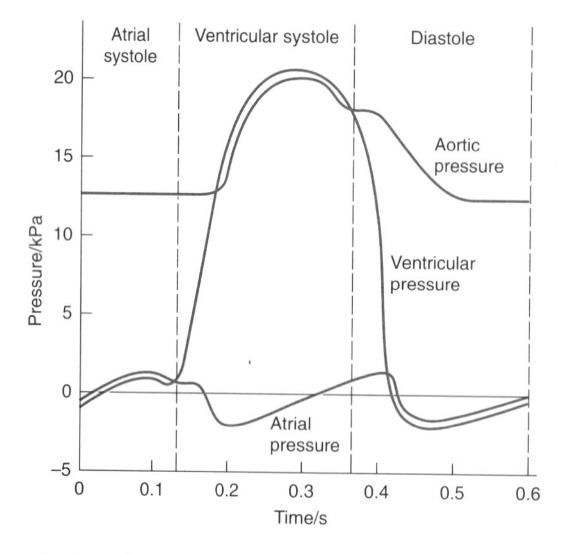

Fig. 7.17 Pressure changes in the atria, ventricles and aorta during one cardiac cycle

Control of heartbeat (cardiac rhythm)

A vertebrate heart continues to beat in a coordinated way even when removed from the body. The heartbeat must, therefore, be initiated from within the heart muscle itself (**myogenic** heart) rather than by separate nervous initiation (**neurogenic** heart). In myogenic hearts the beat is initiated by a 'pacemaker' called the **sinoatrial node** (SA node) – a group of specialised cardiac muscle cells 2 cm by 2 mm, in the right atrium, near the point where the venae cavae enter. The wave of excitation from the SA node spreads outwards causing the atria to contract, emptying their blood into the ventricles. The wave is picked up by a mass of similar tissue in the right atrium near the interatrial septum. This second group of cells is called the **atrioventricular node** (AV node). To ensure that the ventricles contract from the apex upwards and so allow blood to be forced out through the arteries, the wave of excitation is conducted to the apex of the ventricle by the **Purkinje fibres** (bundle of His).

Factors modifying heartbeat (cardiac rhythm)

Chemical control To maintain its rhythm for any length of time the heart requires the correct balance between the ions of calcium, sodium and potassium. The chemicals adrenaline and noradrenaline and drugs such as digitalis increase cardiac output. An increase in the level of carbon dioxide in blood (effectively a fall in pH) also causes an increase in cardiac output. Acetylcholine slows myogenic hearts while accelerating neurogenic ones.

Nervous control Receptors in the aorta and carotid arteries respond to an increase in blood pressure by sending nervous impulses to the cardioregulatory centre in the brain, which in turn sends impulses along the efferent vagus nerve to decrease cardiac output. An increase of pressure in the right atrium causes receptors to send impulses along the afferent vagus nerve to the cardioregulatory centre which then sends impulses to increase cardiac output. The heartbeat is therefore quickened by the sympathetic nervous system and slowed by the parasympathetic nervous system.

The maintenance and control of blood pressure and circulation

In a healthy human at rest the arterial systolic blood pressure is 120 mm Hg and the diastolic pressure is 80 mm Hg. The pressure is created initially by the contraction of the ventricles of the heart. As blood is forced into the arteries, the elastic walls expand causing distension. The recoil of the elastic walls pushes blood away from the heart, creating distension of the artery at a point further away from the heart. (The blood is prevented from returning to the heart by the semilunar valves.) The 'pulses' continue throughout the arterial system and help to maintain blood pressure. The hormone vasopressin increases blood pressure in vertebrates.

The return of blood to the heart in the veins is maintained in a number of ways:

1. There is the residual heart pressure – usually 10 mm Hg or less.
2. The contraction of muscles squeezes veins and forces blood towards the heart; flow in the reverse direction is prevented by pocket valves.
3. Inspiratory movements – when breathing in, the low thoracic pressure helps to draw blood along the major veins towards the heart.
4. Gravity will help return blood from those regions above the heart.

The 5 dm^3 of blood in a human is inadequate to supply the maximum needs of all regions of the body at the same time. Different organs make different demands on the blood in terms of quality and quantity (see Table 7.5).

Table 7.5 *Blood flow to various organs*

Organ	Blood flow in cm^3 per 100 g of organ (when body is at rest)
Heart	80
Liver	90
Brain	55
Muscle	3
Skin	10
Kidney	400
Remainder	2.5

Clearly, during exercise muscles require considerably more blood than they do when at rest. Control of blood flow is achieved by contraction and relaxation of muscles in the smaller artery (arteriole) walls. Control of these muscles is by the autonomic nervous system, with the sympathetic system contracting the vessels and decreasing blood flow and the parasympathetic system dilating them and increasing flow. Apart from changes in blood flow, to meet physiological needs, certain psychological events control flow in humans, e.g. fear, embarrassment and erection of the penis or clitoris due to erotic stimulation.

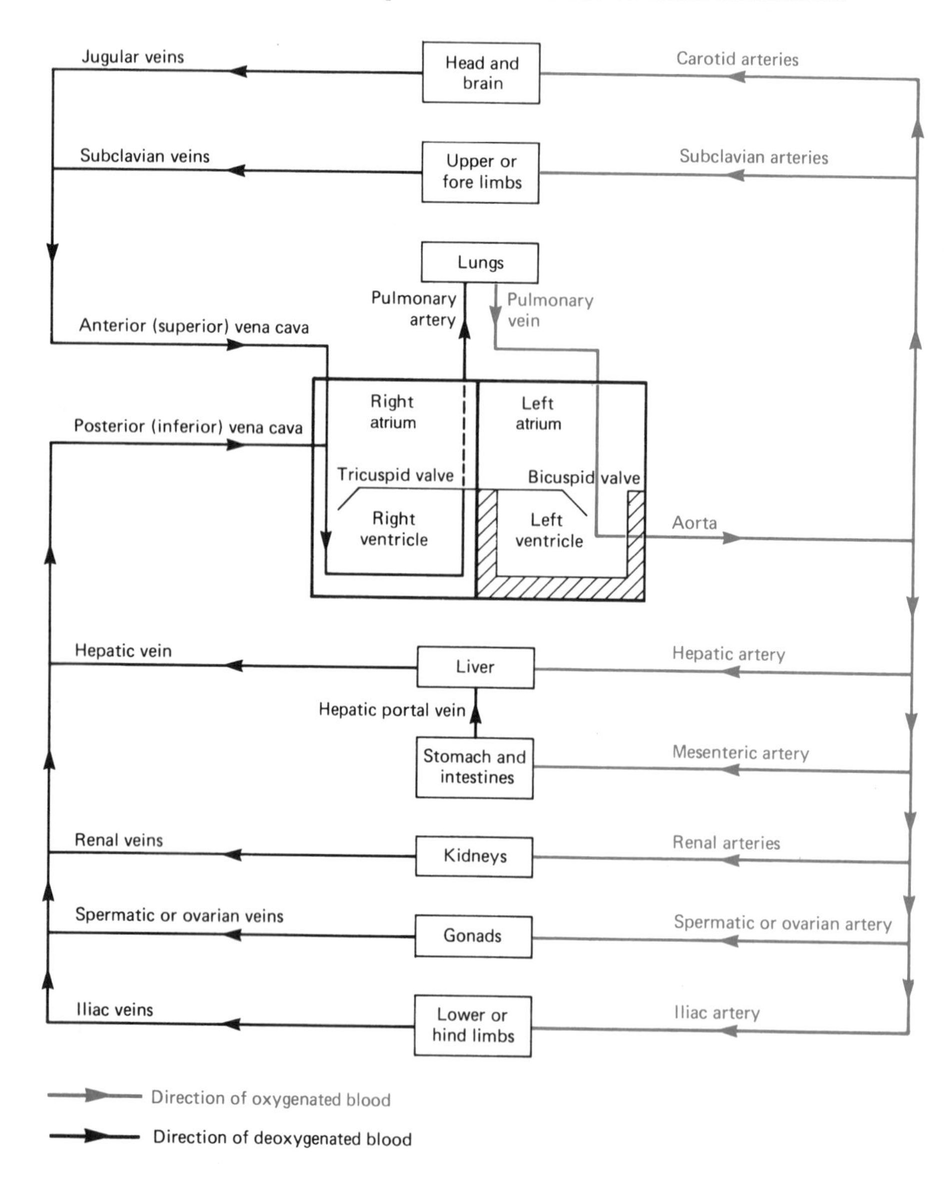

Fig. 7.18 Outline diagram of mammalian circulation

Table 7.6 *Similarities and differences between blood vessels*

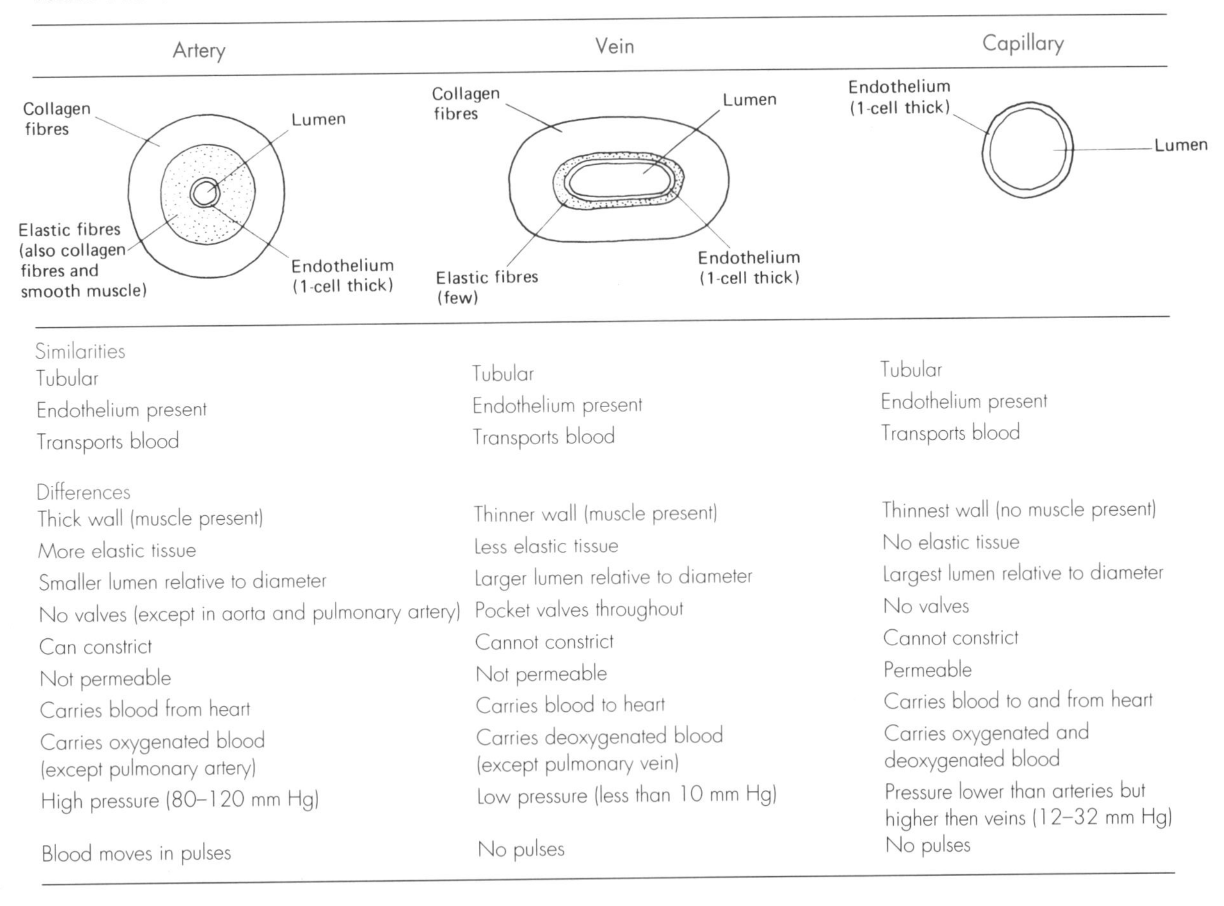

Artery	Vein	Capillary
Similarities		
Tubular	Tubular	Tubular
Endothelium present	Endothelium present	Endothelium present
Transports blood	Transports blood	Transports blood
Differences		
Thick wall (muscle present)	Thinner wall (muscle present)	Thinnest wall (no muscle present)
More elastic tissue	Less elastic tissue	No elastic tissue
Smaller lumen relative to diameter	Larger lumen relative to diameter	Largest lumen relative to diameter
No valves (except in aorta and pulmonary artery)	Pocket valves throughout	No valves
Can constrict	Cannot constrict	Cannot constrict
Not permeable	Not permeable	Permeable
Carries blood from heart	Carries blood to heart	Carries blood to and from heart
Carries oxygenated blood (except pulmonary artery)	Carries deoxygenated blood (except pulmonary vein)	Carries oxygenated and deoxygenated blood
High pressure (80–120 mm Hg)	Low pressure (less than 10 mm Hg)	Pressure lower than arteries but higher then veins (12–32 mm Hg)
Blood moves in pulses	No pulses	No pulses

7.3 TRANSPORT IN PLANTS

Most organisms comprise a variety of different cells grouped into tissues and organs. These tissues and organs have become specialised to perform particular functions. The product of one tissue may be needed by another and consequently a transport system between the two is required.

In addition there are certain advantages for plants in being large, e.g. they can compete more readily for light. Some have become extremely tall (over 100 m) and have thereby obtained competitive advantage. This means that the organs collecting the water, the roots, are some considerable distance from the leaves that require it for photosynthesis. Again this necessitates a transport system between the two structures. The sugars manufactured in the leaves must be transported in the opposite direction to sustain respiration in the roots.

Unlike most animals, plants do not possess contractile cells such as muscle cells. They are therefore dependent to a large degree on passive rather than active mechanisms for transporting material. The evaporation of water from leaves creates a water potential gradient along the leaf mesophyll cells that draws water in from the xylem. As water is drawn from the xylem, the cohesive properties cause it to be pulled up in a continuous column. A water potential gradient in the roots brings water from the soil to the xylem. Only its entry into the xylem involves the expenditure of energy produced by the plant itself.

The downward transport of sugars is less clearly understood. Certain theories (e.g. mass flow theory) favour a totally passive process, while others (e.g. transcellular strand theory) involve energy expenditure.

Before we study these mechanisms of transport in plants in more detail we must look more closely at what is probably the most important biological substance – the water molecule.

THE WATER MOLECULE

Life on earth originated in water, which is the most abundant liquid in the world. It makes up at least half of all living organisms and up to 95% of some species. Abundant it may be, but ordinary it certainly is not. As a result of hydrogen bonds formed between its molecules, water possesses some unusual chemical and physical properties.

Naturally occurring water consists of 99.76% by weight of $^1H_2{}^{16}O$. The remainder consists of various isotopes, e.g. ^2H and ^{18}O. The commonest of these is ^2H (deuterium) which is most often found with the normal hydrogen atom as HDO, though occasionally as D_2O. Both are called 'heavy water' and have a deleterious effect on living organisms.

The molecule of water consists of the two hydrogen atoms bonded to the oxygen atom covalently (by the sharing of electrons). Although the molecule is neutral overall, the oxygen atom retains a slight negative charge and the hydrogen atoms a slight positive one. Such molecules are termed **polar**.

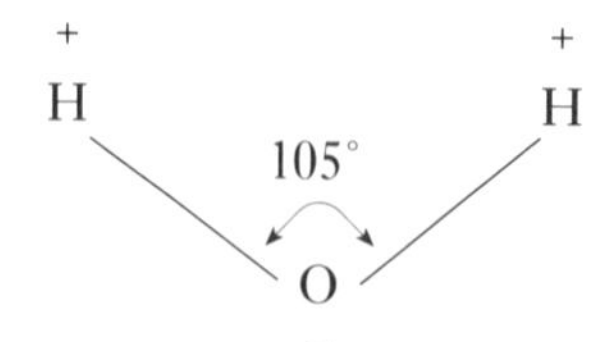

This polarity of molecules causes them to be attracted to each other by their opposite charges. These attractive forces form what are called **hydrogen bonds**. Although very weak and short lived, such bonds collectively constitute an important force which holds water molecules together and makes water a much more stable substance than would otherwise be the case.

PROPERTIES OF WATER

Cohesion and surface tension

Cohesion is the tendency of molecules of one substance to hold together by mutual attraction. The hydrogen bonding of water results in strong cohesive forces. One effect of this is that the surface of a drop of water will assume the smallest possible area, and the drop therefore forms a sphere. The water molecules at the surface are drawn in towards the body of the drop forming a skin-like layer of molecules at the surface. This force is called **surface tension**. Insects walking on the surface of water and the movement of water up plants are two biological processes that can occur as a result of the cohesive properties of water molecules.

Adhesion and capillarity

Adhesion is the attraction of molecules of different compounds to one another. The ability of water to cling readily to other molecules is responsible for the upward movement of water when a small-bore tube is dipped into it. This phenomenon is called **capillarity**. Xylem vessels of a diameter 0.02 mm can, in theory, support a column of water of height 1.5 m by capillarity forces. One of its main biological effects is the upward movement of water in the soil.

Thermal capacity (specific heat)

Another consequence of hydrogen bonding in water is that much heat is needed to cause increased molecular movement and hence gas (steam) formation. The heat energy must first be used to break the hydrogen bonds. For this reason the temperature of water rises only very slowly for a given amount of heat added, when compared with other substances. Similarly it cools more slowly. In all it is thermally stable and so biochemical reactions in a water medium are not subjected to large temperature fluctuations and can take place at a more constant rate. Were it not for hydrogen bonding, water would be a gas at normal environmental temperatures and life as we know it could not exist. For the same reasons much heat is needed to evaporate water and therefore even the evaporation of a small amount of water from the surface of an organism has a large cooling effect, e.g. sweating.

Density

Water has its maximum density at 4 °C. Unlike most other substances it is less dense as a solid than as a liquid. It freezes from the top downwards and the ice that forms at the surface can insulate the warmer water below this layer from the colder temperatures above it. This prevents large bodies of water from freezing solid and has contributed to the survival of aquatic organisms.

Dissociation (ionisation), pH and buffers

There is a slight tendency for water molecules to dissociate into ions according to the equation:

$$2H_2O \rightleftharpoons H_3O^+ + OH^-$$

water molecule — oxonium ion — hydroxide (hydroxyl) ion

It is simpler, however, to consider the dissociation as:

$$H_2O \rightleftharpoons H^+ + OH^-$$

water molecule — oxonium ion — hydroxide (hydroxyl) ion

In a litre of water this dissociation produces 1/10 000 000 (10^{-7}) mole of hydrogen ions. This is equivalent to a pH of 7, which is neutral. If the concentration of hydrogen ions was greater, say 1/1000 (10^{-3}) mole hydrogen ions per litre, the pH would be 3 and the solution would be acidic. Any pH below 7 is acidic, any above is basic. An acid is therefore a substance that donates hydrogen ions and a base is a hydrogen ion acceptor. Note that the pH scale is not linear but logarithmic.

A **buffer solution** is one that retains a constant pH despite the addition of small quantities of acids or bases. Buffers contain both hydrogen ion donors and acceptors. Hydrogencarbonate ions may act as an acceptor

$$HCO_3^- + H^+ \rightleftharpoons H_2CO_3$$

hydrogencarbonate ion — hydrogen ion — carbonic acid

or a donor

$$HCO_3^- + OH^- \rightleftharpoons CO_3^{2-} + H_2O$$

The removal of OH^- allows more water to dissociate and so produce more H^+:

$$H_2O \rightleftharpoons H^+ + OH^-$$

Hydrogencarbonate salts and phosphate salts are responsible for the buffering of blood, maintaining its pH at a constant 7.4. Apart from dissociating itself, water readily causes the dissociation of other substances placed in it. Thus it is an excellent solvent.

Osmosis and animal cells

If a solution is separated from pure water by a partially permeable membrane, the pressure which must be applied to prevent osmosis is called the **osmotic pressure**. As this situation is a hypothetical one, and as a solution does not actually exert any pressure in normal circumstances, the term **osmotic potential** is preferred. As the osmotic potential is the potential of a solution to pull in water, its value is always negative. The more concentrated a solution, the more negative is its osmotic potential.

When two solutions have the same osmotic potential they are said to be **isotonic**. Where one solution has a greater osmotic potential (i.e. is more concentrated) than another it is said to be **hypertonic** to it. The one with the lower osmotic potential (i.e. the less concentrated one) is said to be **hypotonic**.

Osmosis and plant cells

Although the osmotic principles apply equally to plant and animal cells, a different set of terms is currently applied to the osmotic relationship of plant cells. **Water potential** (represented by the Greek letter psi ψ) is a measure of the tendency of water to leave a solution. Pure water is designated a water potential of zero. As the solute molecules in a solution tend to prevent the water molecules leaving it, the solution will have a lower water potential than pure water. Its value will therefore be less than zero, i.e. negative. The more concentrated the solution, the more negative is its water potential.

For practical purposes a plant cell can be considered as a solution of salts and sugars in the vacuole surrounded by a partially permeable membrane (tonoplast, cytoplasm and plasma membrane) and a slightly elastic but completely permeable cell wall. A plant cell therefore has a more negative water potential than pure water and will draw in water when surrounded by it. This entry of water forces the living part of the cell, known as the protoplast, against the cell wall. In effect, the water in the vacuole is being subjected to a pressure from the cell wall. This pressure is referred to as the **pressure potential** (ψ_p). In a turgid plant cell this is a positive value, although in the xylem of a transpiring plant (which is under tension) it is negative. The water potential of a cell is changed by the presence of solute molecules. This change is referred to as the **solute potential** (ψ_s). As solute molecules invariably lower the water potential, its value is always negative. The relationship between these three terms is given as:

$$\underset{\text{water potential}}{\psi} = \underset{\text{solute potential}}{\psi_s} + \underset{\text{pressure potential}}{\psi_p}$$

THE IMPORTANCE OF WATER

Metabolic role

1. All reactions occur in an aqueous medium.
2. Water is essential to all hydrolysis reactions, e.g. polysaccharides to monosaccharides, fats to fatty acids and glycerol and proteins to amino acids.
3. Diffusion requires a moist surface, e.g. respiratory surfaces.
4. Water is a raw material in photosynthesis.

Lubrication

The viscosity of water makes it a useful lubricant, for example:

1. Mucus is used externally by snails and earthworms during locomotion and internally in the mammalian gut and vagina.
2. Synovial fluid lubricates the joints in vertebrates.
3. Pleural fluid lubricates lung movements during breathing.
4. Pericardial fluid lubricates movements of the heart.

Solvent

As water readily dissolves other molecules it is important in the following:

1. Excretory products are removed from the body in water, e.g. ammonia, urea.
2. Transport – substances such as glucose, amino acids, minerals and hormones are transported dissolved in aqueous blood plasma (see Table 7.2). Similarly in plants minerals and sucrose are transported in aqueous solution.
3. Most secretions comprise substances dissolved or suspended in water, e.g. digestive juices.
4. The cytoplasm of all cells contains many dissolved chemicals.

Support functions

Water is not easily compressed, making it a useful structural agent. Its support functions include:

1. the hydrostatic skeleton in earthworms;
2. the coelom of many invertebrates;
3. turgor pressure, which keeps herbaceous plants, and herbaceous parts of woody ones, erect;
4. the aqueous and vitreous humours of the eye;
5. the erection of the penis;
6. the amniotic fluid around the mammalian fetus;
7. as a medium in which many organisms live permanently it provides support for the whole body. .

Other functions

1. Water is a medium for the dispersal of some terrestrial organisms and for the transference of their gametes, e.g. mosses.
2. The evaporation of water is important in the temperature control of organisms, e.g. sweating in mammals.
3. In mammals water makes up much of the endolymph and perilymph, which are important in hearing and balance.
4. Water potential has an effect on the structure of plant cells as shown in Fig. 7.19.

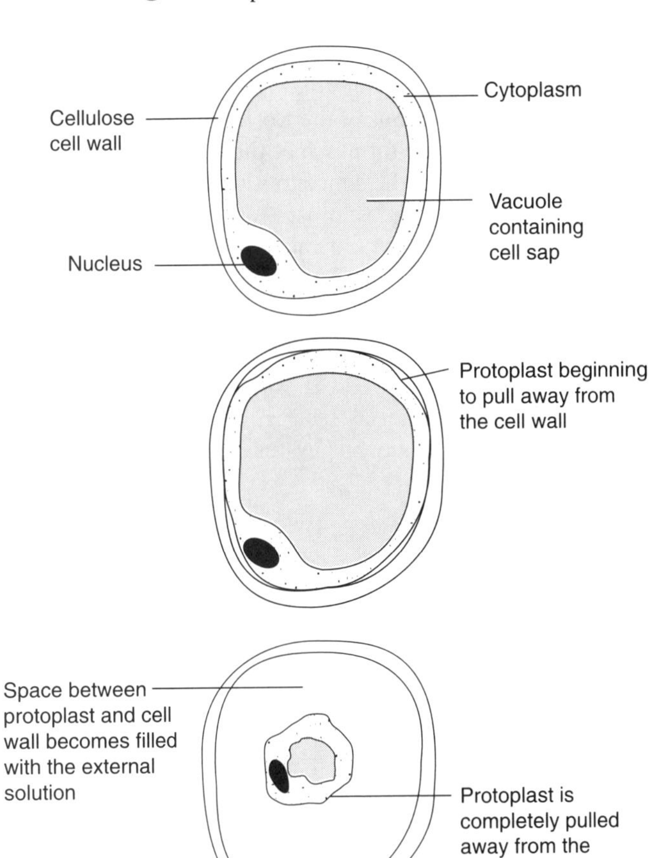

(a) When the water potential (Ψ) of the external solution is less negative (higher) than that of the cell, water enters by osmosis, pushing the cell contents hard against the cell wall. This makes the cell turgid.

(b) When the water potential (Ψ) of the external solution is the same as that of the cell, water neither enters nor leaves. The cell contents, while filling the space within the cell wall, are not pushed up hard against it. Indeed they may in places begin to pull away from it. Such a cell is at incipient plasmolysis.

(c) When the water potential (Ψ) of the external solution is more negative (lower) than that of the cell, water leaves the protoplast and it therefore shrinks away from the cell wall. The cell therefore becomes flaccid.

Fig. 7.19 Effect of differing water potential on individual plant cells

TRANSPIRATION

Transpiration is the evaporation of water from plants. The majority of this water is lost through stomata although some passes directly through the cuticle and, in woody plants, some is lost through areas of loosely packed cells known as **lenticels**. Plants need stomata to permit the exchange of gases necessary to photosynthesis and, to a large degree, transpiration is just the unfortunate consequence of having these openings in leaves. At the same time transpiration contributes significantly to the upward movement of water in large trees.

Factors affecting transpiration

1. **Humidity** The lower the humidity around the plant, the greater is the diffusion gradient between the leaf and the atmosphere. The rate of transpiration is therefore greater when the humidity is low.
2. **Wind speed** An increase in wind speed normally increases the rate of transpiration since saturated air is blown away from the vicinity of the stomatal pore and a large diffusion gradient is maintained. This is counteracted by the cooling effect due to increased evaporation which tends to reduce the transpiration rate slightly.
3. **Temperature** If the temperature of the air and leaf increase simultaneously, the consequent increase in vapour pressure of the leaf increases the diffusion gradient and hence the rate of transpiration.
4. **Light intensity** The stomata of most plants close in the dark and open in the light. Up to a point, therefore, an increase in light intensity will increase the rate of transpiration and vice versa.
5. **Plant structure** Many anatomical features of a plant affect the transpiration rate. A small leaf area, a thick cuticle and fewer stomata can all reduce the rate at which water is lost.

Xerophytes

Xerophytes are plants which are adapted to survive conditions of unfavourable water balance. Xerophytic modifications not only arise in plants in the hot dry regions of the world, but also occur in cold areas where water, though abundant, is frozen into ice for much of the year and is therefore not available to the plant. Plants living in areas where the concentration of salts in the soil is high also find considerable difficulty obtaining water and so must economise on its use. These plants are called **halophytes** and live in areas such as salt marshes fringing the sea and river estuaries. They also show certain of the xerophytic adaptations listed in Table 7.7.

Measurement of the rate of transpiration

The rate of transpiration in a leafy shoot is measured using a **potometer** (see Fig. 7.20). The instrument is set up as shown and, provided it is watertight, the rate at which water evaporates

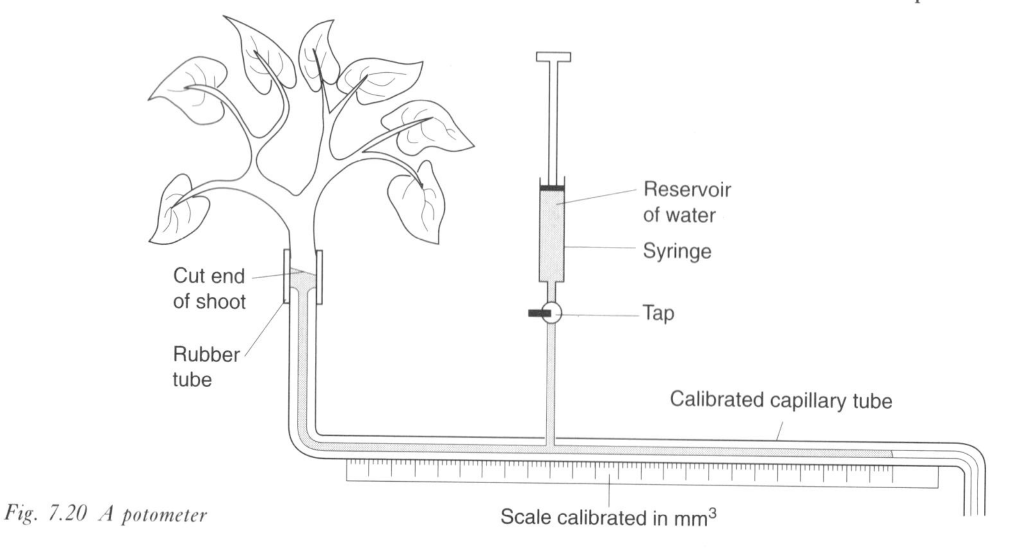

Fig. 7.20 A potometer

Table 7.7 *Adaptations in xerophytes*

Mechanism	Adaptation	Examples	Notes
Reduces transpiration	Thick cuticle	Most evergreens	Reduces cuticular transpiration; often shiny so reflects sun causing leaf temperature and therefore transpiration to fall
	Depression of stomata	*Ilex* (holly), *Pinus*	Lengthens diffusion path and therefore reduces diffusion pressure gradient; may trap still, moist air
	Rolled leaves	*Ammophila* (marram grass), *Calluna*	Folding lengthens diffusion path and traps moist air
	Protective hairs	*Ammophila*	Often accompanies rolling of leaf; traps moist air
	Leaves small or absent	*Pinus*	Small and circular in cross-section to give low surface area/volume ratio and structural support to prevent wilting
		Opuntia	No leaves, flattened stem photosynthesises
	Variations in leaf positions	*Lactuca* sp. (European compass plant)	Leaf positions adjusted so that sun strikes them obliquely; lowers transpiration because lowers temperature
	High solute potential of cell sap	Many xerophytes	Thought to reduce evaporation from cell walls
	Succulent stem and possibly leaves	Cacti	Stores water
Succulence	Diurnal closing of stomata	Many xerophytes	Reduces transpiration; requires metabolic modifications
	Shallow, wide root system	Many xerophytes	Gains maximum benefit from light rain
	Vegetative propagation well developed	Many xerophytes	Seed germination requires water
Extensive root systems		Most xerophytes	Especially developed in the most arid conditions
Resistance to desiccation	Increased lignification (correspondingly reduced leaves) allows resistance to wilting	*Ruscus*	Flattened stem (cladode) takes over photosynthesis
		Acacia	Lamina lost; petiole flattened to form phyllode
		Cacti	Leaves reduced to spines
	Reduced cell size	Many xerophytes	Less likely to wilt than fewer large ones

from the shoot can be measured by following the movement of the column of water along the calibrated capillary tube. Strictly speaking the potometer measures water uptake rather than transpiration as not all the water absorbed will be transpired. Some may be assimilated into the plant while some may be used for metabolic purposes. However, the quantities in both cases will be comparatively small.

The mechanism of stomatal opening and closing

Stomata are found in leaves and herbaceous stems. Each stoma comprises a pair of specialised epidermal cells called guard cells and an elliptical pore – the stomatal aperture (see Fig. 7.21).

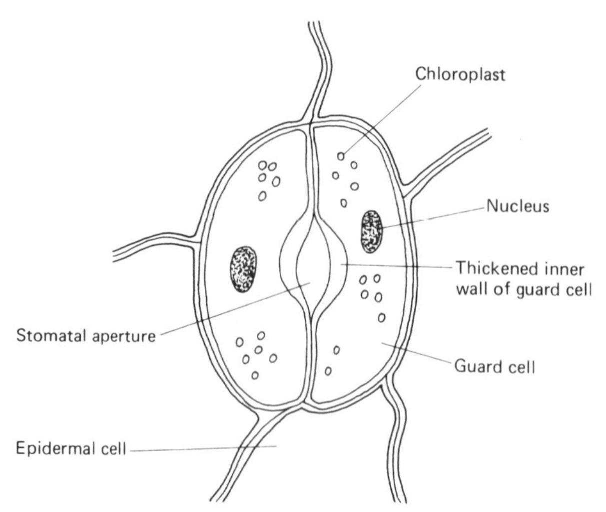

Fig. 7.21 Surface view of a stoma

The frequency of stomata varies with environment and species. They are generally more numerous on the abaxial (under) surface of leaves than on the adaxial (upper) side. Stomata usually open in the light and close in the dark. One theory to explain this states that the necessary change in the water potential of the guard cells in order to cause the opening of the stomata is due to the active transport of ions, in particular potassium ions (K^+), from the surrounding cells. The active pumping of potassium ions causes the water potential of the guard cells to rise. Water is hence drawn in osmotically and the increased turgidity causes the stoma to open. The removal of potassium ions in the dark causes the stoma to close. The ATP required for the active transport of the potassium ions is probably provided by the photosynthetic activity of the chloroplasts in the guard cells. It seems likely that other ions such as chloride ions and/or malate ions are used to maintain the electroneutrality of the guard cells which would otherwise be upset by the movement of the potassium ions.

TRANSPORT OF WATER THROUGH THE PLANT

Transport of water across the leaf

Under normal environmental conditions water vapour will diffuse out from the substomatal air space through open stomatal pores. The water lost from the substomatal air space is replaced by more evaporating from the adjacent spongy mesophyll cells. This draws water from the xylem in the leaf by three routes.

1. **The symplast pathway** As water is lost from the mesophyll cells the water potential of their cytoplasm becomes more negative (lower) compared with adjacent mesophyll cells. Water therefore passes into the first cell from those surrounding it along tiny strands of cytoplasm, called **plasmodesmata,** which link adjacent cells across the cell wall. In this way a water potential gradient is established across the leaf which draws water from the xylem to the substomatal air space (see Fig. 7.22).

2. **The apoplast pathway** Most water travels from cell to cell along the cell wall. As water evaporates from the cell wall into the substomatal air space a tension is set up which pulls water from the xylem along the network of cell walls in the leaf (see Fig. 7.22).

3. **The vacuolar pathway** Some water passes from cell vacuole to cell vacuole of adjacent cells by osmosis. This creates a similar water potential gradient to the one in the symplast pathway which again draws water from the xylem to the substomatal air space (see Fig. 7.22).

Transport of water through the stem

There is much evidence to support the view that water moves up the stem in the xylem:

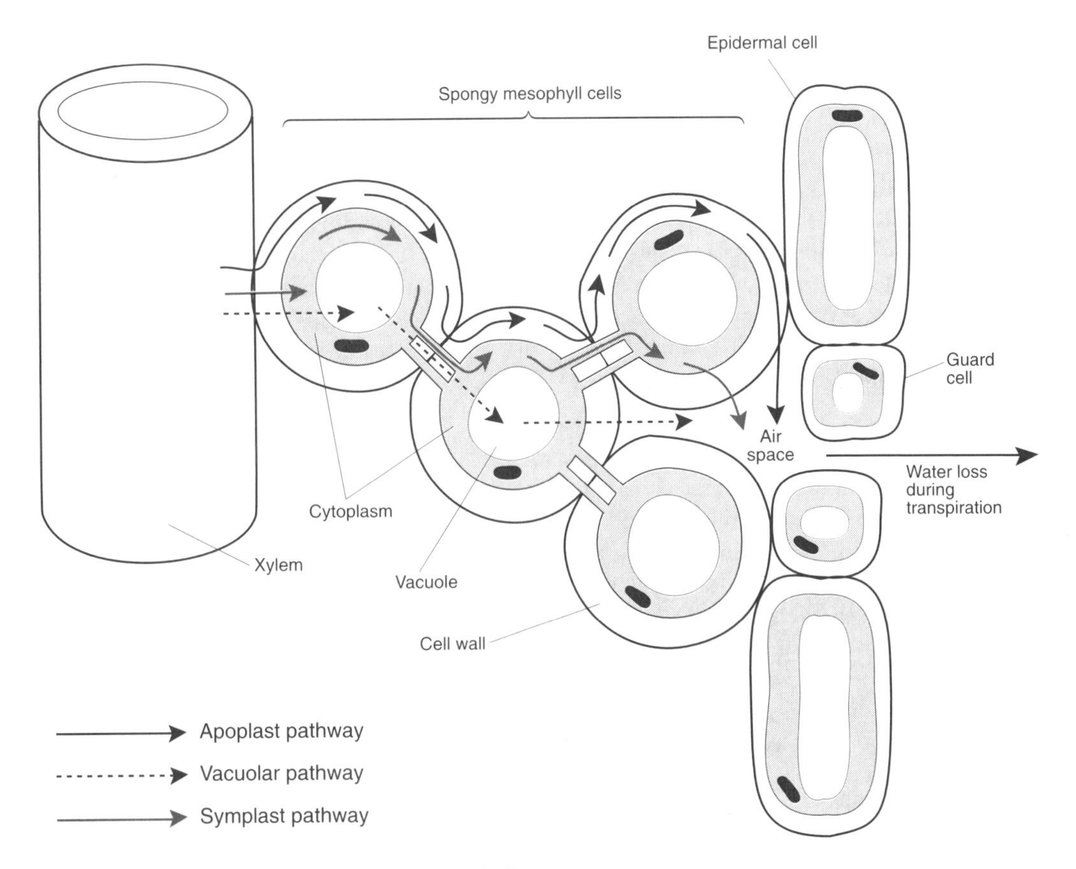

Fig. 7.22 Water transport pathways across the leaf

1. A shoot when cut under water containing a dye takes up the dye. If after a few hours the shoot is cut at various heights up the stem, only the xylem is seen to be stained by the dye.
2. The removal of a ring of tissue from around a woody stem does not impede water flow if only the phloem is cut away. If, however, the outer layers of xylem are removed the leaves wilt.
3. Metabolic poisons do not prevent water movement up the stem. This suggests a passive rather than energy-consuming process is involved in water transport. As xylem vessels are dead, these seem the likely site of a passive process.
4. Artificial blocking of xylem causes plants to wilt.

The generally recognised theory for the movement of water in the xylem is the **cohesion–tension theory**. Water leaving the leaf creates a pull, known as the **transpiration pull**, which draws water up through the xylem. The large cohesive forces between water molecules allow water to be pulled up in a long continuous column. While other forces such as capillarity and root pressure may contribute, it is the transpiration pull which is the major force in drawing water up large trees.

The structure of xylem tissue is discussed in 1.3 and is suited to its function of water transport in many ways:

1. The tissue forms a continuous column – there are no end walls to impede water flow.
2. Xylem comprises long, hollow cells joined end to end.
3. The walls are impregnated with lignin which prevents water escaping except at designated points where there are pits.
4. The lignin in the wall strengthens the tissue and so prevents it collapsing under the tension which the transpiration pull creates.

Water uptake by roots

Available soil water is absorbed from the soil spaces between soil particles and the water film surrounding them. Absorption takes place largely in the younger parts of a root which possess extensions of the epidermal cells known as **root hairs**. Collectively the root hair region is called the **piliferous layer**. The root hairs only remain functional for a few weeks after which the epidermal layer becomes replaced by a layer called the **exodermis**, through which some water absorption still takes place. The younger regions of the roots continue to form new root hairs. The root hairs contain cell sap with a more negative water potential than the available soil water, thereby allowing water to move into them by osmosis. Once into the epidermal cells, water moves across the root cortex by the three routes described earlier, namely the symplast, apoplast and vacuolar pathways.

Movement into the xylem

The vascular cylinder is separated from the cortex by a highly specialised cylinder of cells, **the endodermis**. Each endodermal cell, as well as having a cellulose cell wall, has a **Casparian strip** of suberin which prevents water passing across its radial and horizontal walls. This means that all water passing from the cortex to the xylem must pass through the cytoplasm of an endodermal cell. As the cell ages heavy thickening covers the Casparian strip but some cells remain in the primary condition and are known as passage cells. When a stem is cut just above soil level the cut end exudes water for some time. This is evidence for water being 'pushed' into the stem from the roots by root pressure. Such movement of water from the cortex into the xylem is thought to be an active (energy-requiring) process since the application of a metabolic poison prevents it. The energy for the process may come from the oxidation of starch in the endodermis.

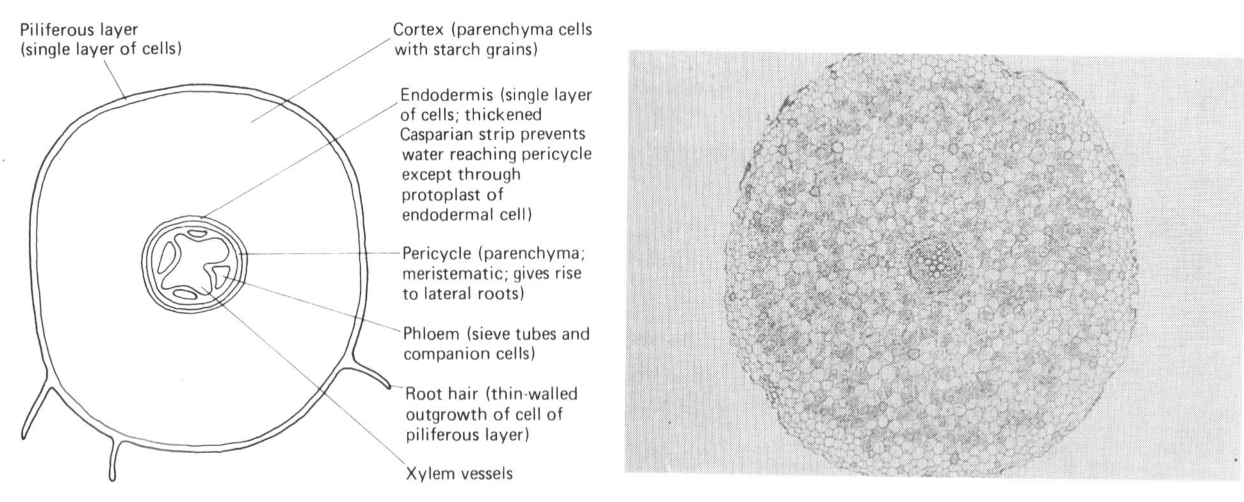

Fig. 7.23(a) TS root of a dicotyledon

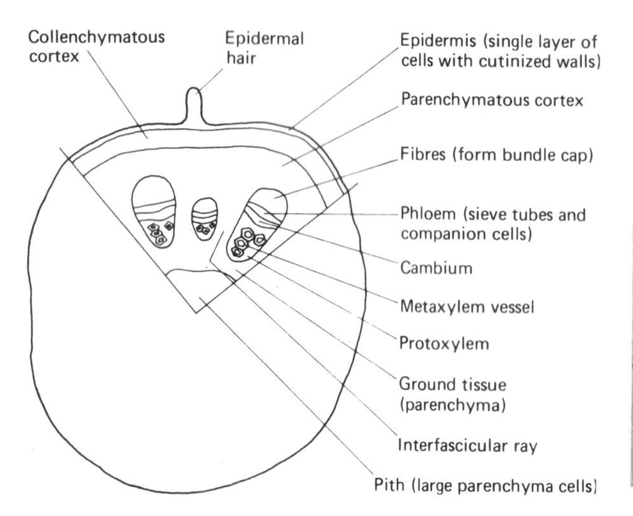

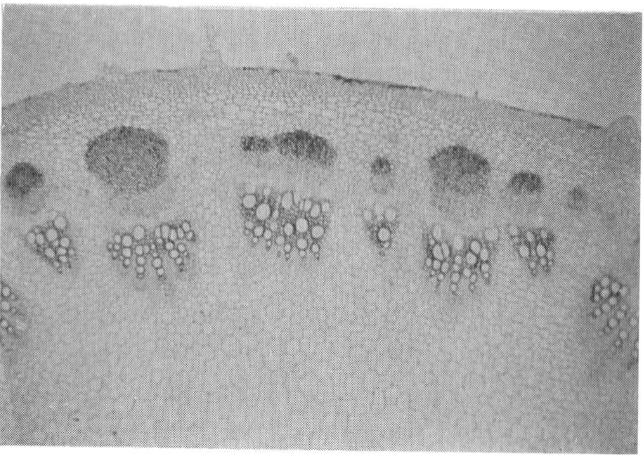

Fig. 7.23(b) TS stem of a dicotyledon

Root pressure

We have seen that the more negative water potential of the epidermal cells causes water to move into them from the soil by osmosis. This water entering the root creates a force known as **root pressure**. This makes a considerable contribution to the movement of water up herbaceous plants, but its effects are less significant in tall woody plants where transpiration pull remains the primary water-drawing force.

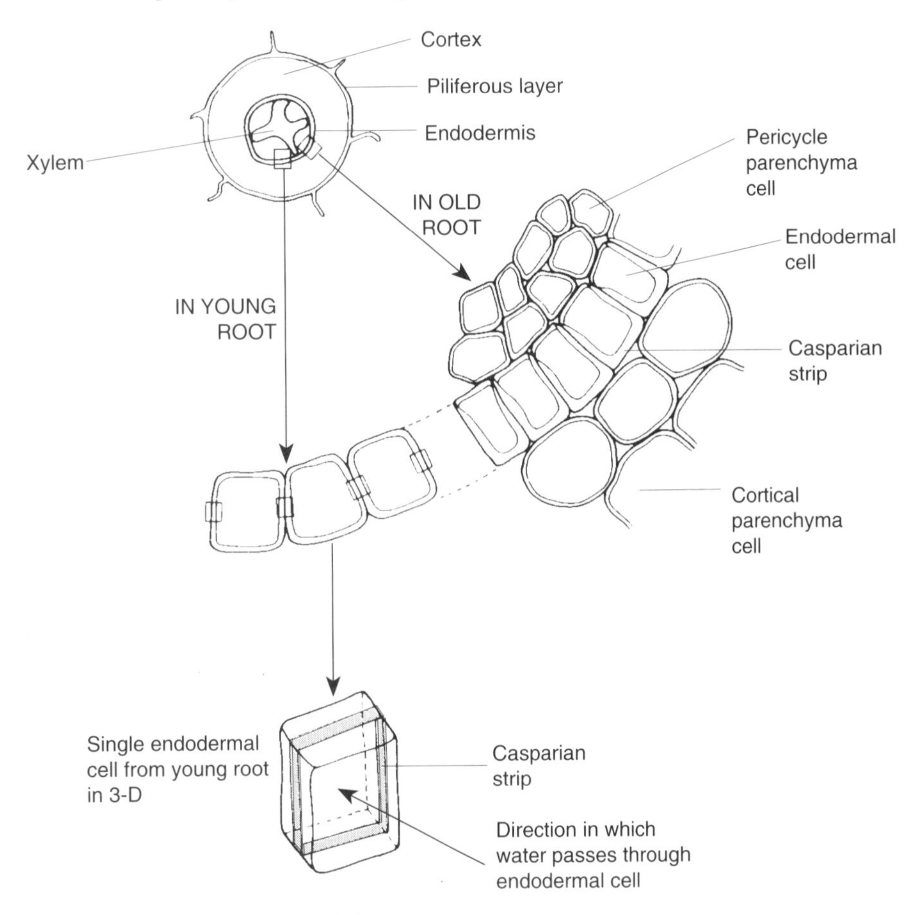

Fig. 7.24 Water transport across the endodermis

UPTAKE AND TRANSPORT OF MINERAL SALTS

Like water, mineral salts (see Table 7.8) are absorbed through the root hairs and other young parts of the root. The application of metabolic poisons prevents their uptake, which indicates that the process is active. There is also evidence that salts are taken up selectively (proportions of minerals inside and outside the plant are different) and against the concentration gradient. The mechanism and route of movement across the cortex to the xylem are not certain. There are a number of theories relating to the mechanism of transport across membranes; most postulate the existence of a carrier molecule, possibly a protein.

Once inside the xylem the ions travel upwards in the transpiration stream. Although xylem is principally responsible, mineral transport also occurs in the phloem and lateral movement from xylem to phloem is possible. When the mineral ions reach the leaves and meristematic areas of the plant they diffuse out of the xylem vessels and are absorbed by the cells for various metabolic functions, e.g. building up proteins and amino acids.

TRANSLOCATION OF ORGANIC MATERIALS

Evidence for movement in the phloem

1. There are diurnal variations of sucrose concentration in the leaf and, after a time lag, these are reflected in the phloem sieve tubes.

Table 7.8 *Mineral elements necessary for plant growth*

	Mineral	Form in which element is absorbed	Function	Effect of deficiency
Macronutrient	Nitrogen	NO_3^-	Component of amino acids, proteins, nucleotides, chlorophyll, some plant hormones	Chlorosis (yellowing of leaves); stunted growth
	Phosphorus	$H_2PO_4^-$	Component of proteins; needed for ATP, nucleic acids and phosphorylation of sugar	Stunted growth; dull, dark green leaves
	Potassium	K^+	Component of enzymes and amino acids; needed for protein synthesis and in cell membranes	Yellow-edged leaves; premature death
	Calcium	Ca^{2+}	Calcium pectate in middle lamella of cell walls; aids translocation of carbohydrates and amino acids; affects permeability of cell membanes	Death of growing points, therefore stunted roots and shoots
	Magnesium	Mg^{2+}	Constituent of chlorophyll; activator of some enzymes	Chlorosis
	Sulphur	SO_4^{2-}	Constituent of some proteins	Chlorosis; poor root development
	Iron	Fe^{2+}	Constituent of cytochromes; needed for synthesis of chlorophyll	Chlorosis
Micronutrient	Boron	BO_3^{3+} or B_4O^{2+}	Aids germination of pollen grains and uptake of Ca^{2+} by roots	Diseases such as heart rot of celery, internal cork of apples
	Zinc	Zn^{2+}	Activator of some enzymes; essential for leaf formation; needed for synthesis of IAA	Malformation of leaves
	Copper	Cu^{2+}	Constituent of some enzyme systems	Growth abnormalities, e.g. dieback of shoots
	Molybdenum	Mo^{4+} or Mo^{5+}	Affects nitrogen reduction through enzyme system	Reduced yield of crop
	Chlorine	Cl^-	In osmosis and important in anion–cation balance of cells	Difficult to demonstrate
	Manganese	Mn^{2+}	Enzyme activator	Type of chlorosis; leaves mottled with grey patches

❷ If a stem is 'ringed' so that phloem is removed, sugars accumulate above the ring. Dyes, however, move normally.

❸ If a plant is given $^{14}CO_2$ and ringed to remove the phloem, the sucrose accumulating above the cut contains ^{14}C.

❹ Aphids use needle-like mouthparts to obtain sugars from phloem sieve tubes. Removal of the aphid, leaving its mouthparts inserted like a pipette in the phloem, allows analysis of the sieve tube contents. Sieve tubes are found to contain a solution of amino acids and sucrose, the latter showing the expected diurnal variations in concentration (see point 1).

❺ The transport in the phloem is normally downwards. However, by darkening or removing the upper leaves, the sugars can be shown, by ringing, to move upwards in the phloem.

Contents of sieve tubes

It has been found using the aphid technique that sieve tubes contain a solution containing

1. up to 30% by weight sucrose
2. up to 1% by weight amino acids
3. traces of
   (a) sugar alcohols
   (b) ionic phosphate and potassium
   (c) hormones
   (d) viruses

Theories of phloem transport

Diffusion is too slow to account for the observed rates of transport; therefore, a number of theories have been proposed:

Mass flow hypothesis (Munch, 1930) Material travels from a region of high concentration (e.g. photosynthesising leaf chloroplast) to a region of low concentration (e.g. storage plastids in a root). In the region of high concentration water is taken up by osmosis and there is a difference in hydrostatic pressure between this region and the storage area. This causes mass flow within the sieve tube lumina. The theory ignores the membrane barrier between the sieve tube and the plastid and assumes an empty sieve tube lumen and fully open sieve plate pores.

Transcellular strands (Thaine) Thaine proposes that protein fibrils surrounding endoplasmic reticulum tubules pass from one end of the sieve tube to the other and has suggested that solutes pass along these fibrils due to the peristaltic action of the protein sheath, in a manner resembling cytoplasmic streaming. This is an active process. Although the existence of these transcellular strands is not above doubt, this is the only theory which accounts for transport of solutes in both directions in one sieve tube.

Electro-osmosis (Spanner) Spanner suggests that the flow of nutrients is produced and maintained by electro-osmotic forces set up across the sieve plates. In particular, this theory proposes that potassium ions are actively transported across sieve plates, carrying with them water and dissolved solute.

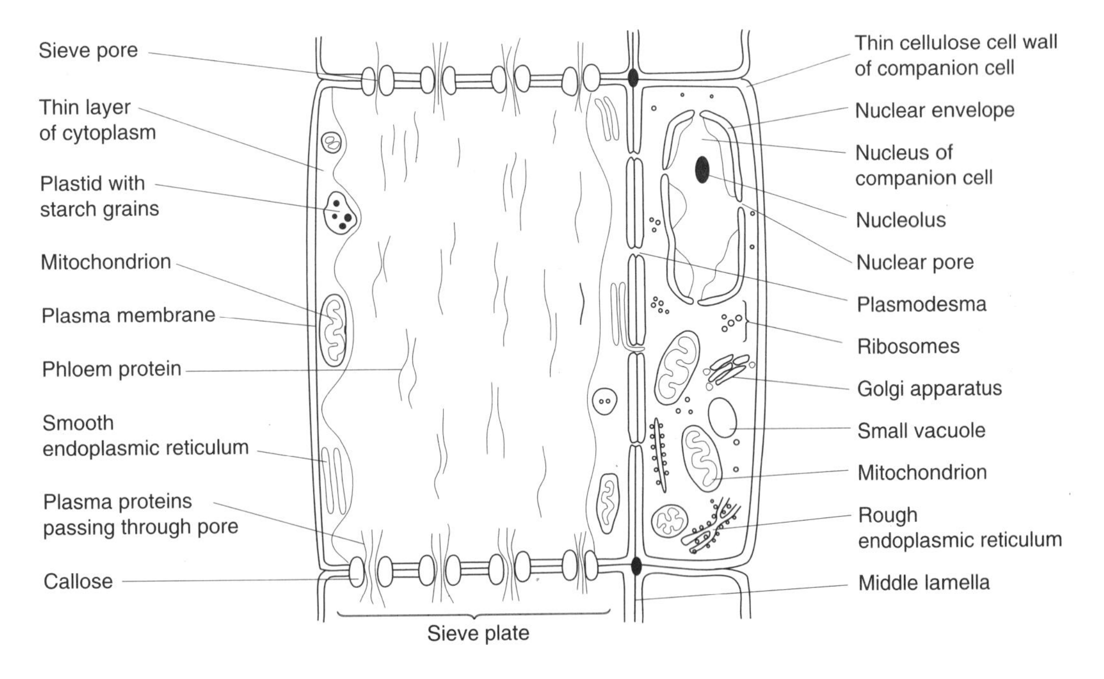

Fig. 7.25 LS of a phloem sieve tube element and companion cell

7.4 OSMOREGULATION AND EXCRETION

You may have noticed how much more you drink on a hot day, even when relatively inactive. This is because terrestrial organisms lose water by evaporation. The hotter, drier and windier the climate, the more they lose. It may, however, surprise you to know that some animals living in water also experience dehydration. Where the solution surrounding an organism is hypertonic to (more concentrated than) its body fluids, water will leave by osmosis. This situation can arise in sea water. Freshwater animals have the opposite problem, as water tends to flood into them by osmosis. In all these cases mechanisms have evolved to ensure that any fluctuations in water balance are corrected. This process is known as **osmoregulation**.

During metabolic activities wastes are produced which are toxic and, if allowed to accumulate, poisonous to the organism. The process by which they are removed is called **excretion**. Many are removed by diffusion, as part of some other process, e.g. carbon dioxide diffuses into the lungs during breathing. In animals, however, nitrogenous wastes resulting from the breakdown of excess amino acids pose particular problems. The ammonia produced is especially toxic. If water is readily available it can be diluted sufficiently and removed. Where water needs to be conserved (e.g. in terrestrial organisms and marine vertebrates) nitrogenous wastes need to be more concentrated before being removed. They must be made less toxic by conversion to urea. Where water is particularly scarce or flight makes the storage of watery urine impractical, the ammonia is converted to uric acid which requires almost no water for its removal. Birds and insects excrete uric acid. By virtue of their autotrophic mode of nutrition plants take in the materials they need and no more. They therefore have almost no complex excretory products – only simple diffusible ones. Their high surface area/volume ratio is suited to the removal of these simple substances, e.g. carbon dioxide (dark), oxygen (light). Any complex excretory material can be removed with the leaves when they are discarded.

EXCRETORY PRODUCTS IN PLANTS

Carbon dioxide When the rate of respiration exceeds the rate of photosynthesis (e.g. at night) carbon dioxide will diffuse out of the green parts of plants. In roots carbon dioxide is constantly diffusing out into the air spaces of the soil.

Oxygen When the rate of photosynthesis exceeds the rate of respiration (e.g. in high light intensities) oxygen diffuses out of the photosynthesising parts of plants.

Complex wastes As plants make their requirements according to demand there are few other wastes. Silicates and tannins that are not required may be stored in leaves and fruits and are therefore removed when these are shed from the plant.

EXCRETORY PRODUCTS IN ANIMALS

Carbon dioxide Produced during cellular respiration, this is carried to the respiratory surface where it simply diffuses out.

Excess water and mineral salts These are lost in a controlled manner usually through a specialised excretory and osmoregulatory organ, e.g. the kidney in vertebrates.

Bile pigments These are the breakdown products of haemoglobin. In mammals they are excreted by the liver in the bile juice and eliminated with the faeces.

Nitrogenous waste This is the result of the breakdown of amino acids. Organisms remove it in three main forms:

1. **Ammonia** is highly toxic and never allowed to accumulate in living cells. It is soluble, readily diffusible and excreted in dilute form by freshwater animals and marine invertebrates. Animals excreting ammonia are said to be **ammoniotelic**.

2. **Urea** is less toxic than ammonia but still needs to be diluted for elimination from the body. It is excreted by some terrestrial organisms and marine ones whose body fluids are hypotonic to sea water. These animals are **ureotelic**. Urea is formed from ammonia by the ornithine cycle:

$$\underset{\text{carbon dioxide}}{CO_2} + \underset{\text{ammonia}}{2NH_3} \longrightarrow \underset{\text{urea}}{CO(NH_2)_2} + \underset{\text{water}}{H_2O}$$

3. **Uric acid** is nontoxic. As it is highly insoluble, little water need be lost in its elimination from the body. It is a common excretory product of animals living under arid conditions. Insects and birds excrete uric acid and are said to be **uricotelic**.

Ornithine (urea) cycle

Excess amino acids cannot be stored. By deamination the amine group is removed from them resulting in the formation of ammonia. Ammonia is also toxic and so it is converted through a series of reactions, the ornithine cycle, to urea. The enzyme arginase catalyses the formation of urea from arginine. The purpose of the cycle is to regenerate arginine using excretory ammonia. In mammals this cycle occurs in the liver cells.

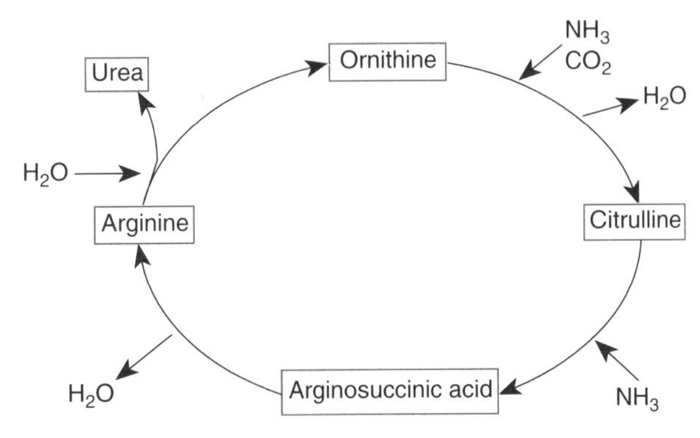

Fig. 7.26 Ornithine (urea) cycle

Osmoregulation in freshwater protozoa

Freshwater protozoa have cell contents which are hypertonic to the medium in which they live. Since the cell membrane is partially permeable water enters by osmosis. To prevent the animal bursting the excess water must be removed. Freshwater protozoa such as *Amoeba* and *Paramecium* have contractile vacuoles which constantly fill and empty at the surface. The water which enters the cytoplasm must move into the contractile vacuole against the concentration gradient. As the process requires energy, there are often aggregations of mitochondria near contractile vacuoles. The contractile vacuoles of freshwater protozoa moved to a medium with a higher salt concentration fill and empty less frequently than in fresh water.

Osmoregulation in fish

Although much of the surface of a freshwater bony fish is impermeable to water there is still a tendency for it to enter across the gills and lining of the buccal cavity and pharynx. The organ of osmoregulation in bony fish is the kidney, which reacts to the influx of water in two main ways:

1. production of a large volume of urine – aided by the many large glomeruli in the kidney;
2. production of very dilute urine – achieved by extensive reabsorption of salts from the renal fluid into the blood stream.

However, a freshwater bony fish still suffers some loss of salts and so this is offset by the active uptake of salts from the environment by special chloride-secreting cells in the gills.

Marine bony fish, which evolved in fresh water and have body fluids slightly hypotonic to sea water, have a tendency to lose water to the environment by osmosis. As in the freshwater bony fish the areas of the body permeable to water are the gills and the lining of the buccal cavity and pharynx. To combat dehydration the kidney:

1. produces a small volume of urine – glomeruli are relatively few and small;
2. produces the nitrogenous excretory product trimethylamine oxide. Freshwater bony fish excrete ammonia since they have plenty of water available for its dilution. Trimethylamine oxide is nontoxic and therefore a more suitable excretory product for marine bony fish since little water is needed for its expulsion.

Marine bony fish also drink sea water; the excess salts ingested are actively removed from the body by the chloride-secreting cells of the gills.

Osmoregulation in a terrestrial insect

All terrestrial organisms are liable to water loss through evaporation. In an insect the loss is reduced in three main ways:

1. Having an impermeable surface – the cuticle is coated with wax. Retention of water can never be complete and there is bound to be some loss of water from respiratory surfaces.
2. Production of uric acid – because this nitrogenous waste product is nontoxic and insoluble, it can be eliminated from the body as a semisolid without the loss of water.
3. Reabsorption of water by the Malpighian tubules and rectal gland. The Malpighian tubules are a bunch of blind-ending tubules at the junction of the mid-gut and rectum. Potassium urate is produced by insect tissues and this enters the Malpighian tubules with water and carbon dioxide. In the tubules they form uric acid and potassium hydrogencarbonate. The potassium hydrogencarbonate is reabsorbed into the haemolymph. Reabsorption of water from the uric acid occurs in the Malpighian tubules and the rectal gland so that the urine eliminated is in a very concentrated form.

Mammalian kidney

The basic unit of nitrogenous excretion in a mammal is the kidney tubule, or nephron. Each human kidney contains over one million nephrons and filters about 120 cm^3 min^{-1}. Over the length of the nephron reabsorption from the glomerular filtrate takes place and results in the formation of urine which is hypertonic to the blood.

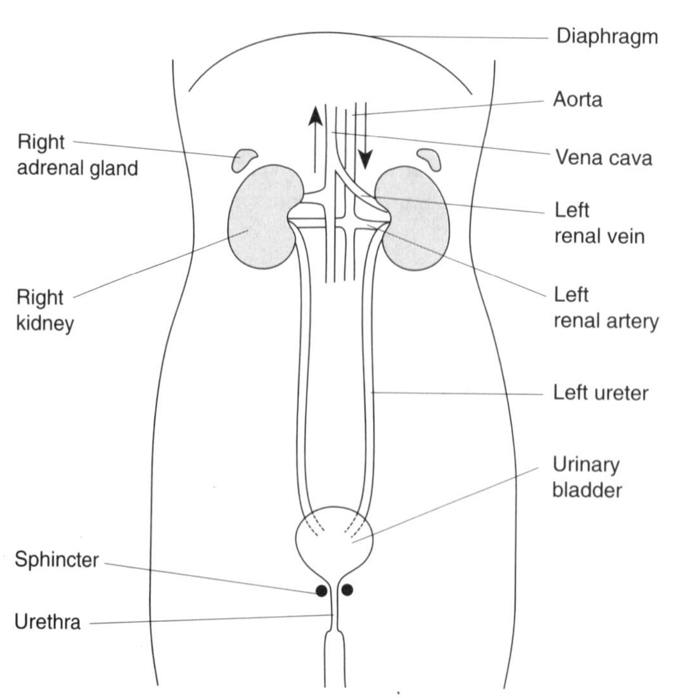

Fig. 7.27 Position of the kidneys in humans

Functions of the mammalian kidney These are as follows:

1. excretion of metabolic wastes, especially urea;
2. osmoregulation;
3. maintaining the acid-base balance of the body, which is mainly brought about by the excretion or retention of H^+ and HCO_3^-; the pH of the urine can vary between 4.5 and 8.0;
4. maintaining the balance of ions, e.g. Na^+, K^+, Ca^{2+}, Mg^{2+}, H^+, Cl^-, HCO_3^-.

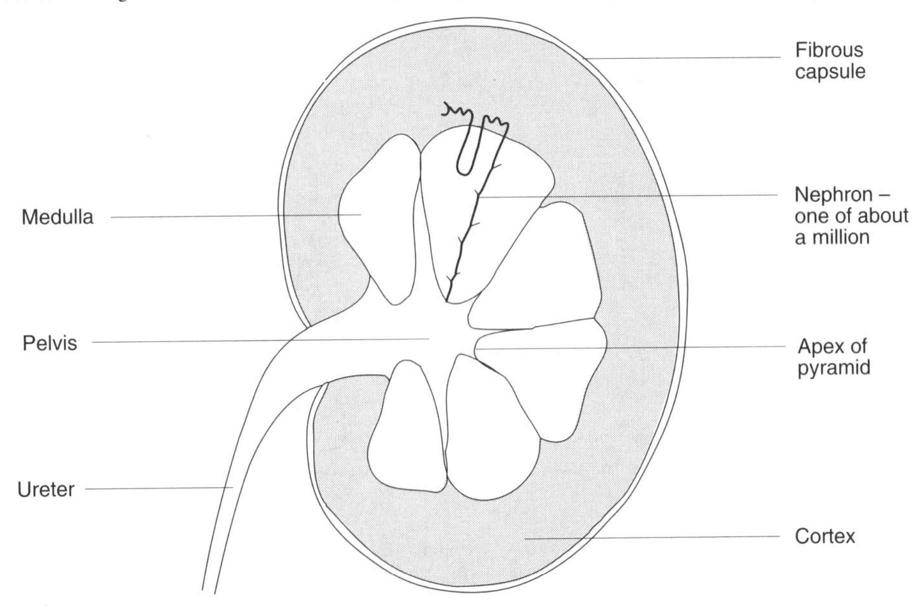

Fig. 7.28 LS mammalian kidney showing position of a nephron

Formation of hypertonic urine The loop of Henle, found in birds and mammals, constitutes a **counter-current multiplier**, the active mechanism being the transport of sodium ions from the ascending to the descending limb. This movement of sodium ions results in a very high concentration developing in and around the apex of the loop of Henle, deep in the medulla. In this region the thin-walled collecting ducts open into the renal pelvis. The high concentration of sodium in the tissues causes water to be drawn out of the collecting ducts by osmosis. Consequently the renal fluid, which has been isotonic or hypotonic to the blood through the length of the nephron, becomes hypertonic in the collecting ducts. It has been estimated that 99% of the fluid filtered by the Bowman's capsule is reabsorbed by the nephron. The length of the loop of Henle is proportional to the concentration of urine, e.g. desert mammals, such as the kangaroo rat, which produce very hypertonic urine, have very long loops of Henle.

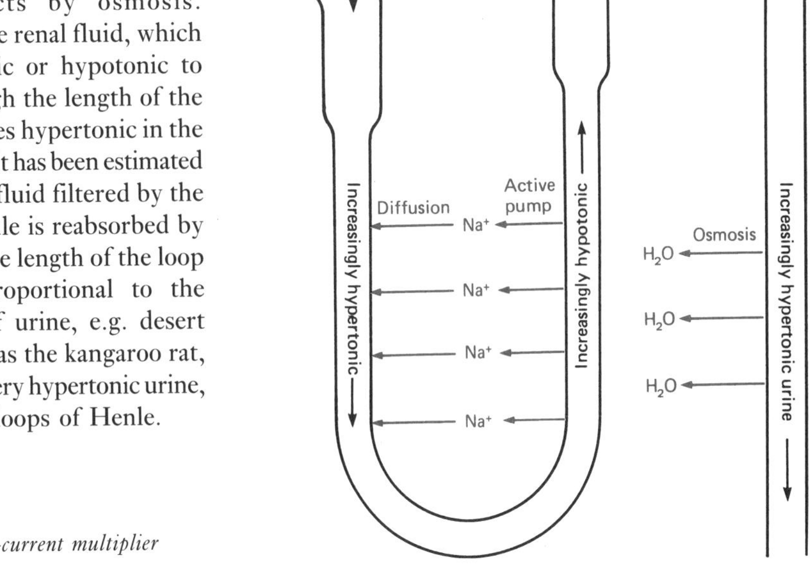

Fig. 7.30 Counter-current multiplier

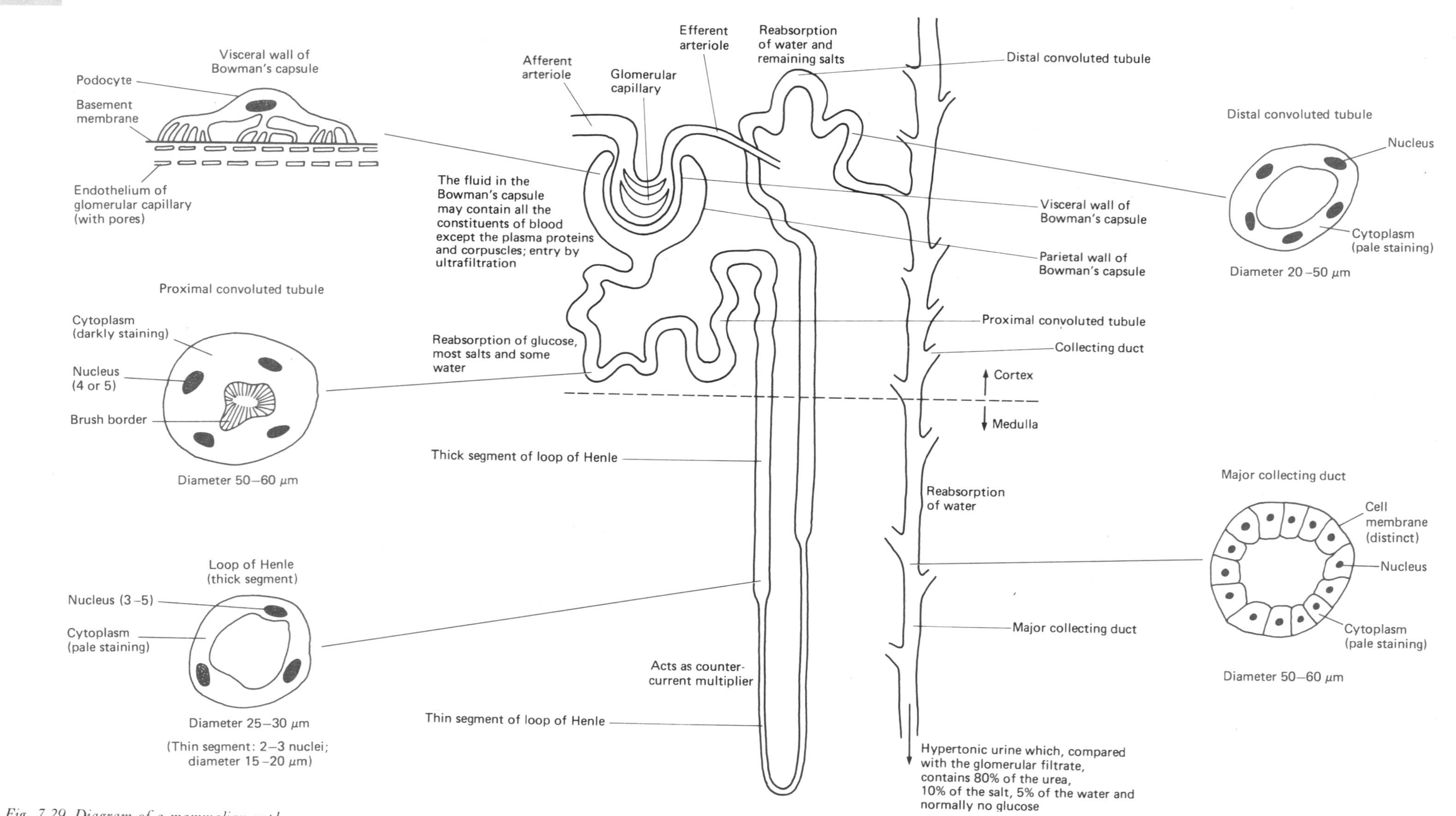

Fig. 7.29 Diagram of a mammalian nephron

Control of water and salt balance in mammals

Antidiuretic hormone (ADH) The permeability of the collecting ducts to water is affected by the hormone ADH, secreted by the posterior lobe of the pituitary. The presence of ADH increases the permeability of the membrane and results in the production of more hypertonic urine. If no ADH is present the membrane permeability is lower and more dilute urine is produced. The production of ADH is controlled by the hypothalamus whose osmoreceptors monitor the solute concentration of the blood. If the solute concentration of the blood is high, ADH secretion is stimulated, the permeability of the membranes of the collecting ducts increases, water is reabsorbed into the blood and very hypertonic urine is produced. ADH secretion may also be triggered, via the hypothalamus, by blood volume receptors in the walls of the heart, the aorta and the carotid arteries.

Aldosterone This hormone produced by the adrenal cortex stimulates the reabsorption of sodium from the kidney. When the sodium concentration in the kidney tubule decreases, the kidney produces the enzyme renin which catalyses the conversion of the blood protein angiotensinogen into angiotensin. Angiotensin stimulates the adrenal cortex to produce more aldosterone.

Chapter roundup

In this chapter we have begun to piece together some of the many and diverse activities which are carried out by organisms and to look at how they may be linked into a unified system. You should by now appreciate that a small, simple organism can carry out many functions without the need for specialised transport systems to link its various parts – passive processes such as diffusion are adequate to allow the exchange of materials. As organisms increase in size and complexity, however, the range of substances to be moved, and the larger distances over which they must be transported, necessitate the development of a specialised transport system.

We have also reviewed the different substances which need to be exchanged and the development of specialised tissues or cells dedicated to transporting a particular material. We have looked at how wastes are removed and how a favourable water balance is maintained. It is not, however, just water which must be maintained at a constant level; other substances such as sugars, amino acids and mineral salts must be kept in strict balance, as indeed must factors such as pH and electroneutrality. We will, in the next chapter, investigate how constancy is achieved through what biologists call homeostasis.

We will also complete our review of how organisms unify their various activities by studying how coordination is achieved by both nervous and chemical control. This will lead us onto the development of complex sense organs such as the eye and ear and to the development of the brain as the central coordinating organ. We will then survey the ways in which this complex coordinating system can lead to varied patterns of behaviour. In conclusion, we will consider a practical example of how many different systems can operate efficiently together by looking at movement in animals.

Illustrative questions and worked answers

1 The lung volume of an adult human was measured by a spirometer during a variety of breathing exercises and the following trace obtained.

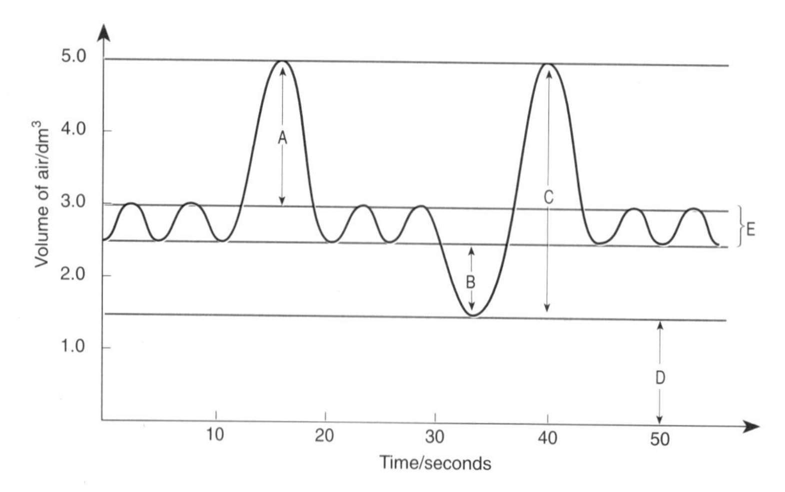

(a) By means of an appropriate letter state which volume represents:
(i) inspiratory reserve volume, (ii) expiratory reserve volume, (iii) tidal volume, (iv) residual volume, (v) vital capacity. (5)

(b) In another experiment a subject was given various gaseous mixtures to breathe and the rate of breathing was measured.
The results are shown graphically below.

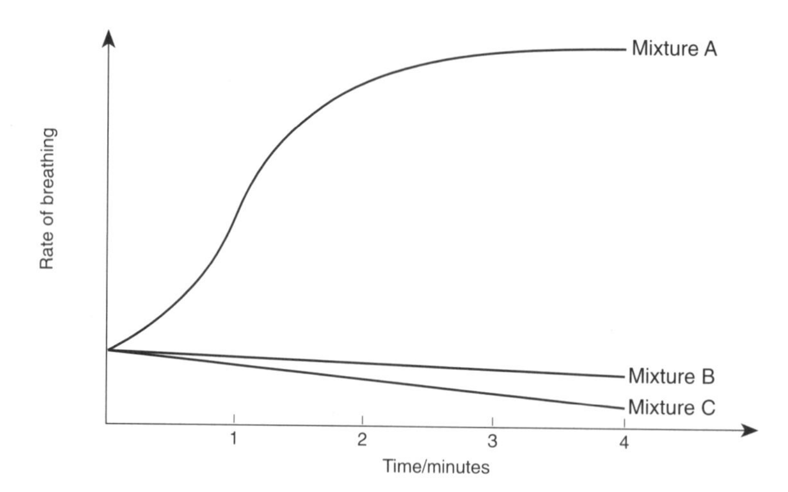

Mixture A = 90% oxygen + 10% carbon dioxide
Mixture B = normal atmospheric air
Mixture C = 100% oxygen

(i) Explain what the data show concerning how the rate of breathing is controlled. (5)

(ii) In the light of the information provided by the graph show why mouth-to-mouth resuscitation is a better means of artificial respiration than pressing on the chest wall. (2)

(c) Why is it more dangerous to rebreathe expired air if it is passed through soda lime? (3)

Time allowed 25 minutes

Tutorial note

Part (a) of this question is straightforward recall and should present no problems if you have revised adequately. In part (b), however, you will need to apply, as well as recall, your knowledge and this is never as easy. It is the phrase 'concerning how the rate of breathing is controlled' in (b)(i) which is important – you must do more than just explain the data.

Normal atmospheric air is 21% oxygen and 0.04% carbon dioxide; therefore mixture A contains more carbon dioxide than mixture B, which in turn contains a little more than mixture C. The graph shows a similar relationship between the three mixtures in terms of the rate of breathing. This suggests that the carbon dioxide concentration of inspired air is the factor controlling breathing rate. You should indicate to the examiner that it cannot be the oxygen concentration that controls breathing since when this rises from 21% (mixture B) to 90% (mixture A) the breathing rate increases, but a further rise to 100% (mixture C) causes a fall. Therefore any relationship between breathing rate and oxygen concentration is not a straightforward one.

You will need to recognise in b(ii) that expired air contains around 4% carbon dioxide (some 100 times more than normal atmospheric air) and link this with your earlier deduction that the higher the carbon dioxide content of inspired air, the faster is the rate of breathing.

Similarly in part (c) you will need to know that soda lime absorbs carbon dioxide and then link this knowledge to the graphical data showing that where no carbon dioxide is present (mixture C) breathing all but ceases.

Suggested answer

(a) (i) A (ii) B (iii) E (iv) D (v) C

(b) (i) Normal atmospheric air contains 21% oxygen and 0.04% carbon dioxide. When the carbon dioxide concentration is increased to 10% (mixture A) the breathing rate increases and when it is decreased to 0.0% (mixture C) the rate decreases. This suggests that it is the concentration of carbon dioxide in inspired air that controls breathing rate. It is not the oxygen concentration in inspired air since when this is higher than normal at 90% the breathing rate increases, but when it is even higher at 100% the rate decreases. There does not therefore seem to be a straightforward relationship between the oxygen concentration in inspired air and breathing rate.

(ii) During mouth-to-mouth resuscitation expired air contains about 4% carbon dioxide and this stimulates an increase in the patient's respiratory rate and aids recovery. Pressing and releasing the chest wall will cause atmospheric air with only 0.04% carbon dioxide to enter the patient's lungs. With its lower carbon dioxide level this air is not as effective in stimulating the patient's own respiratory rate and recovery is therefore slower.

(c) If expired air is rebreathed its oxygen content decreases and the carbon dioxide content progressively increases. The breathing rate is therefore increased due to the rise in the carbon dioxide concentration of the air. This faster breathing rate, to some extent, compensates for the lowering of the oxygen concentration and also acts as a warning to the person because the faster breathing rate causes distress. If the expired air is passed through soda lime, the carbon dioxide is completely absorbed and when rebreathed this air no longer stimulates faster breathing. Thus, although the oxygen concentration of the air continues to fall, there is neither a compensatory increased breathing rate nor a warning of the danger and so unconsciousness and death may follow.

2 The oxygen dissociation curves of human haemoglobin for a normal person at rest at 37 °C and for a human fetus are given overleaf.

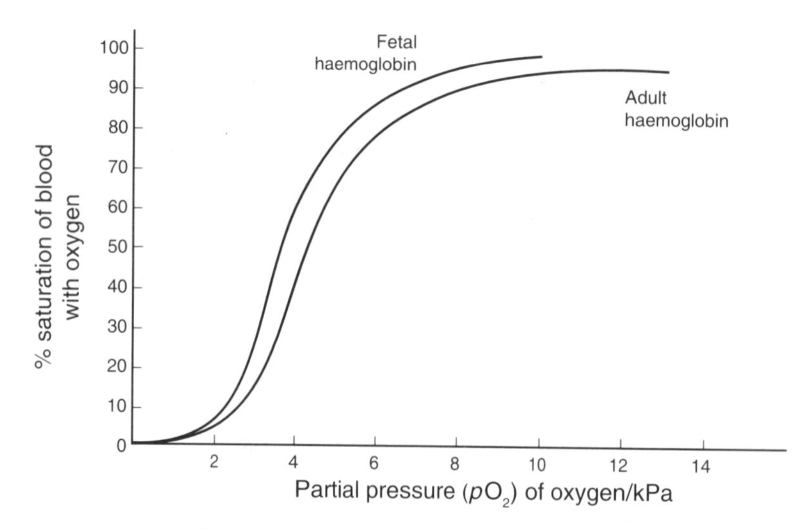

(a) State the % saturation of adult blood with oxygen when pO_2 is:
(i) 4 kPa (ii) 7 kPa (2)

(b) Using the letters X, Y and Z as indicated, mark on the graph for adult haemoglobin the positions of the oxygen tensions you would expect for the blood in: (i) the pulmonary vein (X), (ii) the pulmonary artery (Y), (iii) the femoral artery (Z). (3)

(c) How would the dissociation curve differ if the blood had a high concentration of carbon dioxide? (1)

(d) What is the significance of this difference? (2)

(e) Explain why the oxygen dissociation curve for fetal haemoglobin is to the left of that for adult haemoglobin. (2)

(f) Where would the dissociation curve for myoglobin be in relation to that for adult haemoglobin? Explain your answer. (3)

(g) Where would the dissociation curve for the haemoglobin of a mud-dwelling organism such as *Arenicola* be in relation to that of an adult human? Explain your answer. (3)

Time allowed 25 minutes

Tutorial note

You will require both the ability to use graphical information and an understanding of the principles underlying oxygen dissociation curves in order to answer this question well. Rather than having to rely on learned information about the oxygen dissociation curves of many organisms in many situations, you would be well advised to simply remember the following principle. The more the curve is shifted to the left, the more easily the blood takes up oxygen, but the less easily it releases it. The reverse applies when the curve is shifted to the right. This principle should enable you to tackle parts (e), (f) and (g) provided you think about the other biological principles involved.

Part (a) simply involves reading figures from the graph; always use a ruler or, better still, a set square to ensure accurate readings.

In part (b) you will need to ask yourself about the oxygen content of the blood in the three vessels listed. Where the blood is highly oxygenated the saturation level will be high and where deoxygenated it will be lower, but remember that blood is rarely, if ever, below 50% saturation. A possible careless error you could make would be to mark the points on the 'fetal', and not the 'adult', graph. Do not let the term 'femoral artery' put you off (it in fact serves the legs) as its name is unimportant in this context because any artery, other than the pulmonary artery, should carry highly oxygenated blood.

You will need to have an understanding of the Bohr effect to enable you to answer parts (c) and (d).

Suggested answer

(a) (i) 40% (ii) 85% (Answers determined as shown on graph).

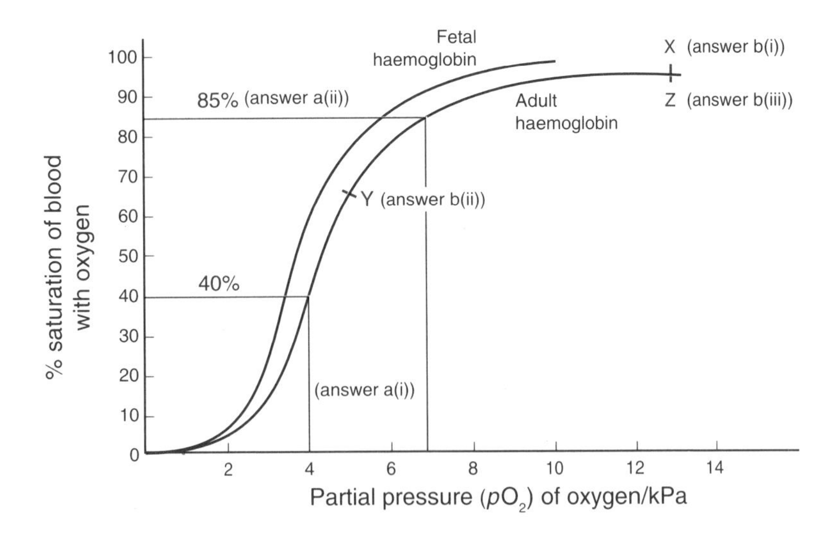

(b) (i) The answer is marked X. The pulmonary vein carries oxygenated blood from the lungs where pO_2 is about 13 kPa and the percentage saturation of haemoglobin is 95%. Since no oxygen is lost on its journey from the lungs to the pulmonary vein the oxygen tensions will be the same in both.

(ii) The answer is marked Y. The pulmonary artery carries blood that has returned from the tissues, via the heart, to the lungs. This blood is deoxygenated and pO_2 is about 5 kPa giving an oxygen saturation of around 66%.

(iii) The answer is marked Z. There should be no loss of oxygen between the lungs and the femoral artery as the blood does not pass through any tissue that absorbs O_2 from it during its journey. The oxygen tension should therefore be the same as that in the lungs (13 kPa).

(c) It would be of a similar shape but displaced to the right of the adult haemoglobin curve.

(d) The displacement to the right means that haemoglobin has a lower affinity for oxygen in the presence of high carbon dioxide concentrations. Since such concentrations occur in respiring tissues haemoglobin readily gives up its oxygen at these tissues. The faster the tissue respires, the faster the haemoglobin gives up its oxygen, which is what is needed by a rapidly respiring tissue.

(e) Being to the left of the adult oxygen dissociation curve the haemoglobin of a fetus has a higher affinity for oxygen than the haemoglobin of its mother. This means oxygen will readily pass from maternal to fetal blood at the placenta.

(f) It would be displaced a long way to the left of the oxygen dissociation curve for adult haemoglobin, which means it has a much higher affinity for oxygen. This is because myoglobin is found in muscle tissue where it can store oxygen for immediate use, and its higher affinity for oxygen allows it to readily absorb oxygen from the haemoglobin in the blood.

(g) It would be displaced to the left. In mud the organism lives in an environment of very low oxygen tension because oxygen is readily used by organisms breaking down organic material in the mud and there is little circulation of water to bring in more oxygen. If the organism did not have haemoglobin with the ability to absorb oxygen at very low oxygen tensions it would be unable to continue respiration aerobically.

3 An experiment to determine which tissue conducts sugar through the stem was set up using the apparatus below.

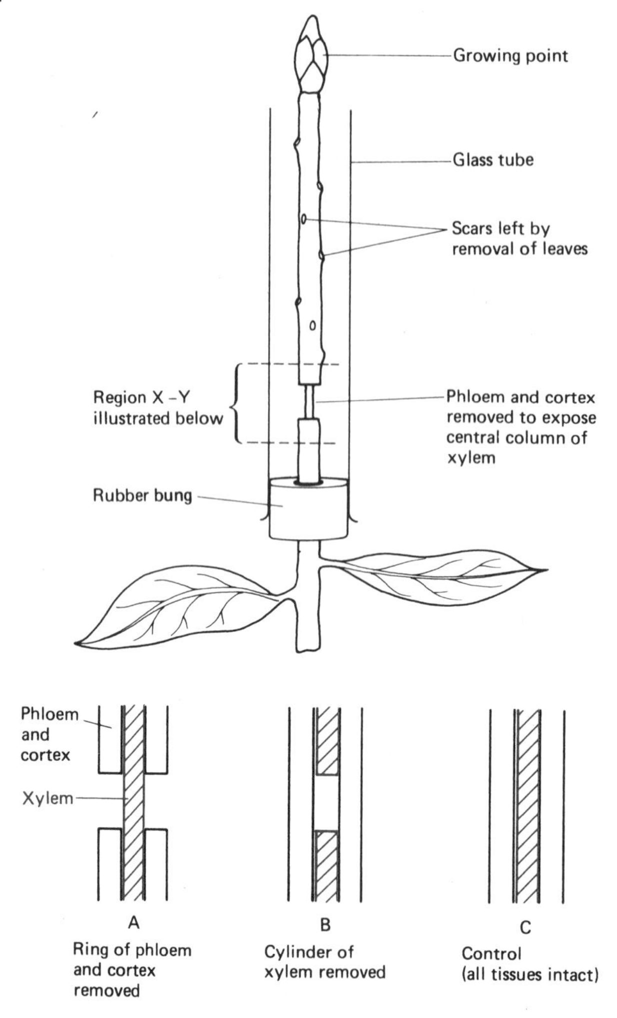

The leaves were removed in the upper region of the plant, as shown. Three sets of apparatus were set up as follows:

A A ring of all tissues outside the xylem was removed at the base of the defoliated region.

B A cylinder of xylem was removed from the stem at the base of the defoliated region. The other tissues were left intact except for a small cut through which the xylem was removed.

C All tissues were left intact (control).

The cylinders were filled with fresh distilled water each day. The results obtained after a week are shown in the table below.

	A Phloem and cortex removed	B Xylem removed	C Control
Increase in stem length in mm	5.3	51.6	65.9
Total sugar content of stem above region X–Y in mg	0.06	6.43	3.36

(a) Why were the following procedures carried out?

(i) The leaves are removed from the upper part of the stem. (1)

(ii) The stems are enclosed in a glass tube containing distilled water. (1)

(b) Explain the difference in the results between
   (i) A and B (2)
   (ii) B and C (2)

(c) Why is an increase in the length of the stem not a particularly good measure of translocation? (2)

Time allowed 15 minutes

Tutorial note

This question requires you to understand an experimental procedure and to be able to interpret results and evaluate their accuracy. As with all questions of this type you must avoid the tendency to attempt answers before having fully appreciated the detail of the experiment. You should therefore spend a little time trying to understand what the experiment is about, how the results are obtained and what they mean.

You must provide detailed answers of A-level quality. In a(i), for example, the removal of the leaves was for a rather more scientific reason than 'to allow the stem to be fitted into the glass tube'! Another common failing is for candidates to write confusing, often muddled responses. You can usually avoid this by careful planning and by not attempting too much at once. This one step at a time approach is illustrated in part (b) where if (i) and (ii) are treated separately you are far less likely to muddle yourself, and the examiner, than if you attempt them together.

Suggested answer

(a) (i) The experiment is designed to indicate which tissue transports sugars. To do this the sugars must be made to move in one direction, in this case upwards, over the experimental region X–Y. If the leaves are removed any sugars present above this region could only have been translocated from below. If the leaves were not removed it would be impossible to determine whether sugars above X–Y were translocated from below or manufactured in the leaves above the region.
(ii) To prevent drying out of the now exposed internal tissues.

(b) (i) In A both the extent of the growth and the amount of the sugar above region X–Y are considerably smaller than in B. This indicates that the phloem and cortex carry much more of the sugar than does the xylem, which transports very little sugar, if any.
(ii) In comparing B and C the growth is greater in the control (C), which shows that the xylem must carry some material necessary for growth. However, the sugar content is less in C than in B. It is unlikely that there is reduced translocation in C because the phloem and cortex are intact. The reduced level of sugar must therefore be due to it being used for the extra growth that occurs in C.

(c) Although sugars are essential for growth in plants, so are many other substances. A lack of growth when the phloem supply is cut only indicates that some factor is not being supplied. Whether or not this missing factor is sugar can only be deduced from consideration of the second set of data obtained.

4 What is the importance of osmotic control in animals? Describe the methods by which the salt and water content of the body is regulated in mammals.

Time allowed 40 minutes

Tutorial note

This is an essay question and therefore demands that you plan your answer carefully. One method might be to 'brainstorm' ideas – write down each word/phrase which occurs to you on the topic. Then try to organise them into some logical form, rejecting those which on further consideration are not relevant. Another method is to think of a number of headings which seem to you to cover the main areas of the question and then try to add further detail under each one. Whatever method you use, you should after about 5 minutes have a clear idea of the content of your essay, the order in which you intend to approach the different aspects and a logical means of connecting together the different parts. A short introduction and conclusion often bring appropriate 'style' marks where these are available.

In discussing the importance of osmotic control in animals you should restrict yourself to general points, although it is always good practice to illustrate your answer with examples which support your case. What you must avoid is describing in detail how osmotic control is achieved in specific animals – 'importance of' are key words.

The regulation of water and salt content in mammals is achieved largely by the kidneys. Your answer should therefore include details of kidney function. A diagram of the nephron is essential and, if appropriately labelled and annotated (notes added to the labels), it can save much time otherwise spent on lengthy descriptions. Take care, however, to avoid the very common mistake of repeating in words all the information displayed on the diagram. To save time you should not shade, colour or in other ways embellish your drawing – keep it neat, clear, accurate and above all well labelled.

The word 'regulated' is the important one and thus, as well as the kidney, the controlling functions of hormones should be emphasised.

Suggested answer

Importance of osmotic control:

1. maintenance of constant body fluid composition in order that metabolic activities can take place properly;
2. the regulation of water and solute concentration essential to the proper functioning of animals;
3. increased environmental independence because animals can survive in places where the water and mineral salt concentration differs from that of their bodies.

Regulation:

1. Kidney functioning – include ultrafiltration, reabsorption of water by osmosis, sodium pump, counter-current multiplier, reabsorption of sodium and chloride ions; role of the proximal and distal convoluted tubules, loop of Henle and collecting duct.
2. Hormonal control – role of the hypothalamus (contains osmoreceptors) and pituitary in the production of the antidiuretic hormone and in its function of altering kidney tubule permeability; role of the adrenal cortex in producing aldosterone and its function in increasing active uptake of sodium ions from the glomerular filtrate; role of renin and angiotensin in stimulating the release of aldosterone.

Question bank

1 Below is a diagram of a simple respirometer, which is used to measure the volume of oxygen taken up by organisms.

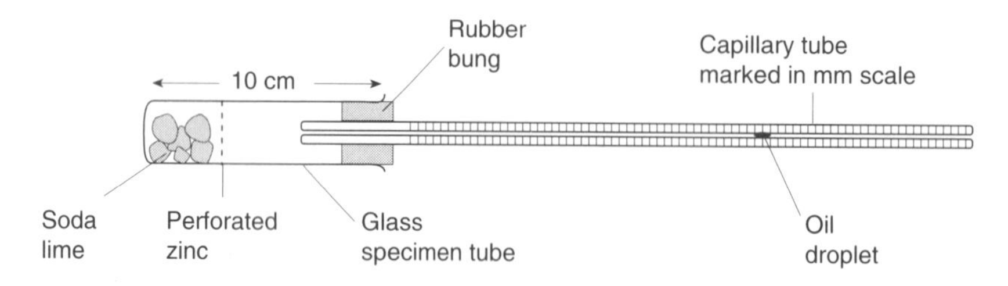

(a) Suggest an organism which could be used in the apparatus. (1)

(b) Briefly describe how the apparatus works and how you would use it. (4)

(c) What control could be set up and how might it be used to modify the results obtained? (2)

(d) List *four* sources of error that could arise when using the apparatus. (4)

(e) State *four* factors that could alter the rate of energy uptake by the organisms in the apparatus. (4)

(f) If, during one experiment, the oil droplet moved 50 mm in 10 minutes and the total mass of the organisms in the glass specimen tube was 4 g, calculate the volume of oxygen taken up per hour, per gram of organism. The capillary tube has a uniform bore of 1.0 mm. (2)

Time allowed 30 minutes

Pitfalls

You should note the scale on the drawing and ensure that your choice of organism fits the tube – mice or rats would be far too big! Give sufficient detail when describing its use – a good rule is to ensure that the procedures could be successfully followed by someone with no knowledge of biology. There is no need for detail in parts (d) and (e) – 'list' and 'state' imply short statements rather than explanations. The most probable error in part (f) is not to convert the linear dimensions of the capillary tube into a volume; you will need to be aware that the volume of a cylinder can be calculated according to the formula $\pi r^2 h$ where r is the radius and h the height (length) of the cylinder.

Points

(a) Choose from: locust, beetle, housefly, earthworm, woodlouse, maggot, bee, germinating pea seeds.

(b) The apparatus should be set up as in the diagram. A sample of one or more of the appropriate organisms should be weighed and put in the specimen tube. By inserting the bung further into the tube the oil droplet should be positioned as far away from the specimen as possible. The position of the droplet in the capillary tube should be recorded from the scale marked on it and a stop clock started. As the animal(s) respire according to the equation

$$6O_2 + C_6H_{12}O_6 \rightarrow 6CO_2 + 6H_2O + \text{energy}$$

the volume of carbon dioxide produced is equal to the volume of oxygen taken up. The carbon dioxide, however, is rapidly absorbed by the soda lime and its volume therefore becomes negligible. The volume of water is also negligible as this too is absorbed by the soda lime. The only measurable volume change is therefore due to the reduction in oxygen volume as it is absorbed by the animals. This reduction in volume causes a reduction in pressure within the specimen tube. Atmospheric pressure now exceeds the internal pressure and the oil droplet is forced towards the specimen tube. The time taken for the droplet to move a set distance (e.g. 15 cm) should be recorded and the process repeated a few times to allow an average time to be obtained. If the distance moved is h then the volume of oxygen consumed can be calculated from the formula $\pi r^2 h$ where r is the radius of the capillary tube. This can be calculated by finding the internal bore diameter (using a travelling microscope) and dividing by two. The volume consumed per minute is calculated by dividing the total volume by the number of minutes taken for the droplet to move the distance h. This figure should then be divided by the weight in grams of the animals to give a final figure of oxygen consumed per minute per gram of organism.

(c) A second set of apparatus, as identical as possible to the experimental one, should be set up in exactly the same way as described in (b), except that the animals should be excluded. This control should be placed as close as possible to the experimental apparatus so that the environmental fluctuations affect both equally. If there is no movement of the droplet in the control apparatus the results of the experimental apparatus require no alteration. If, however, the changes in atmospheric pressure or temperature cause the droplet in the control tube to move, the distance moved should be recorded. This should be added to the distance measured in the experimental tube if the droplet moves away from the specimen tube in the control, or subtracted if it moves towards the specimen tube. In this way the control acts as a thermobarometer.

(d) Choose from:

1. apparatus not air tight
2. temperature changes during the experiment (will alter pressure)
3. pressure changes during the experiment
4. soda lime may be exhausted (i.e. has ceased to absorb more carbon dioxide)
5. capillary tube may not be horizontal (gravity may move droplet)

(e) Choose from:

1. external temperature fluctuations
2. age of the animals used
3. activity of the animals (i.e. resting or moving excitedly)
4. amount of oxygen available in the specimen tube
5. light intensity (may cause increase/decrease in activity depending on the animals used)

(f) The movement of the oil droplet means that the organisms have effectively absorbed a cylinder of oxygen of height 50 mm.
Volume of cylinder = $\pi r^2 h$
Here h = 50 mm

$$r = \frac{1.0 \text{ mm}}{2} \text{ (diameter of cylinder)} = 0.5 \text{ mm}$$

Therefore the total amount of oxygen absorbed = $\pi \times (0.5)^2 \times 50$.
This is for a period of 10 minutes; therefore the amount of oxygen taken up in 1 hour = $\pi \times (0.5)^2 \times 50 \times 6$.
This is for 4 g of organisms; therefore the amount of oxygen taken up

$$= \frac{\pi \times (0.5)^2 \times 50 \times 6}{4} = 58.9 \text{ mm}^3 \text{ per hour, per gram}$$

2 Describe the means by which blood circulation is maintained and controlled in a mammal and review the part played by the blood in ensuring immunity from infectious disease.

Time allowed 30 minutes

Pitfalls

You should plan your answer to this essay carefully (the points below constitute a simple plan). You must limit your answer to a *mammal*. The words 'maintained' and 'controlled' are different but are commonly dealt with by candidates as though they are the same. The maintenance of blood circulation is achieved by the pumping of the heart, the recoil of the elastic artery walls and the squeezing of valved-veins during muscle contraction. Control, on the other hand, involves nervous and hormonal effects on the heart as well as the vasodilation and vasoconstriction of arteries.

The word 'review' suggests a summary of the processes of immunity rather than a highly detailed account. Nonetheless you must include some detail and examples without overrunning the available time.

Points

Maintenance of circulation:

1. mechanism of heartbeat (SA node, AV node, Purkinje fibres, cardiac muscle features)
2. recoil action of arterial wall
3. muscular contraction squeezing veins and the role of pocket valves
4. inspiratory movements drawing blood into the heart
5. gravity where this acts in the same direction as the blood flow

Control of circulation:

Control of heart beat

1. effects of hormones, e.g. adrenaline
2. nervous effects – cardioacceleratory and cardioinhibitory centres in the medulla oblongata of the brain and their influence via the vagus nerve

Control of blood flow

1. vasoconstriction and vasodilation
2. role of the autonomic nervous system

Immunity:

1. recognition of foreign material
2. antigen–antibody response
3. production of antibodies – different types and their functions
4. phagocytosis
5. continued immunity for some period after control of the initial infection

3 Two different species of flowering plants growing in pots were subjected to still air conditions and the rates of transpiration for each were measured four times during the course of one hour. By means of an electric fan the same plants were then subjected to a constant flow of air of 5 m s^{-1} and the transpiration rates were again measured four times during the next hour. The process was repeated in this way for wind speeds of 10, 15 and 20 m s^{-1}. The data obtained are set out below.

Wind velocity	Time (hours)	Rate of transpiration (g^{-1} $hour^{-1}$) Plant A	Plant B
Still air	0.25	200	60
	0.50	180	50
	0.75	200	60
	1.00	190	50
5 m s^{-1}	1.25	290	70
	1.50	300	80
	1.75	210	80
	2.00	290	70
10 m s^{-1}	2.25	400	100
	2.50	410	110
	2.75	400	110
	3.00	420	110
15 m s^{-1}	3.25	540	140
	3.50	530	140
	3.75	530	130
	4.00	460	140
20 m s^{-1}	4.25	420	170
	4.50	300	180
	4.75	210	180
	5.00	120	180

(a) Name the piece of apparatus most commonly used to measure water uptake in plants. (1)

(b) Why is the 'water uptake' measured by this apparatus not necessarily the same as the transpiration rate? (2)

(c) Plot the graph of transpiration rate against time for both plant species on the same axes. (3)

(d) What information is provided by the graph concerning the transpiration rate of the two species during the first three hours of the experiment? (2)

(e) Explain the reduction in transpiration rate for species A after 3.75 hours. (2)

Time allowed 20 minutes

Pitfalls

The required answer in part (b) is often mistakenly thought to require a source of error within the experiment, e.g. a leaky potometer. In fact it is a physiological factor which is being looked for, e.g. some of the water taken up may be used in photosynthesis rather than being transpired.

You should have regard to the following when drawing your graph:

1. Choose the scale to maximise use of the graph paper.
2. Choose an easy-to-use scale and mark it clearly.
3. Name each axis with the parameter being measured and the units.
4. Plot points in pencil so that errors can be erased.
5. Use a cross or encircle the plotted points so that they can be seen easily.
6. Draw a line of best fit through the points as it is reasonable in this case to assume that the intermediate values fall between the points. (However, you are unlikely to be penalised for joining adjacent points.)
7. Distinguish the two lines clearly, e.g. use different colours, and label both or use a key to identify them.
8. Mark the points on the time axis where the wind speed is increased.
9. Give the graph a suitable and complete title.

You may prefer to draw a histogram to a point graph, although this will almost certainly be more time consuming.

In your answer to part (d) you should try to quantify the information in some way, e.g. the initial transpiration rate for species A is almost four times greater than that for species B.

Points

(a) The potometer

(b) Some water taken up will not be transpired. For example, it may be retained in cell vacuoles to give turgidity, used in growth or other metabolic activities such as photosynthesis or secreted by the plant (e.g. nectar). Some water transpired may not have come directly from water taken up during the experiment, e.g. it may be a metabolic product of reactions such as respiration, in which case it comes from stored food material made before the experiment began.

(c) *Transpiration rate against time for two species of flowering plants in different wind velocities*

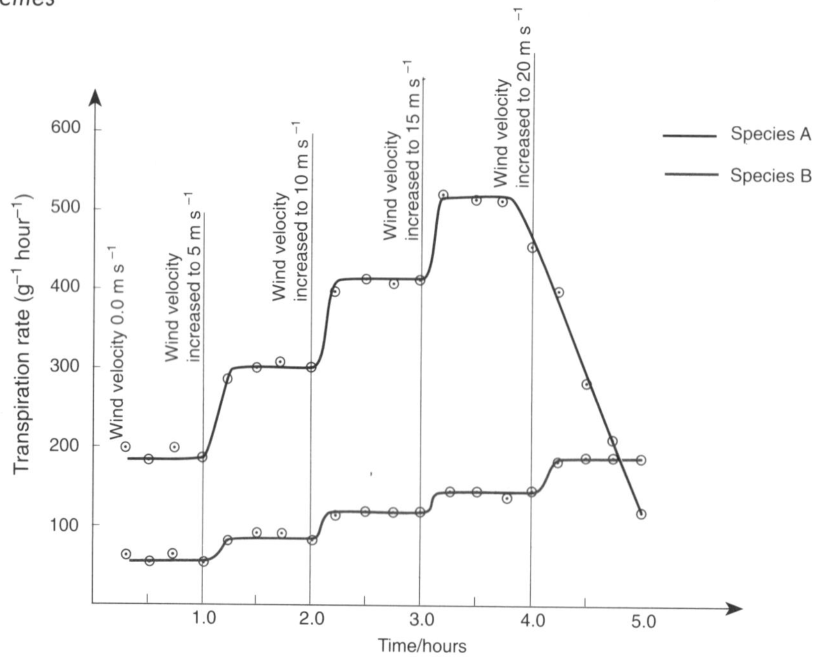

(d) The transpiration rate of both species increases initially as the wind speed increases, but later stabilises for a given wind speed with only small fluctuations. For both species the transpiration rate approximately doubles over the three-hour period. However, the initial rate for species A was almost four times that of species B. The difference in transpiration rate between the two species therefore doubles during the three hours.

(e) At this point the loss of water by transpiration for species A far exceeds the supply of water to the plant. The plant wilts, i.e. the leaves lose turgidity and become flaccid. This reduces their surface area considerably with a consequent reduction in the rate of transpiration. In addition, loss of turgidity in the guard cells causes the stomata to close, considerably reducing the transpiration rate because most water loss occurs through open stomata.

4 Three adjacent plant cells have solute and pressure potential as shown in the table below.

Cell	Pressure potential/kPa	Solute potential/kPa
1	200	−900
2	500	−1000
3	500	−600

Which one of the diagrams below correctly illustrates the direction of water movement as indicated by the arrows?

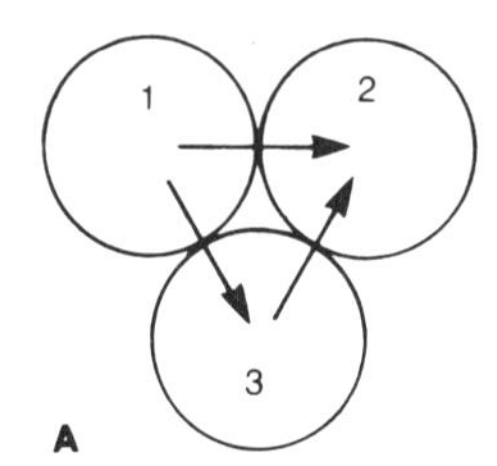

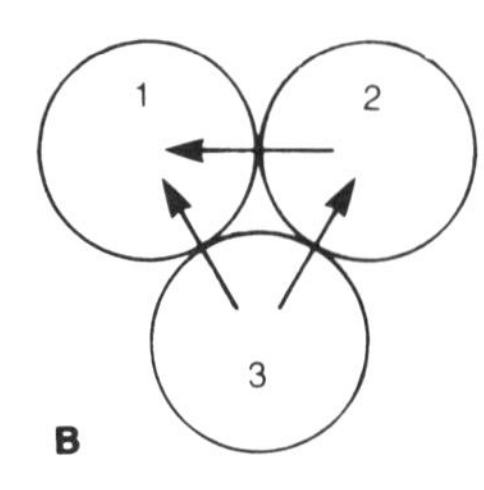

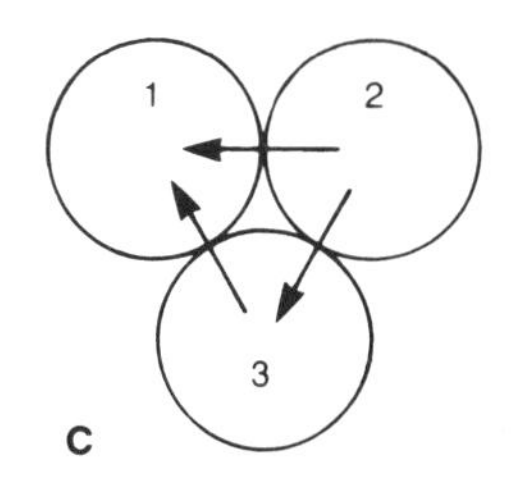

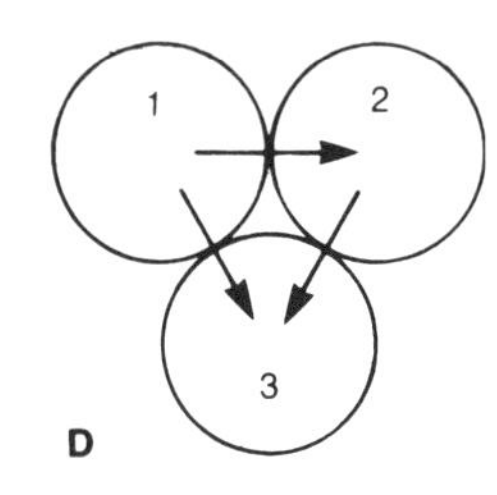

Time allowed 1½ minutes

Points

There are two essential pieces of information required to answer this question. Firstly the following equation:

Water potential = Solute potential + Pressure potential

Secondly, that water flows osmotically from cell to cell from a region with a less negative (higher) water potential to a region with a more negative (lower) water potential.

Using the first piece of information it is necessary to calculate the water potential for each of the cells 1,2 and 3. For example in cell 1:

$$\text{Water potential} = -900 + 200 = -700 \text{ kPa}$$

A similar calculation for cells 2 and 3 gives values of −500 kPa and −100 kPa, respectively. Note that all values for water potential are negative. These values can then be applied to the three cells in the diagram:

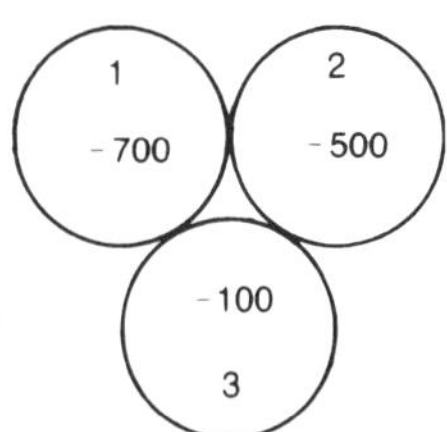

Using the second piece of information (water moves from the less negative to the more negative) it can be seen that water will move from both cells 2 and 3 into cell 1 and at the same time from cell 3 into cell 2, i.e. as shown in diagram B.

CHAPTER 8

COORDINATION, RESPONSE AND CONTROL

Units in this chapter

Chapter objectives

Imagine a factory or hospital in which each individual acted independently, doing what they liked when they chose; imagine a road system where all users ignored the highway code; imagine an organism in which each cell performed its activities with complete disregard to the other cells. The resultant chaos is obvious. Clearly, to operate effectively and efficiently a system, whether a business, public service, transport operation, plant or animal, must work in a controlled and coordinated manner. We saw in Chapter 1 that as larger organisms evolved, cells became specialised to a particular function and in doing so lost the ability to perform others. This led to the dependence of one cell on another and in turn to the need for cells to act in concert. To do otherwise could lead to the efforts of one group of cells being neutralised by the actions of others. In this chapter we will look at how integration and control are achieved in organisms.

We will also see how plants and animals use chemical messages in the form of plant growth substances or hormones to produce a slower, but often more permanent, response than the nervous communication that is additionally used by animals, which produces a quick, but normally reversible, reaction. The variety of sense organs which have developed in animals and the evolution of a central coordinating nervous system – the brain and spinal cord – will be studied.
Finally, we will attempt to put together many of the aspects covered in this book by looking at how a number of systems (respiratory, blood, skeletal, muscular, nervous, hormonal, etc.) can integrate to produce a complex activity such as movement. But first we will study the concept of homeostasis and attempt to unravel the complexities of how, despite a constantly changing external environment, many organisms can maintain a remarkably constant interior.

8.1 HOMEOSTASIS

First coined in 1932, the term homeostasis is used to describe the mechanisms by which a constant internal environment is achieved. The advantage of such constancy is that an organism has greater environmental freedom. It need not be as restricted by temperature, water supply, saline conditions, etc. as it might otherwise be. It has a greater geographical range and therefore better access to food, water and other essentials of life.

TEMPERATURE REGULATION

Before we survey the methods by which organisms control their temperatures let us look at the way they are classified according to temperature:

- **Endotherms** derive their body heat by internal metabolic activities.
- **Ectotherms** derive their body heat predominantly from external sources, e.g. the sun.
- **Homoiotherms** can regulate their body temperatures within very narrow limits, e.g. around 37–38 °C for most mammals and 40 °C for most birds.
- **Poikilotherms** allow their body temperature to vary with the external temperature.

Organisms gain heat in two ways:

1. from the metabolic activities inside their cells;
2. by absorbing solar energy, either by reflection, convection or conduction.

Organisms lose heat in four main ways:

1. conduction to the ground or other matter;
2. convection to the air or water around them;
3. radiation to air, water or earth;
4. by the evaporation of water from their bodies.

While most animals attempt some control of their body temperature, it is in the birds and mammals that thermoregulation is most apparent. Most heat exchange occurs through the skin and so, not surprisingly, this organ plays a major role in controlling temperature. The structure of mammalian skin is illustrated in Fig. 8.1.

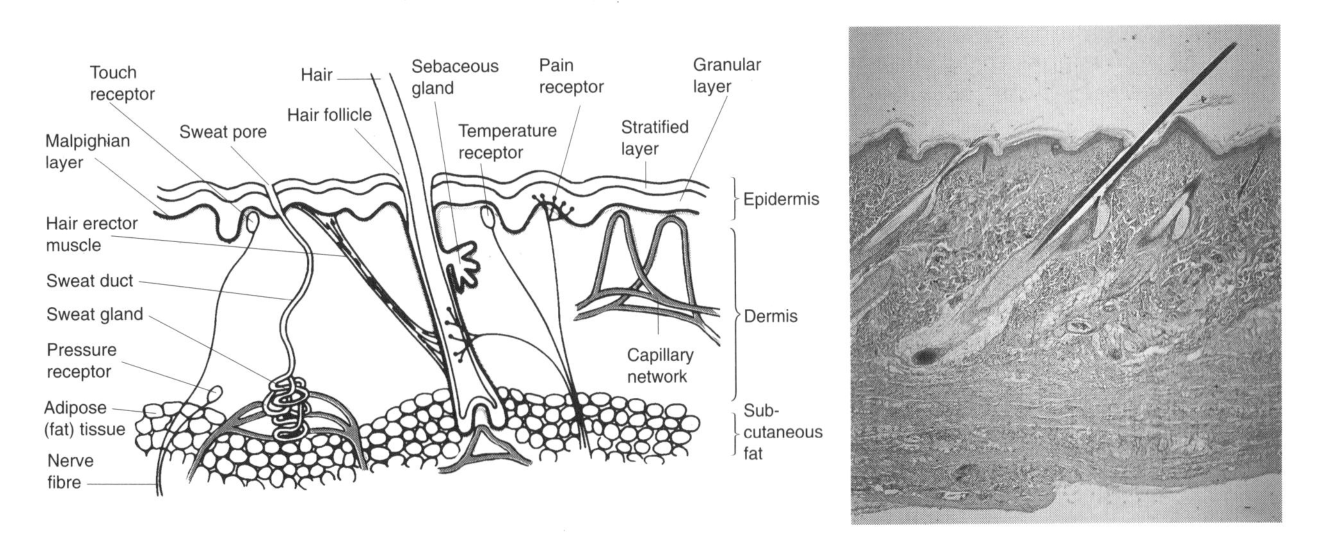

Fig. 8.1 Vertical section through mammalian skin

Mechanisms for losing heat in a warm environment

1. Vasodilation – blood passes close to the skin surface due to the dilation (widening) of arterioles in the skin.
2. Sweating/panting/licking – water evaporating from the skin surface carries away heat.
3. Large surface area to volume ratio, e.g. large ears can be used to radiate body heat.
4. Less insulation – fat and/or fur layers tend to be thinner.
5. Behaviour – animals may shelter and be inactive during the hottest part of the day or even **aestivate** (hibernate) during the summer months.

Mechanisms for retaining or gaining heat in a cold environment

1. Vasoconstriction – blood is kept away from the surface by the constriction of arterioles in the skin.
2. Shivering/increased activity – both increase muscular movement and so increase the production of metabolic heat.
3. Small surface area to volume ratio – compact bodies with shorter extremities reduce the area over which heat is lost.
4. More insulation – fat and/or fur layers tend to be thicker, thus insulating the body from the cold.
5. Increased metabolic rate – animals in colder climates tend to have a higher metabolic rate than their counterparts in warm areas. More heat is therefore generated.
6. Behaviour – animals in cold climates may huddle together to reduce heat loss. They often hibernate over the coldest months, allowing their body temperature to fall and thus reducing the need for food.

CONTROL OF BLOOD SUGAR LEVEL

Cells are sensitive to changes in sugar level and it is therefore important to maintain 90 mg of glucose in each 100 cm^3 of blood. The liver plays a vital role in this control through:

1. glycogenolysis – the breakdown of glycogen
2. gluconeogenesis – the conversion of protein to glucose
3. glycogenesis – the conversion of blood glucose into glycogen for storage

It is the interconversion of glucose and glycogen which maintains a constant blood sugar level. This is under the control of two hormones both produced by special pancreatic cells called **islets of Langerhans**. The cells are of two types:

1. **α-cells** produce the hormone **glucagon**, which is involved in the conversion of glycogen to glucose and so raises the blood sugar level when it falls below normal.
2. **β-cells** produce the hormone **insulin**, which is involved in the conversion of glucose to glycogen and so lowers the blood sugar level when it exceeds the normal.

Under the control of the **adrenocorticotrophic hormone (ACTH)** from the pituitary gland, the adrenal glands will release glucocorticoid hormones such as cortisol when the liver's supply of glycogen is exhausted. This causes amino acids and glycerol to be converted into glucose to supplement supplies. **Adrenaline** from the adrenal glands will raise the blood sugar level in times of stress by increasing the production of glucose from the glycogen supply in the liver.

THE LIVER

Many homeostatic functions are performed by the liver. The most important ones include:

1. **maintenance** of a steady blood sugar level by the conversion of glucose to glycogen and vice versa, under the influence of insulin and glucagon

❷ **breakdown** or removal of excess or used materials such as
(a) old red blood cells by phagocytosis
(b) excess lipids
(c) excess cholesterol
(d) sex hormones and adrenaline
(e) excess amino acids by deamination
(f) poisons by detoxification

❸ **manufacture** of

(a) plasma proteins such as fibrinogen

(b) cholesterol

(c) bile

❹ **storage** of iron, copper, potassium and vitamins A, D and B_{12}

❺ **production** of heat

To be able to carry out these activities, the liver has a very rich blood supply. As well as oxygenated blood from the aorta it receives blood containing digested food in the hepatic portal vein, coming directly from the small intestine.

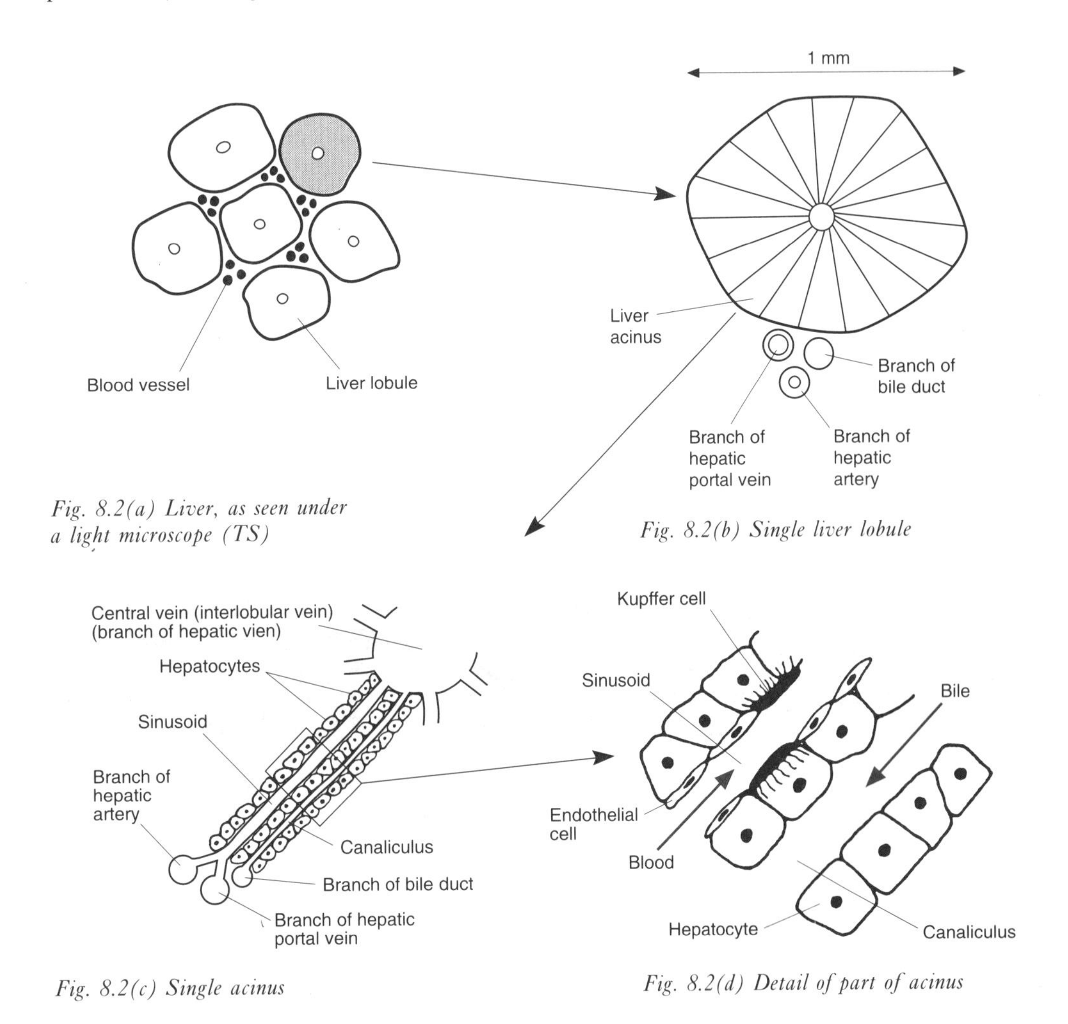

Fig. 8.2(a) Liver, as seen under a light microscope (TS)

Fig. 8.2(b) Single liver lobule

Fig. 8.2(c) Single acinus

Fig. 8.2(d) Detail of part of acinus

We have looked at only a few homeostatic mechanisms here. Others include the control of salt and water balance, which is dealt with in 7.4, and more examples will be referred to in the following. Whatever the factor being controlled, however, there is always a principle that underlies all homeostatic processes – feedback.

FEEDBACK MECHANISMS

All homeostatic processes involve a **detector** (or sensor) which monitors the factor being controlled. When the detector senses a change from the normal it informs a **controller**, a coordinating system which decides upon an appropriate method of correcting the deviation. The controller communicates with one or more **effectors** which carry out the corrective procedures. Once the correction is made and the factor returned to normal, information is fed back to the detector which then 'switches off'. The system is summarised in Fig. 8.3.

In most biological control systems the feedback loop causes the detector to 'switch off', i.e. the controller is no longer alerted to the deviation from the normal. This is called **negative feedback**. An example of negative feedback controlling the osmotic concentration of the blood is given in Fig. 8.4.

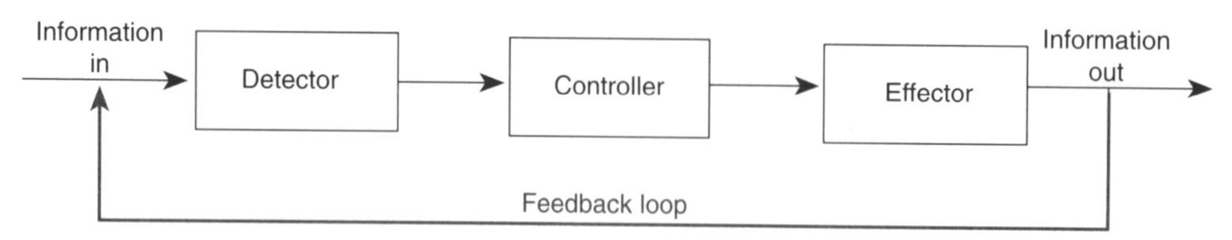

Fig. 8.3 Principles of homeostatic control

Occasionally the feedback loop causes the detector to increase, not decrease, its message to the controller. This is called **positive feedback** and an example occurs when the hormone oxytocin is released during childbirth. It stimulates contractions of the uterus which in turn stimulate more oxytocin production leading to more rapid and violent contractions until the baby is born.

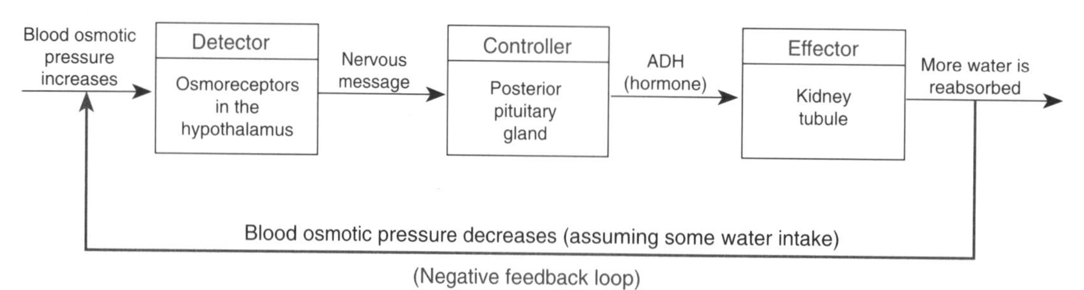

Fig. 8.4 Simplified diagram showing homeostatic control of blood osmotic pressure

8.2 CONTROL AND COORDINATION IN PLANTS

You might think that plants have little need for coordinating activities since they are, with the exception of certain algae, unable to move from place to place without outside assistance. It is true that they have no contractile tissue (e.g. muscle) which allows rapid response to stimuli. Nevertheless their survival depends upon being able to move to or from stimuli such as light and water. Such movements take place as a result of sudden osmotic changes or as part of the growth process. Any such movements must be controlled and coordinated if they are to be of value, and in plants this is achieved largely by chemicals called **plant growth substances**, not least because plants do not possess a nervous system.

PLANT GROWTH SUBSTANCES

Charles Darwin was one of the first researchers to investigate the effects of plant growth substances. His experiments and those of his contemporaries are summarised in Fig. 8.5.

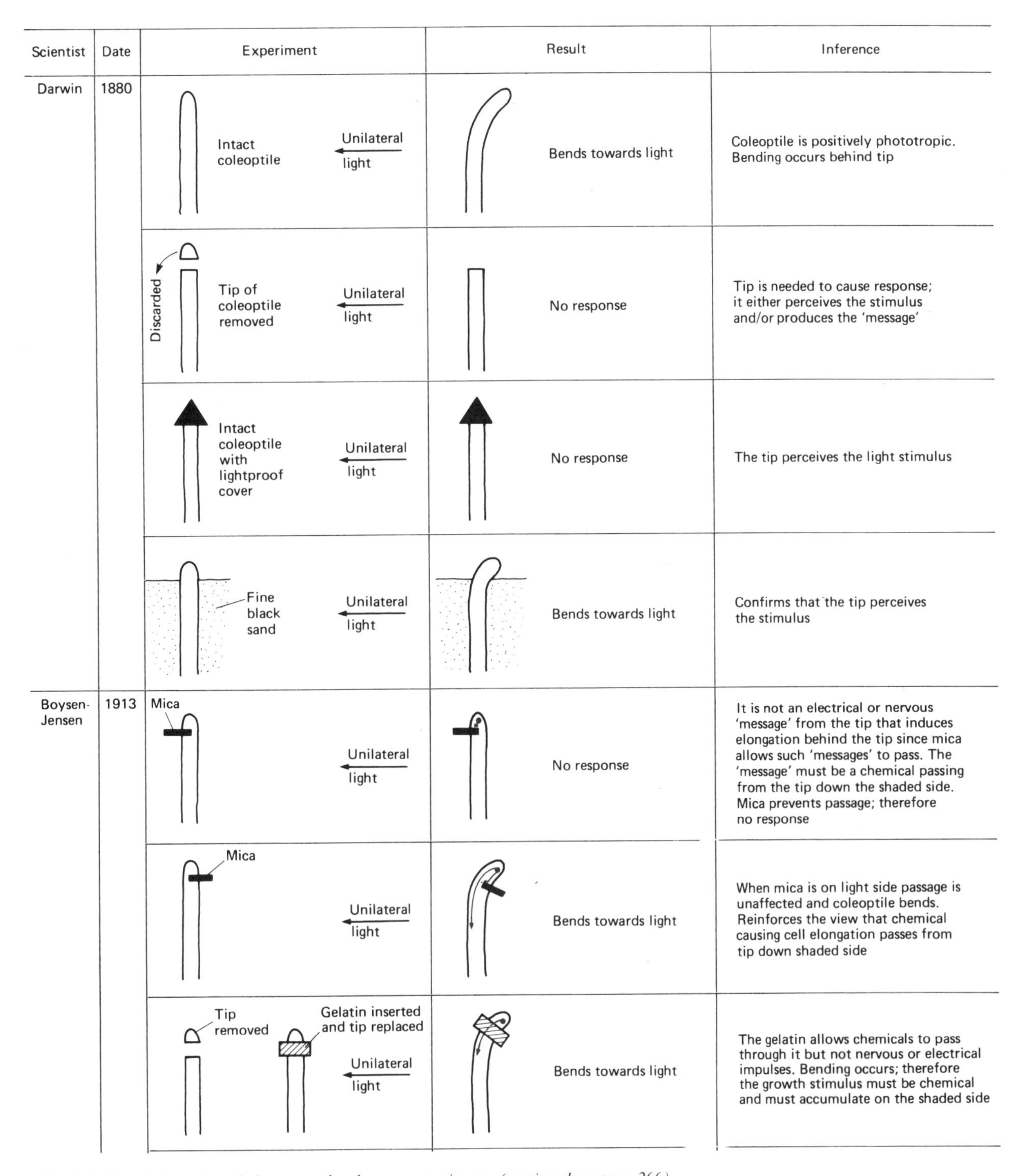

Scientist	Date	Experiment	Result	Inference
Darwin	1880	Intact coleoptile; Unilateral light	Bends towards light	Coleoptile is positively phototropic. Bending occurs behind tip
		Tip of coleoptile removed (Discarded); Unilateral light	No response	Tip is needed to cause response; it either perceives the stimulus and/or produces the 'message'
		Intact coleoptile with lightproof cover; Unilateral light	No response	The tip perceives the light stimulus
		Fine black sand; Unilateral light	Bends towards light	Confirms that the tip perceives the stimulus
Boysen-Jensen	1913	Mica; Unilateral light	No response	It is not an electrical or nervous 'message' from the tip that induces elongation behind the tip since mica allows such 'messages' to pass. The 'message' must be a chemical passing from the tip down the shaded side. Mica prevents passage; therefore no response
		Mica; Unilateral light	Bends towards light	When mica is on light side passage is unaffected and coleoptile bends. Reinforces the view that chemical causing cell elongation passes from tip down shaded side
		Tip removed; Gelatin inserted and tip replaced; Unilateral light	Bends towards light	The gelatin allows chemicals to pass through it but not nervous or electrical impulses. Bending occurs; therefore the growth stimulus must be chemical and must accumulate on the shaded side

Fig. 8.5 Historical review of plant growth substance experiments *(continued on page 266)*

Plant growth substances are chemicals produced in very small quantities in one part of the plant and transported to another where they promote, inhibit or in some way modify growth. There are five main classes of plant growth substances: auxins, gibberellins, cytokinins, inhibitors and ethene. A summary of the effects of each group is given in Table 8.1.

Phytochromes

A number of plant growth responses are influenced differently by light of different wavelengths. For light to have an effect it must be absorbed by a photoreceptor substance. In about 1960 the pigment phytochrome was isolated; this exists in two interconvertible forms:

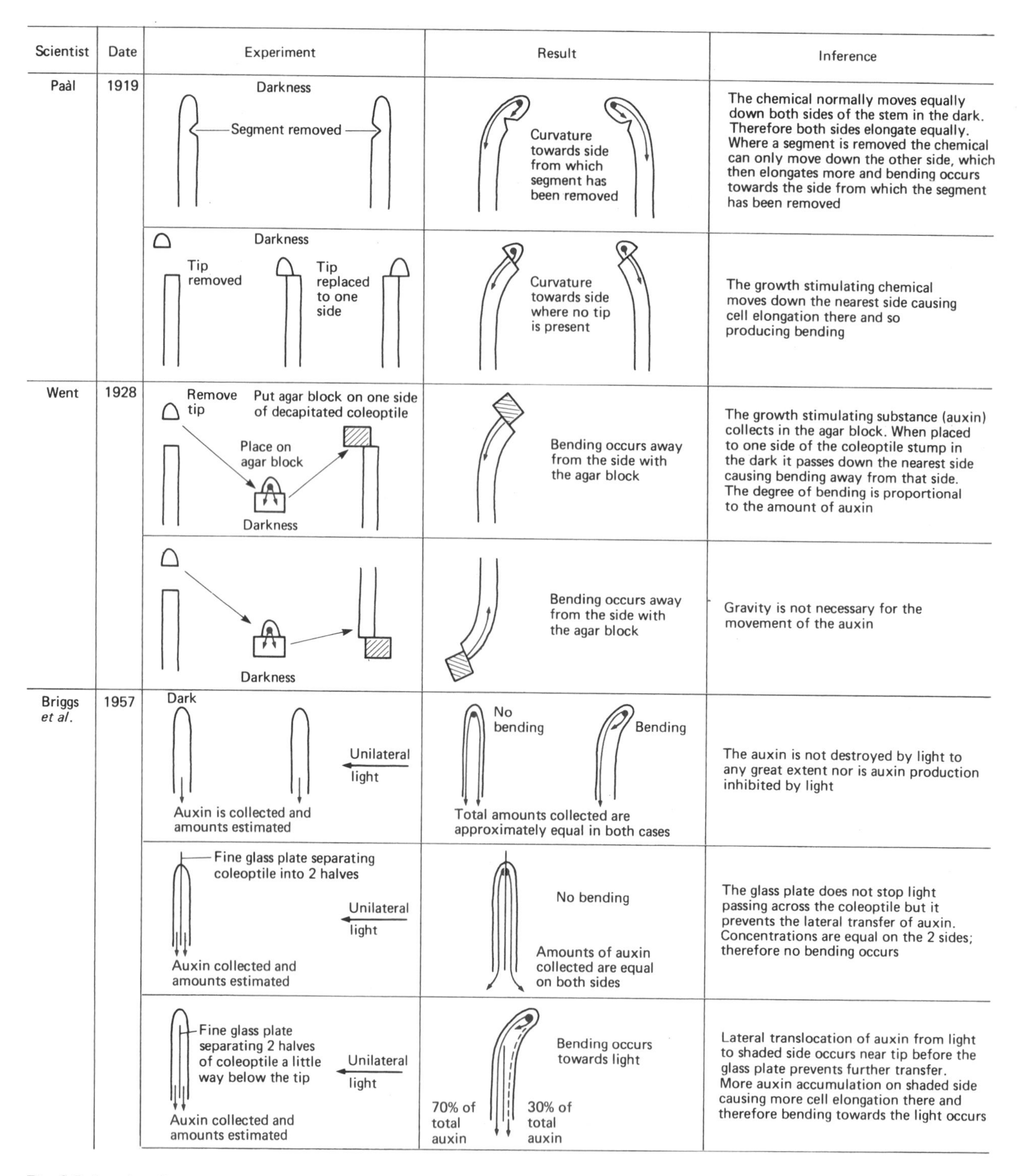

Scientist	Date	Experiment	Result	Inference
Paàl	1919	Darkness; Segment removed	Curvature towards side from which segment has been removed	The chemical normally moves equally down both sides of the stem in the dark. Therefore both sides elongate equally. Where a segment is removed the chemical can only move down the other side, which then elongates more and bending occurs towards the side from which the segment has been removed
		Darkness; Tip removed; Tip replaced to one side	Curvature towards side where no tip is present	The growth stimulating chemical moves down the nearest side causing cell elongation there and so producing bending
Went	1928	Remove tip; Place on agar block; Put agar block on one side of decapitated coleoptile; Darkness	Bending occurs away from the side with the agar block	The growth stimulating substance (auxin) collects in the agar block. When placed to one side of the coleoptile stump in the dark it passes down the nearest side causing bending away from that side. The degree of bending is proportional to the amount of auxin
		Darkness	Bending occurs away from the side with the agar block	Gravity is not necessary for the movement of the auxin
Briggs *et al.*	1957	Dark; Unilateral light; Auxin is collected and amounts estimated	No bending; Bending; Total amounts collected are approximately equal in both cases	The auxin is not destroyed by light to any great extent nor is auxin production inhibited by light
		Fine glass plate separating coleoptile into 2 halves; Unilateral light; Auxin collected and amounts estimated	No bending; Amounts of auxin collected are equal on both sides	The glass plate does not stop light passing across the coleoptile but it prevents the lateral transfer of auxin. Concentrations are equal on the 2 sides; therefore no bending occurs
		Fine glass plate separating 2 halves of coleoptile a little way below the tip; Unilateral light; Auxin collected and amounts estimated	Bending occurs towards light; 70% of total auxin; 30% of total auxin	Lateral translocation of auxin from light to shaded side occurs near tip before the glass plate prevents further transfer. More auxin accumulation on shaded side causing more cell elongation there and therefore bending towards the light occurs

Fig. 8.5 (continued)

one which absorbs red light with a peak at about 660 nm (phytochrome 660 or P_{660}) and the other which absorbs far-red light with a peak at about 730 nm (P_{730}).

$$P_{660} \underset{\text{far-red light = rapid conversion; dark = slow conversion}}{\overset{\text{red light (daylight)}}{\rightleftharpoons}} P_{730}$$

Phytochrome is present in the leaves. After absorbing the appropriate wavelength of light, it causes the conversion of a hormone precursor to a hormone which then affects growth.

Table 8.1 *Plant growth substances*

Growth substance	Effects		Examples
Auxins Example is indole acetic acid. Synthetic auxins include 2-4-dichlorophenoxyacetic acid (2-4-D) and 2-4-5-trichlorophenoxyacetic acid (2-4-5-T)	Cause cell elongation	Phototropism	Oat coleoptiles bend towards light
		Geotropism	Roots grow downwards into the soil
	Cause cell division	Stimulate cambial activity	Development of wound tissue (calluses)
		Initiate root development	Root powders initiate root growth from cuttings
		Stimulate fruit growth and parthenocarpic fruit development	Some crops, e.g. apples, are sprayed with synthetic auxins to cause fruit development without fertilisation
	Maintain the structure of cell walls	Inhibit leaf abscission	Leaves do not fall when auxin from leaf exceeds that from stem
		Inhibit fruit abscission	Fruits do not fall when auxin from fruit exceeds that from stem
	Inhibit growth in high concentrations	Inhibit development of lateral buds	The dominance of apical buds is due to the auxin they produce inhibiting lateral ones; removal of apical buds therefore leads to branching
		Kill plants by disrupting growth	Synthetic auxins are used as selective weedkillers
Gibberellins Related to gibberellic acid which is a metabolic by-product of the fungus *Gibberella fujikuroi*. There are a number of different gibberellins. They affect cell elongation in stems, may increase the leaf area of some plants, but have no effect on roots	Promote cell elongation		Cause elongation of plant stems
	Reverse some types of genetic dwarfism		Dwarf varieties of many plants, e.g. peas, *Chrysanthemum*, can be made to grow to normal size when gibberellins are applied
	Promote germination of seeds		Gibberellins promote germination of many seeds, e.g. oats (*Avena*)
	End dormancy of buds		The natural dormancy of many buds, e.g. birch, is broken when gibberellins are applied
	Affect leaf expansion and shape		*Eucalyptus* leaves are transformed from juvenile to mature shape when gibberellins are present; reverse occurs in ivy
	Aid setting of fruit after fertilisation		Some species of cherry, apricot and peach (*Prunus*) readily set fruit after treatment with gibberellins
	Remove the need for cold treatment in vernalisation		Carrots (*Daucus*) normally flower only after a period of exposure to cold; they can be made to flower without this by application of gibberellins
	Affect flowering		The application of gibberellins to henbane (*Hyoscyamus*) kept in short-day conditions will induce flowering
Cytokinins These are derived from adenine (a purine). Interact with auxins to promote cell division in cultures. Example is kinetin.	Increase rate of cell division		Aid the growth of many plants, e.g. sunflower (*Helianthus*)
	Stimulate bud development		Cause buds to develop, e.g. on leaf cuttings of African violet (*Saintpaulia*)
	Increase rate of cell elongation in leaves		The size of fronds of duckweed (*Lemna*) increases in the dark when cytokinins are added

Growth substance	Effects	Examples
Inhibitors These are a group of substances that inhibit growth. Example is abscisic acid	Retard growth	Inhibit the growth of many plant parts, e.g. hypocotyls, radicles and leaves
	Induce dormancy in buds	The growing apex of some plants, e.g. birch (*Betula*), can be transformed into a dormant bud by addition of abscisic acid
	Inhibit germination	Some seeds have their germination inhibited by abscisic acid, e.g. rose (*Rosa*)
Ethene This is a product of plant metabolism	Involved in many auxin-induced responses	Ethene production is frequently stimulated by auxin
	Causes leaf senescence and abscission	The leaves of some species, e.g. *Euonymus japonica*, die earlier and drop from the plant when treated with ethene
	Ripens fruits	Many fruits, e.g. oranges and lemons, ripen much more rapidly in the presence of ethene

Photoperiodism

Flowering is regulated by the length of day. Three basic groups of plants exist, although all intermediates between them may be found:

1. Long-day plants, e.g. clover, barley, radish, petunia. These only flower when the light period in a 24-hour cycle exceeds about 10 hours. In temperate regions these plants flower in the summer.
2. Short-day plants, e.g. tobacco, cocklebur, poinsettia, chrysanthemum. These only flower when the light period is shorter than about 14 hours. In temperate regions they generally flower in the spring or autumn.
3. Day-neutral plants, e.g. carrot, violet, begonia, cucumber. These are indifferent to the length of day.

Long-day plants are thought to flower when the presence of red light, or a long period of sunlight, causes a sufficient accumulation of P_{730} and thus a low level of P_{660}.

Short-day plants flower when far-red light, or a long period of darkness, causes a sufficient accumulation of P_{660} and thus a low level of P_{730}.

A summary of the effects on plants of red and far-red light is given in Table 8.2.

Table 8.2 *Effects of red light and far-red light*

Red light effects	Far-red light effects
P_{660} changes to P_{730}	P_{730} changes to P_{660}
Stimulates germination of some seeds, e.g. lettuce (*Lactuca*)	Inhibits germination of some seeds, e.g. lettuce (*Lactuca*)
Induces formation of anthocyanins (plant pigments)	Inhibits formation of anthocyanins
Stimulates flowering in long-day plants	Inhibits flowering in long-day plants
Inhibits flowering in short-day plants	Stimulates flowering in short-day plants
Elongation of internodes is inhibited	Elongation of internodes is promoted
Induces increase in leaf area	Prevents increase in leaf area
Causes epicotyl (plumule) hook to unbend	Maintains epicotyl (plumule) hook bent

The flowering of long- and short-day plants can be sumarised as follows:

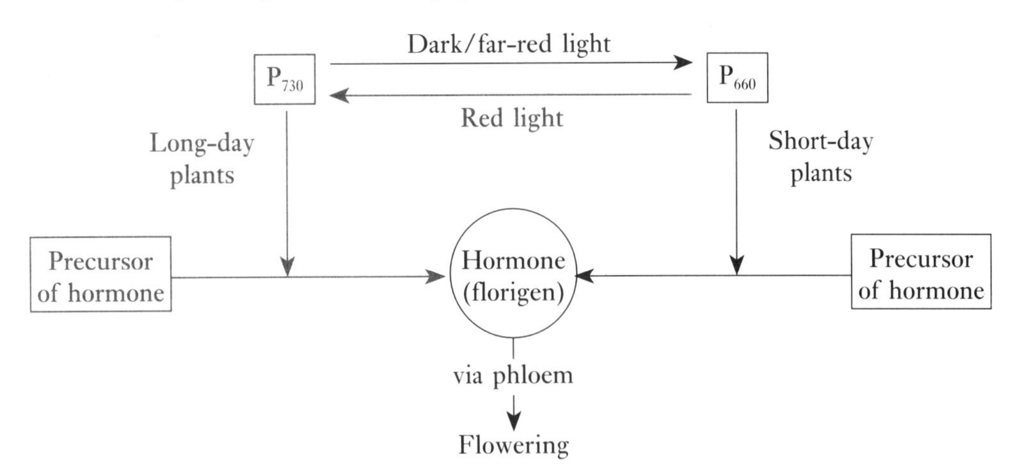

PLANT RESPONSES

Plant responses fall into three broad categories:

1. **Tropisms** are responses to a directional stimulus in which *part* of a plant moves towards or away from the direction of the stimulus.
2. **Taxes** are responses to a directional stimulus in which the *whole* of the plant (or a freely motile part of it) moves towards or away from the direction of the stimulus.
3. **Nasties** are responses of part of a plant to a stimulus in which the direction of the response is unrelated to that of the stimulus.

The nature of the stimulus is used as a prefix in describing responses, e.g. phototropism is a tropic response to light. Where appropriate the direction of the movement is described as positive, if the movement is towards the stimulus, or negative, if the movement is away from the stimulus. Positive hydrotropism in roots is therefore the movement of roots towards water. A summary of the main plant responses is given in Table 8.3.

8.3 THE ENDOCRINE SYSTEM

Animals control and coordinate their various body systems in two main ways: by the use of hormones produced in the endocrine system and by the more rapidly responding nervous system. Both are interlinked and intercommunicate in order to work in close association. We will look at the nervous system in detail in 8.4 but first let us survey the endocrine system.

The term endocrine refers to the fact that the secretory glands have no ducts, their alternative name being **ductless glands**. In the absence of ducts, the hormones produced by endocrine glands are transported to their destinations, called **target organs**, in the bloodstream. Most endocrine glands work under the direction of a master gland – the **pituitary gland** at the base of the brain. The pituitary gland works in conjunction with the nearby **hypothalamus**, which also acts as an important link between the endocrine and nervous systems. Fig. 8.6 details the positions of the major endocrine glands in a human. These glands are not always separate structures; they may be a group of cells within another organ.

Nature of hormones

Hormones do not belong to a single chemical group but are varied in their composition. They are medium-sized molecules and may be polypeptides, proteins, amines or steroids. The

Table 8.3 *Plant responses*

Type of movement and definition	Stimulus	Name of response	Examples
Tropic A growth movement of part of a plant in response to a directional stimulus. The direction of the response is related to the direction of the stimulus, e.g. towards it, away from it	Light	Phototropism	In almost all plants, stems bend towards a directional light source (positive phototropism), roots bend away (negative phototropism) and leaves become positioned at right angles
	Gravity	Geotropism	In almost all plants, stems bend away from gravity (negative geotropism), roots bend towards it (positive geotropism) and leaves become positioned at right angles
	Water	Hydrotropism	In almost all plants, roots are positively hydrotropic and stems and leaves show no directional response
	Chemicals	Chemotropism	Growth of pollen tube towards chemicals from the micropyle. Growth of fungal hyphae away from the products of their metabolism
	Touch	Thigmotropism	Twining of pea (*Pisum*) tendrils around supports. Spiralling of bean (*Phaseolus*) shoots around supports
Tactic The movement of a freely motile organism (or a freely motile part of an organism) in response to a directional stimulus. The direction of the response is related to the direction of the stimulus	Light	Phototaxis	Unicellular green algae such as *Chlamydomonas* will move to regions of optimum light intensity
	Temperature	Thermotaxis	Unicellular green algae such as *Chlamydomonas* will move to regions of optimum temperature
	Chemicals	Chemotaxis	The antherozooids (sperm) of mosses, liverworts and ferns are attracted to chemicals (e.g. malic acid) produced by the archegonium (female part)
Nastic The movement of part of a plant in response to a stimulus. The direction of the response is *not* related to the direction of the stimulus	Light	Photonasty	The leaves of many leguminous plants, e.g. French bean (*Phaseolus*), are lowered in the dark and raised in the light. Many daisies (*Oxalis*) close their flowers in the dark and open them in the light
	Temperature	Thermonasty	Some plants, e.g. *Crocus* and tulip (*Tulipa*), open their flowers at relatively high temperatures (16 °C) and close them at lower temperatures
	Touch	Thigmonasty	The leaves of the Venus flytrap (*Dionaea*) close together rapidly when touched, e.g. by an insect. The leaves of the sensitive plant (*Mimosa*) collapse when touched

chemistry of a particular hormone is very similar in all species, although its effects may vary from animal to animal. Table 8.4 is a summary of the chemical nature of different hormones.

Table 8.4 *Chemical nature of hormones*

Chemical group	Hormones
Polypeptides (less than 100 amino acids)	Oxytocin, vasopressin, insulin, glucagon
Protein	Prolactin, follicle stimulating hormone, luteinising hormone, thyroid stimulating hormone, adrenocorticotrophic hormone, growth hormone
Amines (derivatives of amino acids)	Adrenaline, noradrenaline, thyroxine
Steroids (derivatives of lipids)	Oestrogen, progesterone, testosterone, cortisone, aldosterone

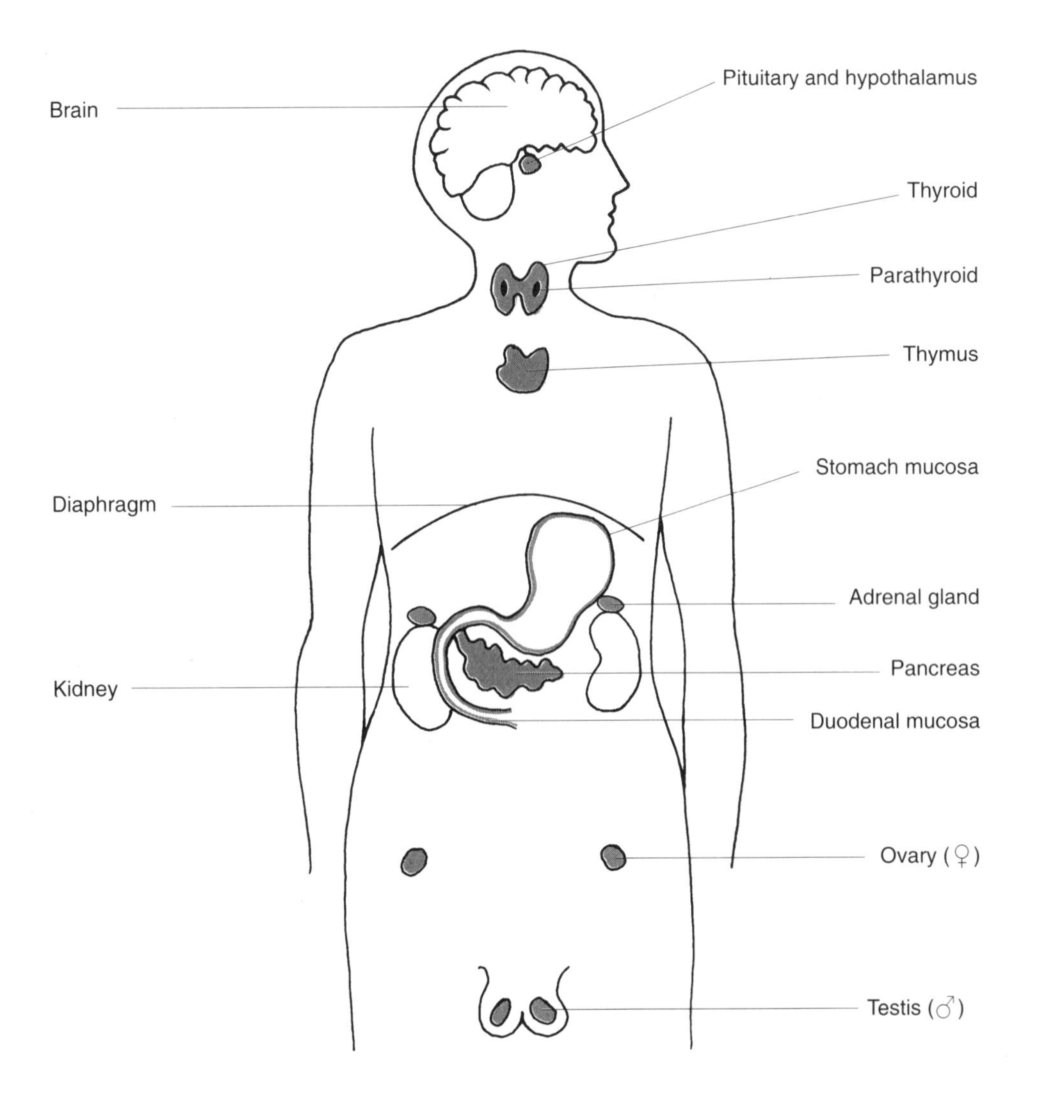

Fig. 8.6 Major human endocrine glands

Action of hormones

Hormones influence molecular reactions in cells. They achieve this in four ways:

❶ affecting transcription of genetic information (e.g. oestrogen)

❷ affecting protein synthesis (e.g. growth hormone)

❸ altering enzyme activity (e.g. adrenaline)

❹ changing the permeability of cell membranes (e.g. insulin)

Hormones only affect specific cells, namely those which possess the relevant **receptor molecules** on their surface. The hormone molecule and the receptor molecule fit one another exactly in a similar way to how enzyme and substrate molecules combine. Two mechanisms of hormone action are illustrated in Fig. 8.7.

Effects of hormones

The effects of hormones are various and highly specific. Some of these have been discussed in some detail earlier, e.g. reproductive hormones were discussed in 4.2. Table 8.5 summarises the effects of all the main mammalian hormones.

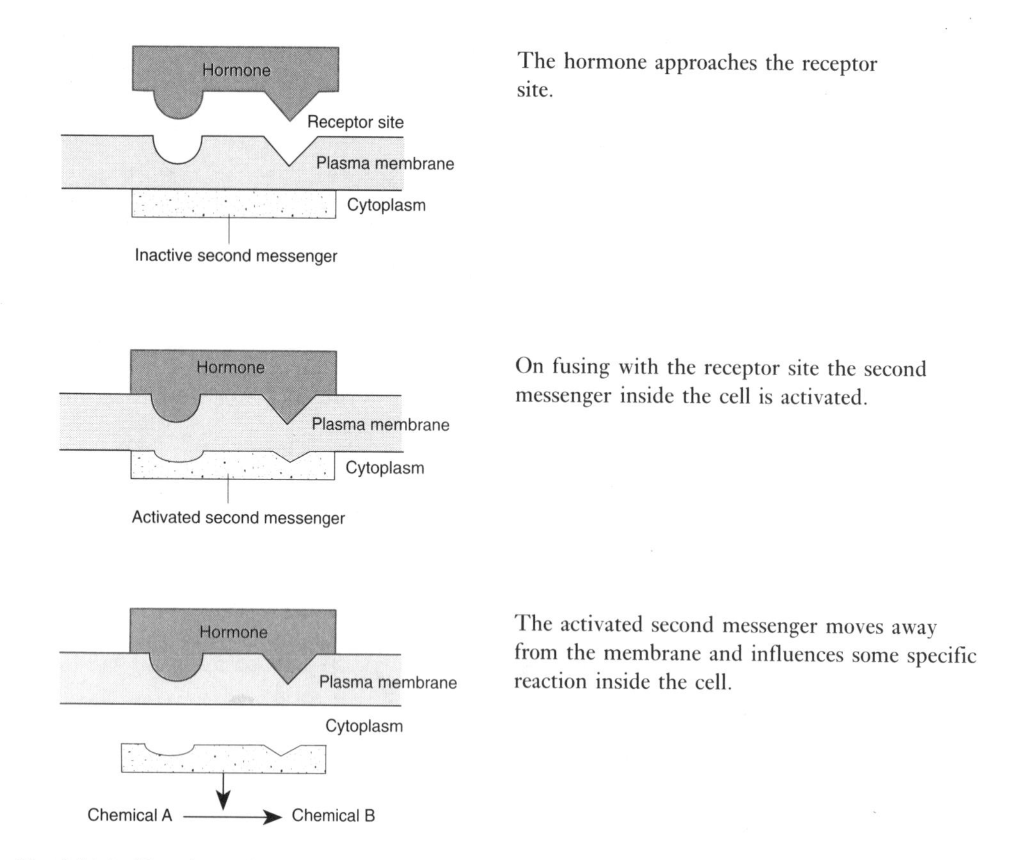

Fig. 8.7(a) Use of second messenger (protein and polypeptide hormones)

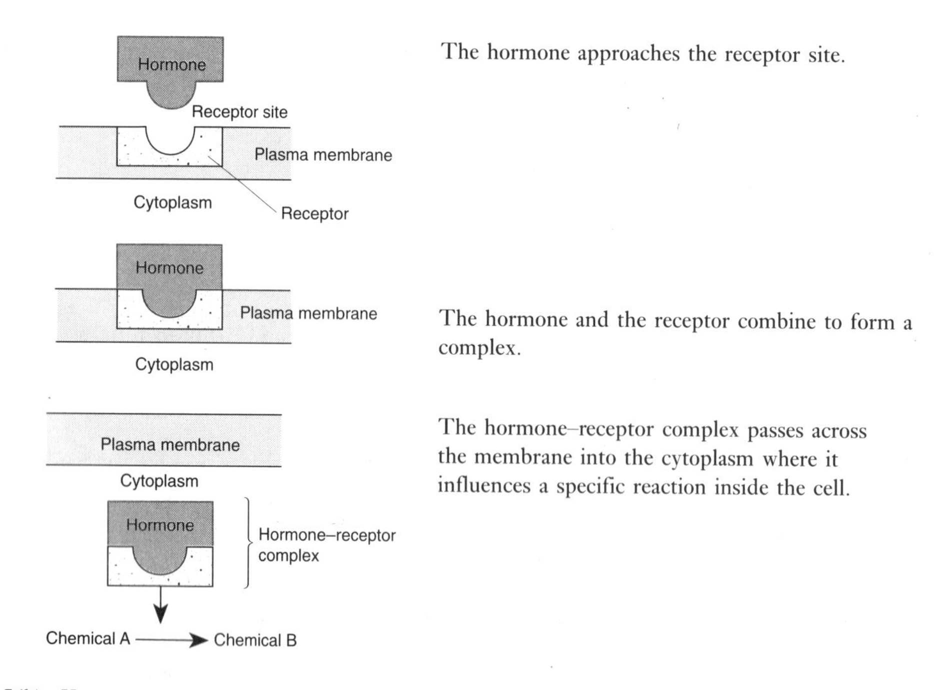

Fig. 8.7(b) Hormone–internal receptor complex (steroid hormones)

Table 8.5 *Summary of mammalian hormones and their effects*

Gland	Hormone	Effects
Pituitary, anterior lobe	Somatotrophin (growth hormone; GH)	Increases growth rate in young animal and maintains size of body parts in adult
	Thyroid stimulating hormone (TSH)	Acts on thyroid gland, thereby controlling metabolic rate
	Adrenocorticotrophic hormone (ACTH)	Controls activity of adrenal cortex
	Follicle stimulating hormone (FSH)	Initiates development of Graafian follicles in females and sperm production in males
	Luteinising hormone (LH), or interstitial cell stimulating hormone (ICSH)	Causes release of the ovum from the ovary and consequent development of the corpus luteum in females; in males it stimulates secretion of testosterone from the testes
	Prolactin, or luteotrophic hormone (LTH)	Mammary gland development and milk production; maternal behaviour in birds
Pituitary, posterior lobe	Antidiuretic hormone (ADH) or vasopressin	Stimulates water reabsorption from the kidney tubules; raises blood pressure by constricting arterioles
	Oxytocin	Causes contraction of smooth muscles in uterus of a pregnant female and the release of milk during suckling
Thyroid	Thyroxine	Increases metabolic rate; induces metamorphosis in frogs
	Calcitonin	Lowers plasma calcium level
Parathyroids	Parathormone	Raises plasma calcium level
Pancreas (islets of Langerhans)	Insulin (from β cells)	Lowers blood glucose level
	Glucagon (from α cells)	Raises blood glucose level
Adrenal cortex	Aldosterone	Stimulates sodium reabsorption from the kidney tubules
	Cortisol	Helps body resist stress, partly by raising blood glucose level
Adrenal medulla	Adrenaline and noradrenaline	Prepare body for activity
Ovaries	Oestrogen	Produces female secondary sex characteristics
	Progesterone	Inhibits ovulation and maintains pregnancy
Testes	Testosterone	Produces male secondary sex characteristics
Stomach mucosa	Gastrin	Stimulates production of gastric juice
Duodenal mucosa	Secretin	Stimulates production of bile by the liver and mineral salts by the pancreas
	Cholecystokinin and pancreozymin	Causes contraction of the gall bladder and stimulates pancreas to produce enzymes
Placenta (during pregnancy only)	Chorionic gonadotrophin	Maintains the corpus luteum in the ovary

8.4 THE NERVOUS SYSTEM

The nervous system permits the rapid passage of information from one part of the body to another, so that suitable responses to stimuli can be made at once. In its simplest form it comprises a network of nerves connecting each part to every other part. With an increase in the number of parts to be connected, this system proved inadequate and a **central nervous system (CNS)** developed. Each part was connected to this central 'switchboard' which then connected incoming messages to the appropriate effector organs. In bilaterally symmetrical animals it was appropriate to have the CNS running along the length of the body with paired nerves branching from it to each segment. As animals developed a definite head region which encountered stimuli before other parts, it became the main location for sense organs. To deal with this increased volume of nervous information at the anterior end, the CNS in this region swelled. The swelling or cerebral ganglion later developed into a brain whose function evolved beyond simple coordination to include storage of learned information and intelligent behaviour.

THE NERVE IMPULSE

The nerve impulse is transmitted along cells known as neurones (Fig. 8.8).

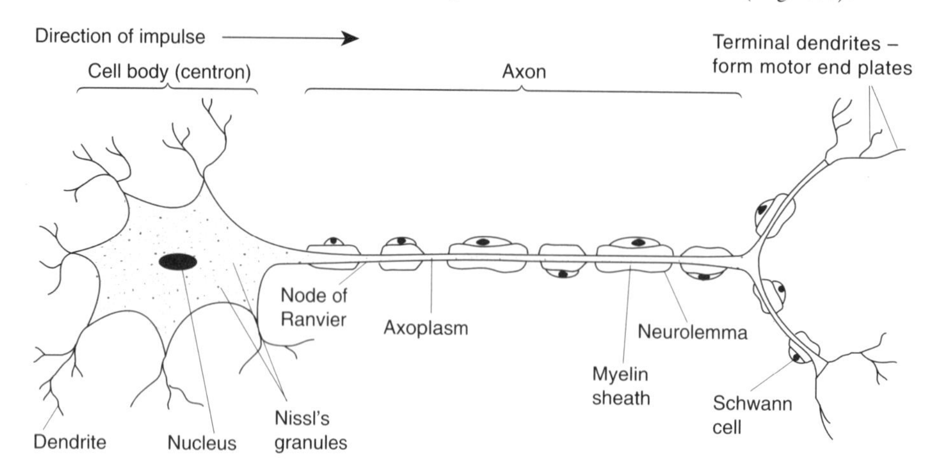

Fig. 8.8 Effector (motor) neurone

The **cell body** of the neurone contains the nucleus. A long process called the **axon** extends from the cell body and carries the nerve impulse away from it. Extensions which carry impulses towards the cell body are called **dendrons** and these may divide into smaller units called **dendrites**. Axons and dendrons are often sheathed with special cells called **Schwann cells,** in which case they are said to be **myelinated**. Between adjacent Schwann cells are gaps known as **nodes of Ranvier**.

Resting potential

At its resting potential the inside of a neurone membrane is negatively charged compared with the outside, the potential difference being around 70 mV. In this state the membrane is said to be **polarised**. This potential difference is the result of the uneven distribution of four ions: sodium (Na^+), potassium (K^+), chloride (Cl^-) and organic anions (COO^-). In particular, the concentration of sodium and chloride ions is greater outside, while that of potassium and organic ions is greater inside. These differences in the concentration of ions across the membrane are maintained by the active transport of these ions against their concentration gradients. This is particularly the case for sodium and potassium ions, which are transported by a **sodium pump** and a **potassium pump** respectively, often collectively referred to as a **cation pump**.

Action potential

If suitably stimulated, the charge on the neurone can be reversed, i.e. the inside of the membrane becomes positive compared with the outside by about +40 mV. This is called the **action potential** and the membrane is said to be **depolarised**. Within 2 ms the membrane returns to its resting potential – it has been repolarised.

The reasons for these changes largely concern the movement of the sodium and potassium ions. Upon stimulation the neurone membrane suddenly becomes very permeable to sodium ions and the resultant sudden influx of these ions begins to depolarise the membrane. This depolarisation in turn increases the membrane's permeability to sodium ions and the rate of influx increases. This is an example of positive feedback. As the inside of the membrane becomes increasingly positive due to these sodium ions, so its permeability to them decreases. At the same time as the sodium ions begin to move inwards, the potassium ions begin to move in the opposite direction. The potassium ions, however, move much less rapidly. These events are summarised in Fig. 8.9.

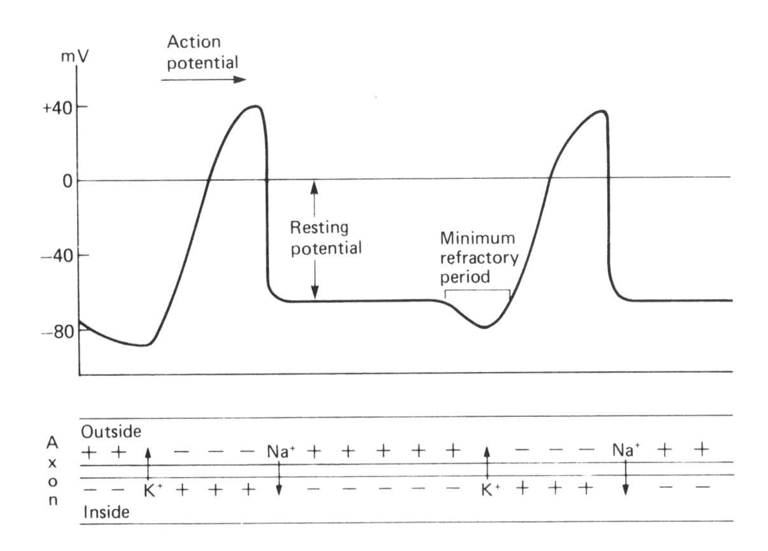

Fig. 8.9 Transmission of action potential

Propagation of the nerve impulse

Nerve impulses are **all-or-nothing** phenomena, i.e. provided the stimulus exceeds a certain minimum value, called the **threshold value**, an action potential of a fixed value proceeds along the whole length of the neurone. There are no intermediate types – the potential is either the full value or it is nothing.

1 At resting potential there is a high concentration of sodium ions outside and a high concentration of potassium ions inside the neurone.

2 When the neurone is stimulated sodium ions rush into the axon along a concentration gradient. This causes depolarisation of the membrane.

3 Localised electrical circuits are established which cause a further influx of sodium ions and thus progression of the impulse. Behind the impulse, potassium ions begin to leave the axon along a concentration gradient.

4 As the impulse progresses, the outflux of potassium ions causes the neurone to become repolarised behind the impulse.

5 After the impulse has passed and the neurone is repolarised, sodium is once again actively expelled in order to increase the external concentration and so allow the passage of another impulse.

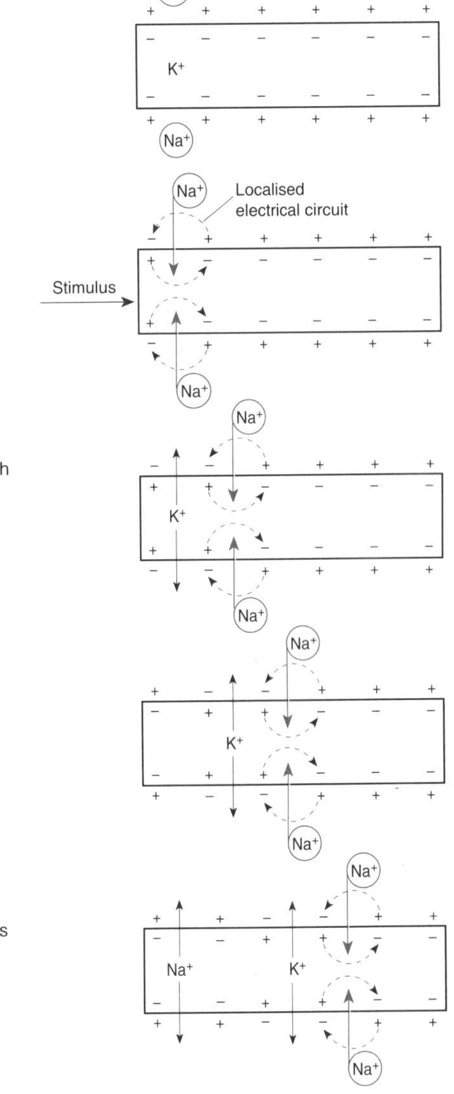

Fig. 8.10 Propagation of a nerve impulse

Following an action potential there is a period of 1 ms during which no further action potential can be generated. This is known as the **refractory period**. It serves two functions:

1. It prevents the action potential being propagated in both directions – it can only pass forwards.
2. It separates one action potential from the next so preventing them from merging.

The propagation of a nerve impulse is due to localised circuits being set up which lead to further depolarisation of the membrane ahead of them and repolarisation behind them. The process is illustrated in Fig. 8.10.

The speed of transmission of a nerve impulse varies from 0.5 m to over 100 m in one ms. Transmission is faster where the axon has a large diameter and where the neurone is myelinated.

THE SYNAPSE

Neurones are not in direct contact with each other but are separated by tiny gaps known as synapses. The structure of a synapse is shown in Fig. 8.11.

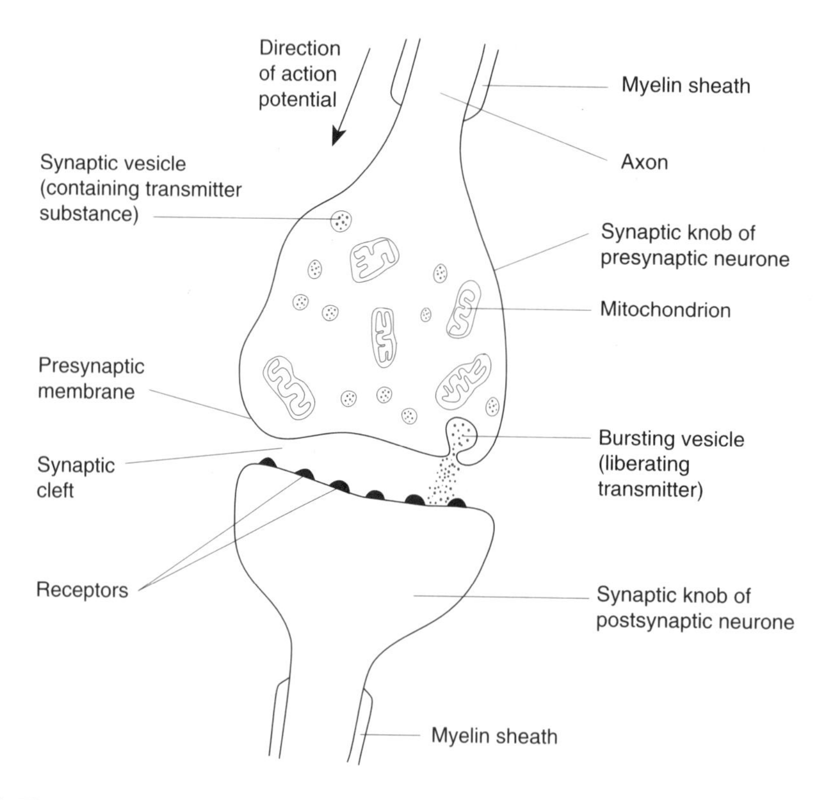

Fig. 8.11 The synapse

Transmission across a synapse

On arrival at a synapse the synaptic knob of the presynaptic neurone liberates a **neurotransmitter substance**. There are two main neurotransmitters **acetylcholine** and **noradrenaline**. The neurotransmitter diffuses across the synaptic cleft and fuses with receptor molecules on the postsynaptic neurone. Upon fusing the permeability of the membrane of the postsynaptic neurone is altered, allowing an influx of sodium ions. This influx creates a new action potential which is then propagated along the postsynaptic neurone to the next synapse. The neurotransmitter is quickly hydrolysed by enzymes on the postsynaptic neurone to prevent it from indefinitely propagating new action potentials. Details of these processes are illustrated in Fig. 8.12.

1 The arrival of the impulses at the synaptic knob alters its permeability, allowing calcium ions to enter.

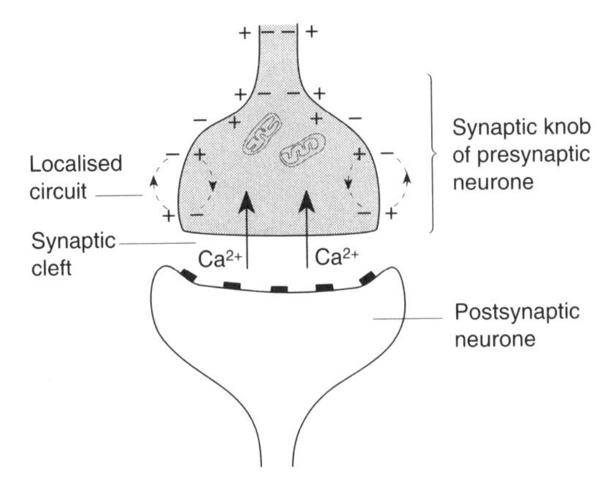

2 The influx of calcium ions causes the synaptic vesicle to fuse with the presynaptic membrane so releasing acetylcholine into the synaptic cleft.

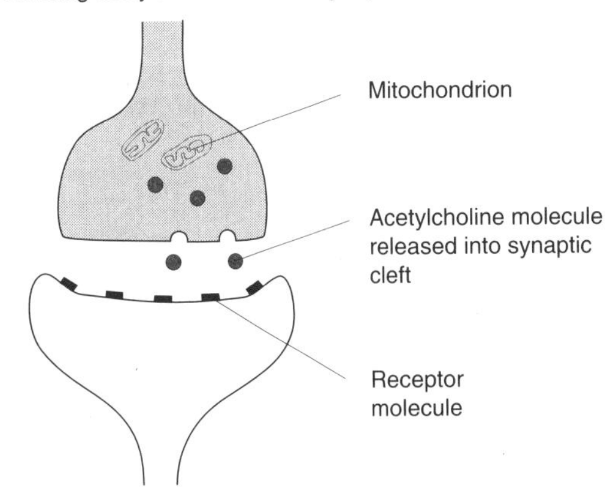

3 Acetylcholine fuses with receptor molecules on the postsynaptic membrane. This alters its permeability, allowing sodium ions to rush in.

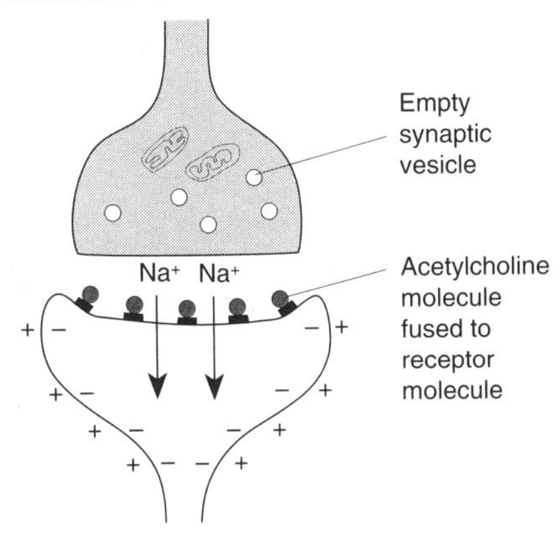

4 The influx of sodium ions generates a new impulse in the postsynaptic neurone.

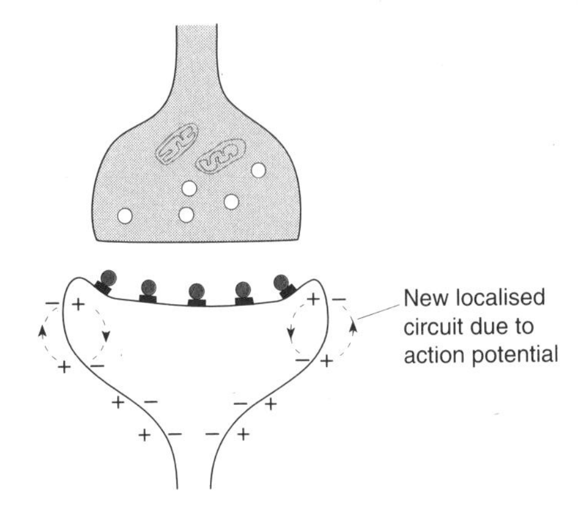

5 Acetylcholinesterase on the postsynaptic membrane hydrolyses acetylcholine into choline and ethanoic acid (acetyl). These two components then diffuse back across the synaptic cleft into the presynaptic neurone.

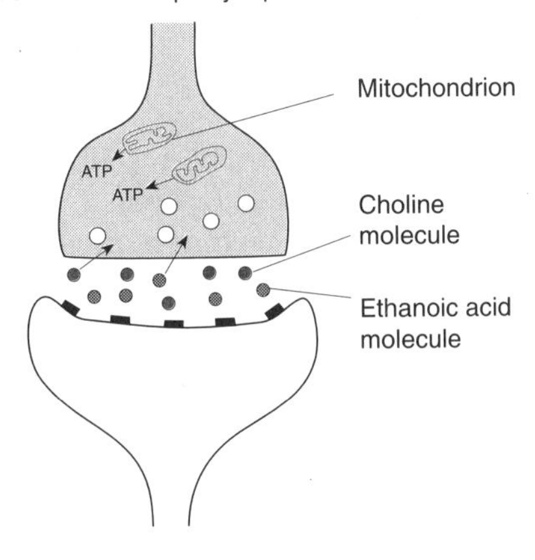

6 ATP released by the mitochondria is used to recombine choline and ethanoic acid (acetyl) molecules to form acetylcholine. This is stored in synaptic vesicles for future use.

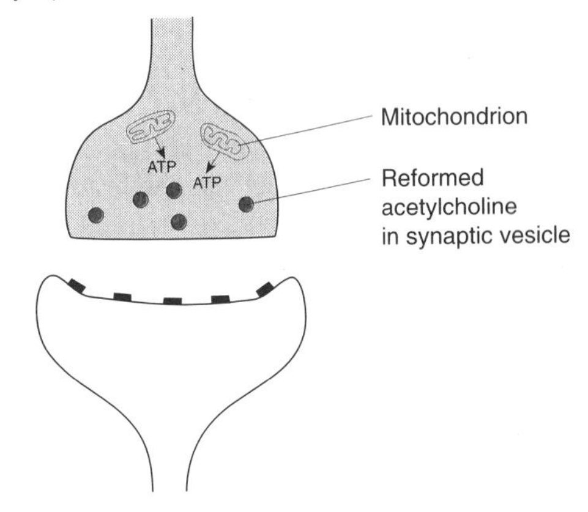

Fig. 8.12 Synaptic transmission

Functions of the synapse

1. To transmit information between neurones.
2. To act as junctions allowing impulses to be divided up along many neurones or merge into one.
3. To act as valves in that they ensure that impulses pass across them in one direction.
4. To filter out low frequency impulses.

Reflex arc

A reflex is an automatic response to a stimulus. These involuntary responses are important in a wide range of bodily functions such as breathing, maintenance of blood pressure, posture, dilation and constriction of the pupil of the eye. Figs 8.13 and 8.14 illustrate the general principles of a typical reflex response.

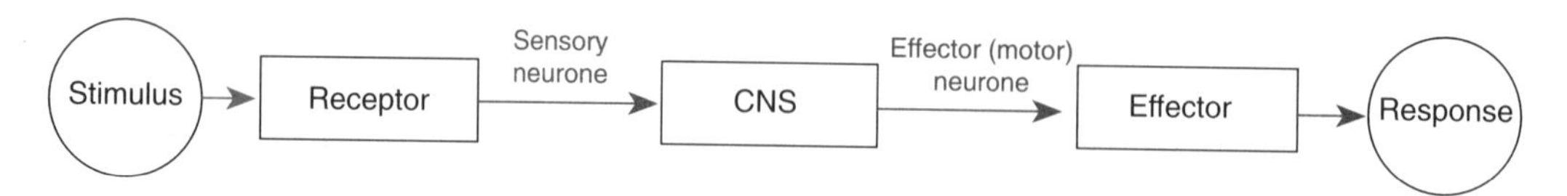

Fig. 8.13 Stages of a reflex response

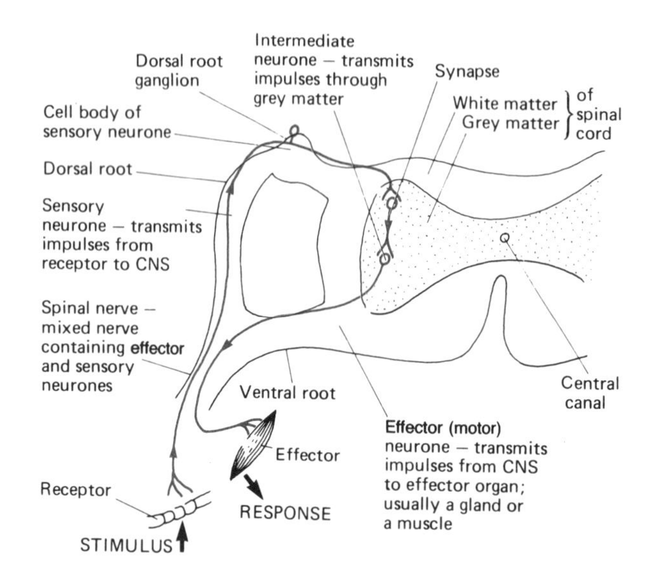

Fig. 8.14 Nerve pathways of a reflex arc

THE CENTRAL NERVOUS SYSTEM (CNS)

In vertebrates the brain developed as a swelling at the anterior end of the dorsal nerve cord, the rest remaining as the spinal cord.

The **spinal cord** consists of tracts of ascending (sensory) and descending (effector) fibres which carry information between the brain, and sense receptors, muscles and glands.

The **brain** may be subdivided into the following main regions each with their own specific functions.

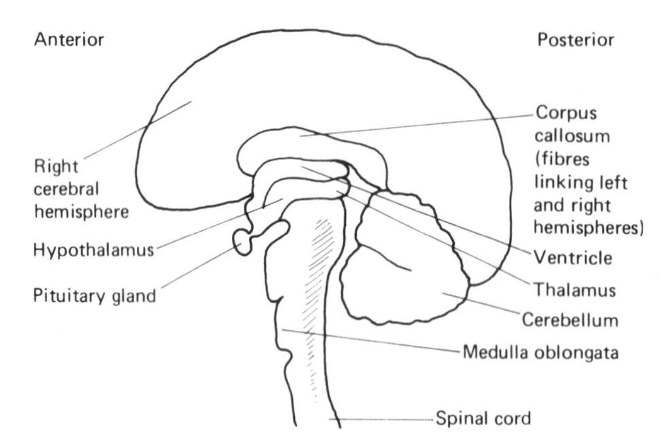

Fig. 8.15 VS human brain

Stimulation of brain cells

Neurones in the brain are normally stimulated by transmitter substances released at the synapse by a neighbouring cell. However, chemical or physical changes in the body can sometimes affect brain cells directly. For example:

1. Osmoreceptor cells in the hypothalamus will stimulate the pituitary to release antidiuretic hormone when the osmotic pressure of body fluids is low.
2. When carbon dioxide levels in the blood are high, reflex respiratory centres in the medulla increase breathing rate and depth.
3. Local temperature changes in the hypothalamus initiate the required responses to raise or lower body temperature.

Table 8.6 *Regions of the brain*

Region	Function
Forebrain	
Olfactory tracts	Concerned with sense of smell; found deep in forebrain
Cerebral hemispheres	Very enlarged and important area of human brain; electrical stimulation has led to 'mapping' of the most superficial area, the cerebral cortex
'Tweenbrain	
Thalamus	Contains ascending and descending tracts linking forebrain with spinal cord
Hypothalamus	Largely controls the pituitary gland; seat of basic emotions or 'drives' such as hunger, thirst, fear, rage and sex
Midbrain	
Corpora quadrigemina (optic lobes of lower vertebrates)	Concerned with sense of sight
Red nucleus	Helps control movement and posture
Hindbrain	
Medulla oblongata	Controls voluntary movement and also has control centres for activities such as swallowing, salivation, heartbeat, vascular constriction and dilation, and respiration
Cerebellum	Regulation and coordination of muscular activities

The autonomic nervous system

This regulates the body's involuntary activities, and is divided into the sympathetic and parasympathetic systems. The sympathetic system has comparable effects to adrenaline.

Table 8.7 *Comparison of the sympathetic and parasympathetic systems*

Sympathetic system	Parasympathetic system
Prepares body for action	Prepares body for relaxation
Increases heartbeat	Slows heartbeat
Dilates arteries in skeletal muscles	Dilates arteries in gut
Slows gut movements	Speeds gut movements
Dilates bronchioles	Constricts bronchioles
Dilates pupil	Constricts pupil
Causes sweat glands to secrete	Causes tear and salivary glands to secrete
Causes hairs to stand erect	Hair is not erect
Bladder and anal sphincters contract	Bladder and anal sphincter relax

BEHAVIOUR

It is difficult to categorise behaviour patterns rigidly, but many of them involve elements of reflex action, orientation and learning.

Reflex action

Reflex actions have the advantage of being fast and automatic, and they are important in escape and avoidance reactions in all animals.

Orientation

Orientation often involves reflex actions, but they are normally developed into behaviour patterns. There are two main forms of orientation found in animals:

1. **Kinesis** This is an increase in random movement under unfavourable conditions. Woodlice move rapidly and turn frequently in dry conditions, but tend to congregate under humid conditions as their rate of movement slows.
2. **Taxis** This is the directional movement of the whole organism in direct response to a stimulus. Such stimuli may be light (phototaxis) or chemicals (chemotaxis), e.g. movement of planarians towards food; location of the ovum by the sperm of lower animals.

Learning

This is the change in behaviour based on experience and it is the most adaptable form of behaviour. Five subdivisions are often recognised:

1. **Habituation** This is the diminishing of a response as a result of repeated stimulation. It is specific to a particular stimulus, e.g. snail (*Helix*) withdraws its tentacles in response to mechanical stimulation, but ceases to do so after repeated stimulation.
2. **Associative learning** In this the animal learns to associate a reward or punishment with a particular form of behaviour or stimulus. Such a pattern may be established either as a result of **classical conditioning** (the work of Pavlov on the salivation of dogs) or through **trial and error**.
3. **Imprinting** Young animals tend to follow, or imprint on, their parents. Experiments have shown that they may follow the first moving thing they see.
4. **Exploratory learning** Even without a specific reward or punishment an animal learns features of its environment which may benefit it later.
5. **Insight learning** This involves the production of a new adaptive response as a result of 'insight'. It is a difficult subdivision to define and any definition of it results in controversy. It is thought to apply only to man.

Table 8.8 *Comparison of learned and instinctive behaviour*

Instinctive behaviour	Learned behaviour
Inborn and not acquired during an animal's lifetime	Acquired during an animal's lifetime
Fixed and not adaptable although some minor modification over a long period may be possible	May be easily and rapidly adapted to suit changing circumstances
Similar amongst all members of a species	Varies considerably amongst different members of the same species
Unintelligent and there is no appreciation of the functions of the behaviour	May be intelligent and the animal often appreciates the function of a particular action
Often comprises a chain of actions in which the completion of one acts as the trigger for the start of the next	No fixed sequence of actions and the completion of one need not necessarily affect which action should follow
Apart from minor modifications the behaviour is permanent	Usually a temporary and short-lived form of behaviour, although it may be reinforced, thus making it more or less permanent
Although there are many forms of instinct the basic form of this behaviour is the same for all organisms	Wide range of learned behaviour ranging from simple taxis or imprinting to complex forms of intelligence and reasoning

Other factors affecting behaviour

- Hormones may modify behaviour patterns; testosterone, for example may increase the incidence of aggressive behaviour, and prolactin releases nest-building responses in birds.
- Animals may show rhythmic changes in behaviour according to the time of day or the season.
- Characteristic behaviour patterns may be released when an animal meets another member of its own species. A male bird, meeting a female on his territory in spring, may begin a courtship display. However, should he meet another male he will threaten him until the intruder withdraws, so that competition for food or a mate is removed. Intraspecific interactions are especially important to humans, who owe their success largely to their ability to live in cooperative groups.

Comparison of endocrine and nervous systems

While both systems operate together to a large degree, they also function separately at times, which inevitably leads to some differences between them. These are outlined in Table 8.9.

Table 8.9 *Comparison of the endocrine and nervous systems*

	Hormonal communication	Nervous communication
Origin of stimulus	Gland	Sense receptor
Nature of stimulus	Hormone	Nervous impulse
Means of transmission	Bloodstream	Nerve fibre
Destination of stimulus	All over body	To a specific point
Receptor	Target organ	Effector (muscle or gland)
Speed of transmission	Usually slow	Rapid
Effects	May be widespread	Localised
Duration	Usually long-lasting	Usually brief

SENSE ORGANS

Organisms have evolved specialised sense receptor cells to provide them with information about changes in both their internal and external environment. It is the function of these receptor cells to convert whatever form of energy they are sensitive to into a nervous impulse, i.e. they act as **biological transducers**. Groups of receptor cells collected together and operating in conjunction are called **sense organs**. Receptors can be classified according to their positions:

- **Exterioreceptors** receive information from the external environment.
- **Interioreceptors** receive information from the internal environment.
- **Proprioreceptors** provide information on the relative position of muscles.

They can also be classified according to the form of energy the stimulus takes:

- **Chemoreceptors** detect chemical stimuli, e.g. taste, blood pH
- **Electroreceptors** detect electrical fields, e.g. in certain fish.
- **Mechanoreceptors** detect mechanical stimuli, e.g. pressure, gravity, movements.
- **Thermoreceptors** detect temperature changes, e.g. heat and cold receptors in the skin.
- **Photoreceptors** detect electromagnetic radiation, e.g. light.

PHOTORECEPTION – THE MAMMALIAN EYE

The eye detects that part of the electromagnetic spectrum which lies in the range 400–700 nm. It lies in a bony socket called the **orbit** within which it can move by use of the **rectus muscles** attached to the skull. Fig. 8.16 illustrates the structure of the eye and the functions of each of its components.

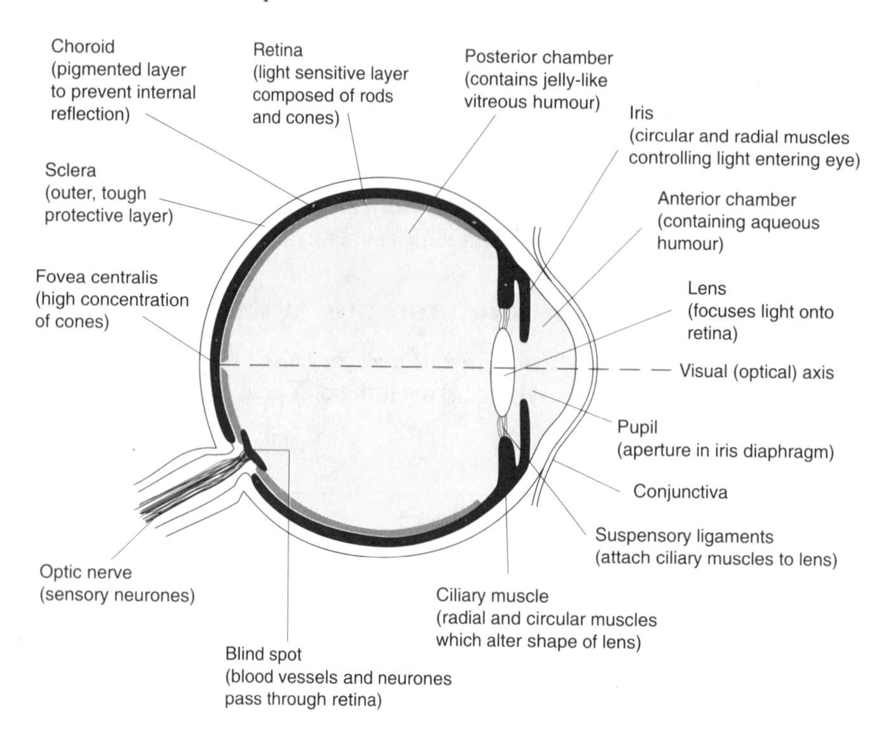

Fig. 8.16 Vertical section through the human eye

Control of light entering the eye

The functioning of the eye depends on the appropriate light intensity reaching the retina. Control of the amount of light entering the eye is achieved by the circular and radial muscles of the iris. Photoreceptor cells in the retina act as the stimulus for initiating movements of the iris muscles which alter the diameter of the pupil (Table 8.10).

Table 8.10 *Action of the iris muscles*

	Dim light	Bright light
Circular muscles of iris	Relax	Contract
Radial muscles of iris	Contract	Relax
Pupil	Dilates	Constricts

Focusing of light rays on the retina

To provide a clear image it is necessary for light rays to be focused onto the retina. This involves the bending (refraction) of these rays, the degree of bending being determined by the distance from the eye of the object being viewed. The closer the eye is to the object the greater the degree of refraction required. While much refraction of the light is achieved by the cornea, it is only the elastic lens which can alter its shape and so adjust the focusing when viewing near or distant objects. The lens shape is altered by means of **ciliary muscles** which are attached to it by **suspensory ligaments**. The changes of shape of the lens to allow focusing are referred to as **accommodation** and this process is illustrated and explained in Fig. 8.17 and Table 8.11.

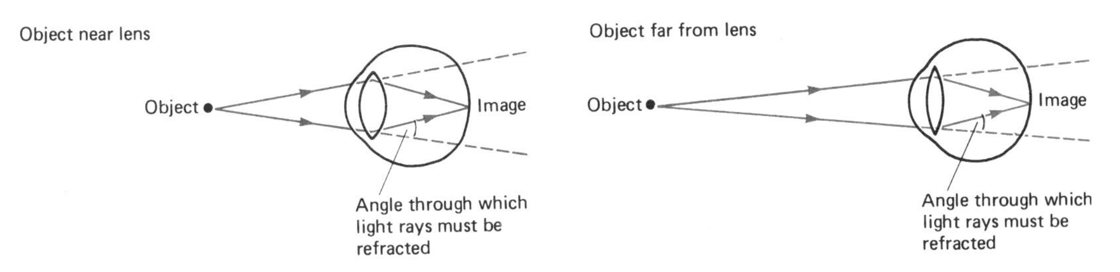

Fig. 8.17 Accommodation

Table 8.11 *Accommodation of the eye*

	Object near eye	Object far from eye
Light must be refracted	a lot	a little
Lens must be	thick	thin
Ciliary muscles	contract	relax
Suspensory ligaments	slack	tense

Structure of the retina

The retina possesses two types of cells: rods and cones. The cellular structure of the retina is shown in Fig. 8.18 and a rod cell is shown in Fig. 8.19. Table 8.12 compares rod and cone cells.

Each retinal photoreceptor cell contains a photosensitive pigment. In the case of a rod cell this is **rhodopsin** or **visual purple**. Light absorbed by the rod cell causes rhodopsin to split into its constituent parts, **opsin** and **retinal**, in a process called bleaching. It is this chemical change which initiates an action potential along the neurone joining the rod cell to the brain.

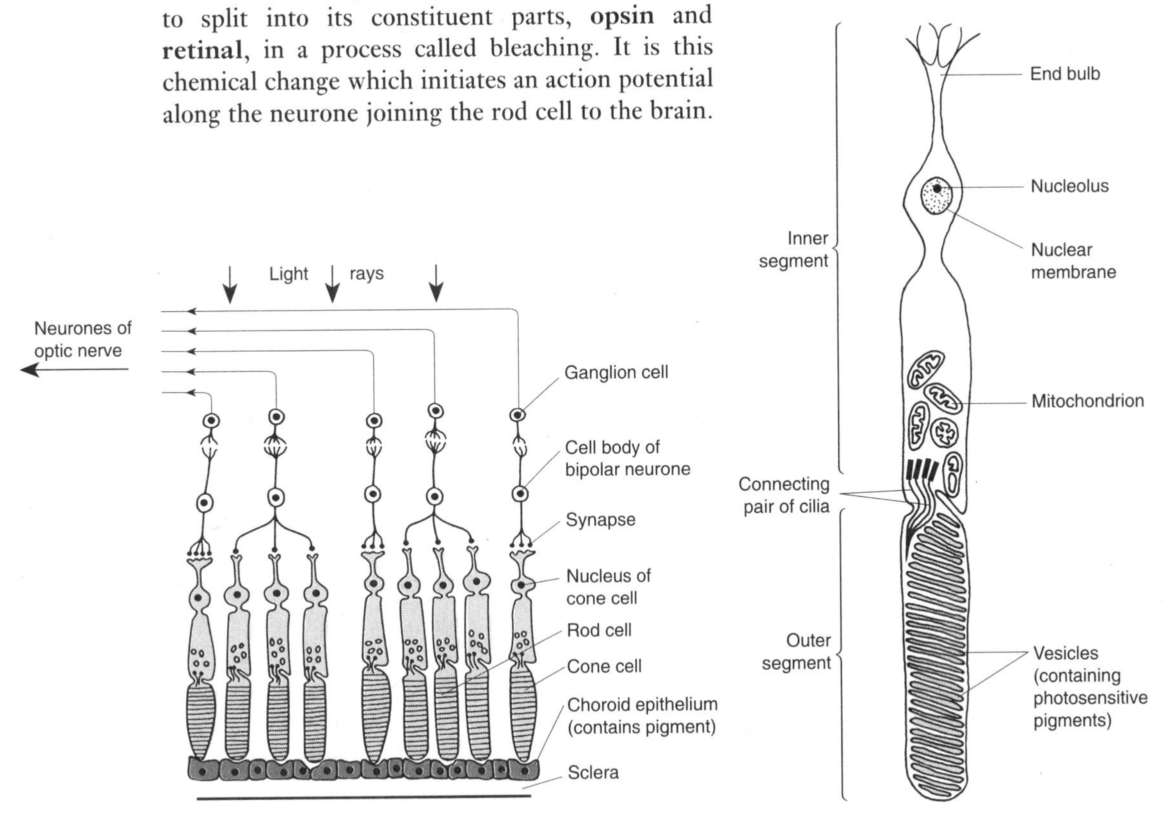

Fig. 8.18 Cellular structure of the retina

Fig. 8.19 Structure of a rod cell

Colour vision

It is only cone cells which are sensitive to different colours. It is thought they are of three types, any one being sensitive to either blue, green or red light. It is the degree of stimulation of each of these three types which provides the full range of colours visible to organisms with colour vision. This theory is called the **trichromatic theory** of colour vision.

Table 8.12 *Comparison of rods and cones*

Rods	Cones
More numerous	Less numerous
Usually around the periphery of retina	Usually located in centre of retina
Arranged in functional units served by one bipolar neurone, therefore acuity low	Each cone served by its own bipolar neurone, therefore acuity high
Very sensitive to low levels of illumination	Only stimulated by bright light
One type of rod only, stimulated by most wavelengths of visible light except red	Three types of cone, each selectively responsive to different wavelengths, therefore allowing colour perception
Rapid regeneration of light-sensitive pigment, therefore can perceive flicker well	Slower regeneration of light-sensitive pigment, therefore less responsive to flicker

8.5 MOVEMENT AND SUPPORT IN ANIMALS

Organisms initially evolved in water, which gave them support. Even so, skeletons were useful as a means of protection and a rigid framework for the attachment of muscles. When organisms became terrestrial it became necessary to use the skeleton for support. For a plant the support needs to be substantial because competition for light has led to the evolution of trees over 100 m in height. Woody plants are supported by their xylem. As xylem is a dead tissue it makes no energy demands on the organism and may therefore be massive without any disadvantage to the plant. Herbaceous plants and woody plants with herbaceous parts rely on the hydrostatic pressure created by water entering cells by osmosis. This pressure is termed turgor. Hydrostatic pressure is also utilised by animals. Earthworms, for instance, take advantage of the fact that the liquid in their coelom, like all liquids, is virtually incompressible and can act as a skeleton. Other animals use specialised tissues which, unlike those of plants, are living and require energy. For this reason, and because most animals move from place to place, animals and their skeletons are much smaller. Where a very large size has been attained the organisms have returned to water in order to gain support for their bodies, e.g. blue whale.

Locomotion is a feature of animals and some algae. The mode of nutrition in plants makes movement unnecessary because the essentials of water, carbon dioxide and light can all be obtained while remaining in one place. Indeed, to move from place to place would involve being relatively small and this could be a disadvantage in the competition for light. Being sessile, however, creates certain problems especially in the transfer of gametes during sexual reproduction. In animals locomotion is necessary in order to obtain food, although some aquatic organisms can filter food from the surrounding water. Although cilia and flagella can be used to achieve locomotion in unicellular organisms, the remainder are dependent on some musculoskeletal system. The actual arrangement of this system is dictated by the mode of locomotion, e.g. burrowing, walking, crawling, jumping, climbing, gliding, flying or swimming.

SUPPORT

As an animal increases in size the soft tissues of the body require support. On the whole, aquatic animals need less extensive skeletal support than terrestrial ones. There are three basic types of skeleton: exoskeleton, endoskeleton and hydrostatic skeleton.

Exoskeleton (external skeleton), e.g. arthropod cuticle

An exoskeleton provides more-or-less complete protection for the internal organs and a large area for the attachment of muscles. In arthropods the exoskeleton is composed of a three-layered cuticle secreted by the underlying epidermal cells. The outer layer is a thin, waxy, waterproof layer (epicuticle). Immediately above the epidermis is a flexible layer made of **chitin**, which is a polymer of glucosamide and acetic acid (endocuticle). Between the two is a rigid layer of chitin impregnated with tanned proteins (exocuticle). The proteins together with lipids help to make the chitin impermeable to water. Crustaceans also have calcium carbonate deposited in this layer to give additional strength.

The inflexible plates of the exoskeleton are separated by flexible regions where the rigid exocuticle is absent; this permits movement. Sensory hairs project through the exoskeleton which is also interrupted by the openings of various glands and the digestive, respiratory and reproductive systems. The most serious limitation imposed by an exoskeleton is on growth. Chitinous plates do not grow but must be periodically shed by a process known as **ecdysis (moulting)**.

Endoskeleton (internal skeleton)

The endoskeletons found in vertebrates are cellular although the bulk of the substance is a noncellular matrix secreted by the cells. **Cartilage** forms the first skeleton of mammalian embryos, but later most of it is converted to bone. Cartilage combines rigidity with a degree of flexibility. Most vertebrates are supported by a bony endoskeleton, bone providing a strong, rigid framework requiring the presence of joints if movement is to take place.

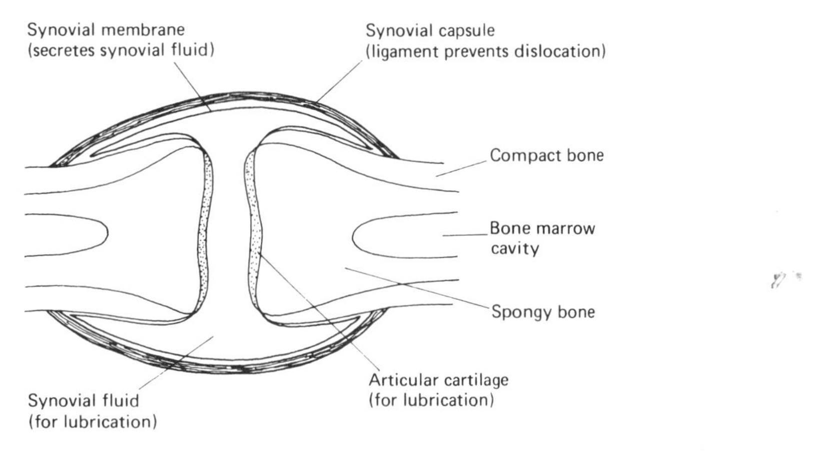

Fig. 8.20 Generalised synovial joint

Hydrostatic skeleton

This is found in certain invertebrates (e.g. earthworm) where the coelomic fluid forms an incompressible but flexible material to aid locomotion. The muscles are arranged segmentally and act on the coelomic fluid. Contraction of the circular muscles makes the body longer and thinner while contraction of the longitudinal muscle makes the body shorter and thicker. Chaetae on each segment anchor the appropriate part of the body so that these alternate contractions bring about movement. Not all the circular or longitudinal muscles contract at one time; waves of contraction pass back along the body.

Muscular movements

Sliding-filament theory of muscular contraction Basically, this supposes that the muscle filaments do not shorten when the muscle contracts but that they slide between one another. The strongest evidence for this theory comes from electron microscope studies of muscles fixed at different degrees of tension. These show (see Fig. 8.21):

❶ the lengths of the actin and myosin filaments remain unchanged at different tensions;

2. during contraction the isotropic band (I-band) shortens;
3. during contraction the length of the anisotropic band (A-band) remains more or less constant.

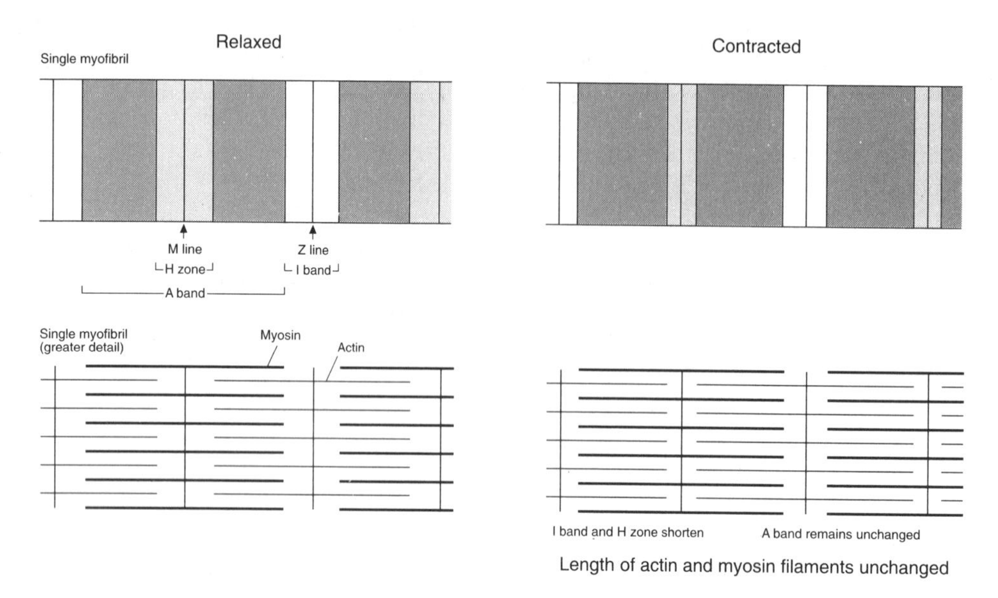

Fig. 8.21 Relaxation and contraction in skeletal muscle

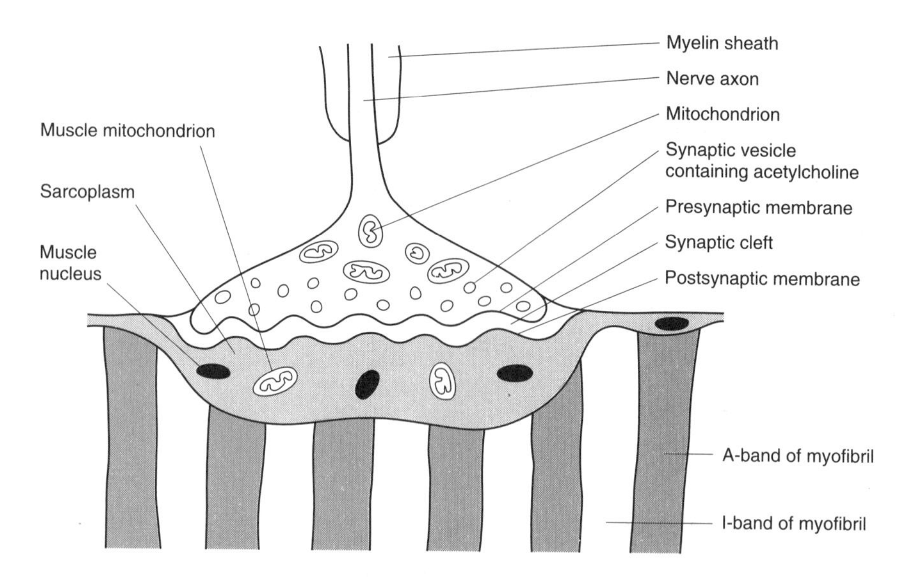

Fig. 8.22 Neuromuscular junction – the end plate

The process suggested by this theory is as follows. When an impulse reaches the neuromuscular junction (see Fig. 8.22) acetylcholine is released, which causes the sarcolemma of the muscle to be depolarised. Calcium ions released from the sacroplasmic reticulum bind to a protein molecule called **troponin** (see Fig. 8.24), changing its shape. The troponin now displaces a filamentous protein, **tropomyosin**, which has been preventing extensions of the myosin filament attaching to the actin filament. A change in position of these extensions causes the actin filament to slide past the stationary myosin one. ATP causes the separation of the myosin extension from the actin filament, and its hydrolysis to ADP releases the energy for the extension to reposition itself further along the actin filament. By a continuation of this ratchet-type movement, the muscle shortens. The need for ATP explains the occurrence of numerous mitochondria in muscles.

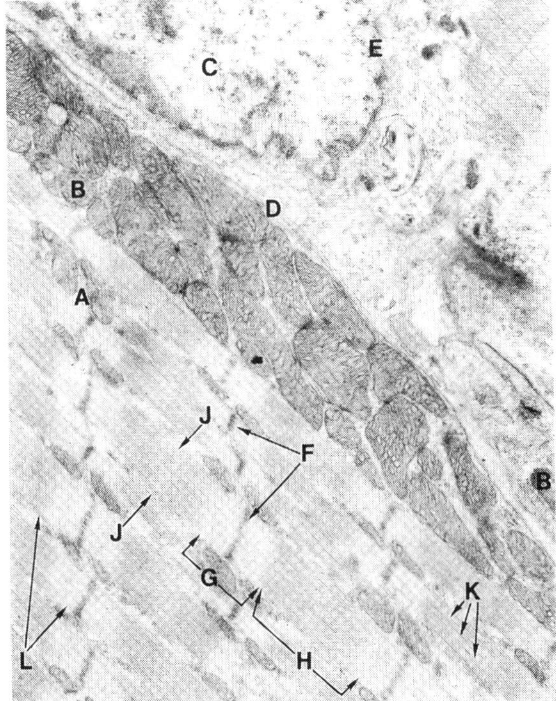

Key
Biceps (striated) muscle of mouse, magnification 25 000 ×

- **A** Sarcoplasmic reticulum
- **B** Mitochondrion
- **C** Nucleoplasm
- **D** Sarcolemma
- **E** Nuclear envelope
- **F** Z line
- **G** I band
- **H** A band
- **J** H zone
- **K** Muscle filaments
- **L** Myofibrils

Fig. 8.23 Electron micrograph of skeletal muscle

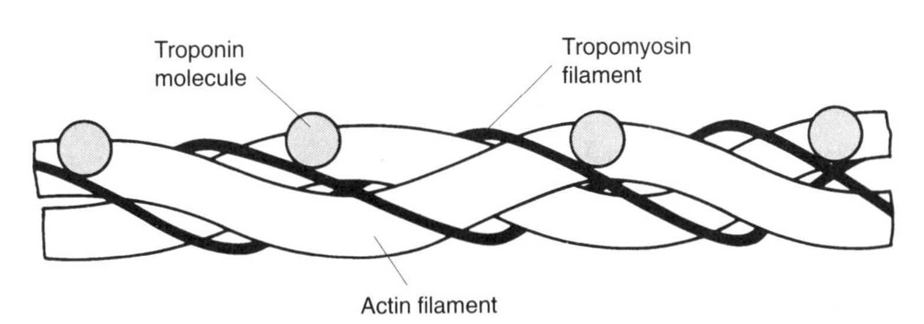

Fig. 8.24 Relationship of tropomyosin and troponin to the actin filament

Chapter roundup

In this chapter we have seen that in order for organisms to attain a degree of independence from their surroundings they need to maintain a constant internal environment. To achieve this requires careful monitoring of both internal and external conditions. Organisms therefore possess various receptors which provide information. This is then used to make the required adjustments to prevent any marked deviation from normal.

Plants, with their photosynthetic mode of nutrition, do not need to move from place to place in order to obtain nutrients. This means that they possess neither a muscular system nor any form of nervous control. They are nevertheless capable of movement – not the rapid muscular type of many animals, but the slower growth variety. Both plants and animals use chemical forms of coordination to integrate their activities; animals also utilise nervous coordination to enable them to respond rapidly to stimuli. Such stimuli are perceived by a range of internal and external receptors and are coordinated in higher animals by a central nervous system. This system includes, in many animals, a complex brain which permits, among other things, intelligent behaviour.

It is, in a sense, the culmination of millions of years of evolution which permits you to read this book, memorise (and we hope understand) the information, and then apply what you have learnt in order to achieve A-level success.

Illustrative questions and worked answers

1 Two people drank a solution which contained 100 g of glucose. The blood sugar level of each person was measured during the next 3 hours and the results are shown in the table.

(a) Using the same axes plot the blood sugar levels against time for X and Y.
(b) Summarise the information provided by the graphs.
(c) Suggest explanations for the changes in blood sugar levels of X and Y.

	Blood sugar level (mg per100 cm^3 blood)	
Time (mins)	X	Y
0 (glucose drunk)	81	90
20	136	131
40	181	142
60	213	89
90	204	79
120	147	74
150	129	86
180	113	89

Time allowed 30 minutes

Tutorial note

In plotting the graph it is essential to:

1. Plot the independent variable (in this case 'time') on the abscissa or *x*-axis (horizontal axis) and the dependent variable (in this case 'blood sugar level') on the ordinate or *y*-axis (vertical axis).
2. Choose a scale which makes maximum use of the available paper and yet is easy to use.
3. Label each axis clearly with the parameter being measured and the units used.
4. Plot the points clearly (in pencil to allow alteration) preferably using × for one set and ● for the other, to avoid any confusion.
5. Draw a smooth curve through the points – preferably distinguishing between the two lines, e.g. use different colours and label them or use a key.
6. Put an appropriate title on the graph.

In summarising the graph for part (b) make general points but try to support them with actual figures. Bearing in mind the 30 minutes allowed for the question you will need to give a reasonable degree of detail for part (c), relating your answers to the effects of insulin and glucagon on blood sugar levels.

Suggested answer

(a) *Graph of blood sugar level against time for two individuals*

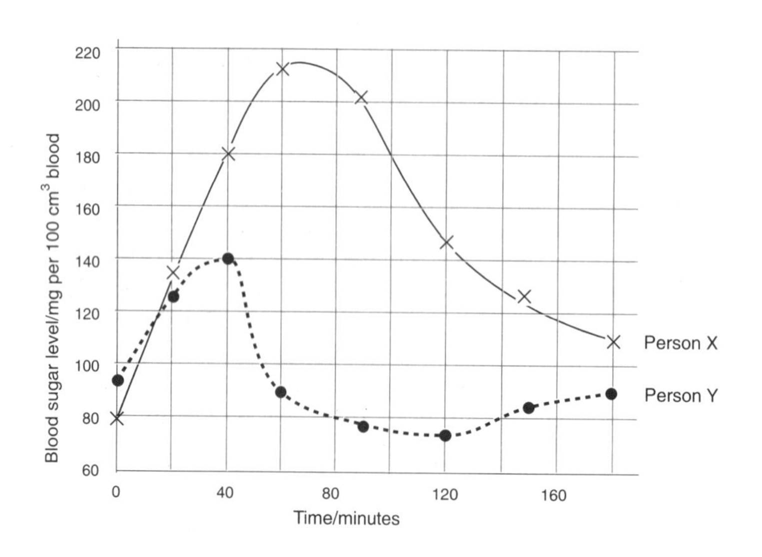

(b) Upon drinking a solution of glucose the blood sugar level of both X and Y rises immediately, person X showing the greater rise from 81 mg per 100 cm^3 blood to a peak of around 215 mg per 100 cm^3 blood after 70 minutes. This is followed by a decrease to 113 mg per 100 cm^3 blood after 3 hours. The *rate* of decrease slows over this period. Person Y appears to control blood sugar level to some extent. The rise is less pronounced (from 90 to 142 mg per 100 cm^3 blood) and the fall in level begins earlier (after 40 minutes). The fall continues until the level is below the starting point after 2 hours, suggesting that person Y has overcompensated for the rise in blood sugar level. The level rises, until after 3 hours it reaches the starting level again.

(c) The blood sugar level of person X rises almost immediately the glucose is swallowed. This is a result of absorption of the glucose into the blood by the wall of the stomach. As more and more glucose is absorbed, the blood sugar level rises steadily. After just over 1 hour most of the glucose has been absorbed. Throughout this period some of the glucose is used up in respiration and some is excreted by the kidney when the blood sugar level becomes very high. These factors all combine to reduce the blood sugar level, rapidly at first, but more slowly as the level falls until the level approaches that before the glucose was swallowed. It seems unlikely from the data that person X produces much, if any, insulin and is therefore diabetic.

The blood sugar level of person Y rises rapidly after the glucose is swallowed again due to absorption by the stomach wall. After 40 minutes the level falls due to the rise in blood sugar level stimulating the secretion of the hormone **insulin** from the islets of Langerhans in the pancreas. This hormone causes blood glucose to be converted to glycogen, which is stored in the liver. Some use of glucose during respiration will also cause the blood sugar level to fall. As the level falls the secretion of insulin decreases. Due to the short time lag between the two events, by the time insulin production has ceased the blood sugar level has fallen below the original value. At this point the pancreas may produce **glucagon** which reconverts some of the glycogen back to glucose and the blood sugar level rises again. An equilibrium between the two hormones is established and the blood sugar level returns to the original value. Person Y shows a normal reaction in the circumstances.

2 Discuss the reasons why animals have to move from place to place. What problems are associated with large size in organisms? How have these been overcome?

Time allowed 40 minutes

Tutorial note

As you will see from the suggested answer there are a large number of points to make in answering this question. The danger is that you may content yourself with only the first three or four on the list of why animals move from place to place. While these are clearly the more important reasons, some discussion of the others will be needed to obtain high marks. Equally most, if not all, of the seven problems of large size listed are necessary for good answers.

A useful way of planning answers to questions of this type is to use the seven descriptive characteristics of life – nutrition, excretion, sensitivity, movement, growth and repair, respiration and reproduction (including genetics and evolution) – as a guideline. Try to think of reasons why animals move under each of the seven headings. You must take care not to limit your response to 'mammals'. The question says 'animals' and examiners will be looking for a wide range of examples, both vertebrate and invertebrate. Use the animal classification groups in 2.5 as a guide and attempt to give an example from each phylum.

In discussing the problems of large size it would be wise to deal with the solution at the same time, i.e. discuss each problem and how it is overcome before moving to the next problem, rather than dealing with all the problems together followed by all the possible solutions. Again a wide range of specific examples from a wide range of organisms is needed. Note the word 'organisms'. It is sadly all too common for answers to totally ignore plants and so miss out on available marks. Our advice, as always, is to read the question carefully and re-read it at least once as you answer to ensure that you are not deviating from it.

Suggested answer

Reasons for locomotion:

1. **To obtain food:** the food requirements of most animals cannot be supplied in their immediate vicinity
2. **To capture food:** apart from the food supply being inadequate for the animal's needs, carnivores must often chase and capture their prey; without locomotion this type of animal cannot survive
3. **Escape from predators:** essential to survival
4. **To find a mate:** essential to the survival of the species
5. **Distribution of individuals:** each individual has a different genotype; movement to new areas allows this variation to be exploited and its evolutionary potential to be realised
6. **Reduction of competition:** prevents overcrowding and intraspecific competition
7. **To find shelter:** from both biotic and abiotic factors
8. **To maintain position:** paradoxically sharks must swim to stay still (this involves movement from place to place because the shark moves horizontally to maintain a vertical position)
9. **Reduced vulnerability to disease:** a scattered population is less likely to suffer epidemics of disease
10. **Escape from waste products:** these are toxic and may carry disease

Problems of large size:

Problem	Solution
More support needed	Bone, cartilage, chitin, xylem, become aquatic
Greater food requirements	Make centre hollow or fill it with dead tissue, e.g. xylem
Movement is difficult	Become sessile or slow moving, return to water, e.g. whale
Surface area/volume ratio	Develop internal or external surfaces to increase surface area, e.g. gills, lungs, long gut
Transport between surface and centre	Develop blood system, circulate fluids, develop transport mechanisms, e.g. respiratory pigments. Phloem and xylem in plants
More waste to dispose of	Develop excretory organs
Difficult to shelter, e.g. from predators	Large size may itself deter predators although camouflage and protective mechanisms will also help

3 In the following question, consider the four statements, decide which are correct and then give a single answer according to the following key:

Answer A if statements 1, 2 and 3 are correct

Answer B if statements 1 and 3 only are correct

Answer C if statements 3 and 4 only are correct

Answer D if statement 2 only is correct

Answer E if statement 4 only is correct

The speed at which an impulse is transmitted along a neurone is altered by:

1 whether the neurone is afferent (sensory) or efferent (motor)
2 the intensity of the generator potential
3 whether the neurone is myelinated or not
4 the axon diameter

Tutorial note

You may encounter this type of multiple choice, depending on your syllabus. They are less straightforward than many multiple-choice questions and the scope for error is therefore greater. There is, for example, the possibility of drawing the correct biological conclusions and then selecting the wrong answer from the key. It always pays to double check.

You should consider each statement in turn and decide whether it is correct; if necessary make a ✓ or ✗ next to each – use a ? where you are unsure. It could be that the key will provide you with the answer to your unknowns. For example, if you have a ✗ next to 1 and 3 and a ✓ next to 2, but are unsure of 4, then since 2 and 4 being correct is not an option on the key, you can safely assume 4 is not correct. A warning however – only use this as a last resort. It is far better to rely on sound biological knowledge than good fortune.

Suggested answer

- Statement 1 is incorrect. All other factors being equal, impulses travel equally fast along afferent and efferent neurones.
- Statement 2 is incorrect. Transmission speed is independent of generator potential; it is an 'all-or-nothing' response.
- Statement 3 is correct. The action potential jumps from one node of Ranvier to the next in myelinated nerves and this accelerates its progress.
- Statement 4 is correct. A large-diameter axon has a large surface area across which ions can move; this leads to a faster impulse.

Statements 3 and 4 are correct, which from the key gives answer C.

Question bank

1 Different plants require varying periods of light to make them flower. These plants fall into three photoperiodic groups: short-day, long-day and day-neutral plants.

(a) (i) Explain what is meant by 'day-neutral plants'.

(ii) An experiment was undertaken to determine the relationship between the number of hours of daylight and flowering in duckweed. State two important considerations in carrying out such an experiment.

(iii) From the results shown in the following graph state, with reasons, the photoperiodic group to which duckweed belongs.

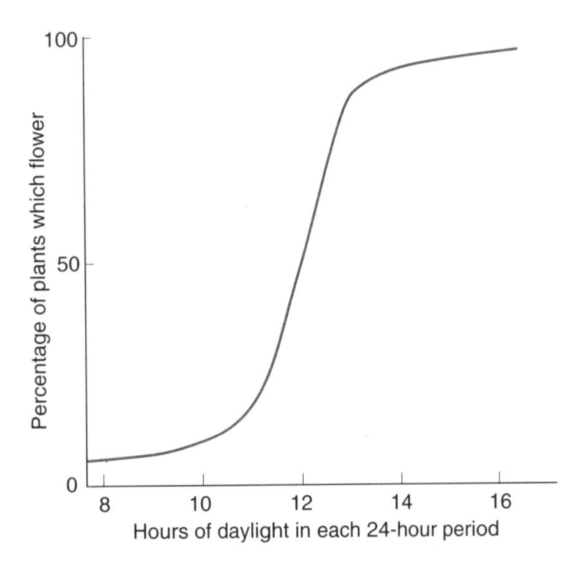

(b) Another species of plant was exposed to various periods of light (unshaded bars) and dark (shaded bars). In some cases the dark period was interrupted by periods of light. The effect on flowering is shown in each case.

(i) To which photoperiodic group does the plant belong?

(ii) By reference to the information provided explain whether it is the length of the light or dark period which is important in stimulating flowering.

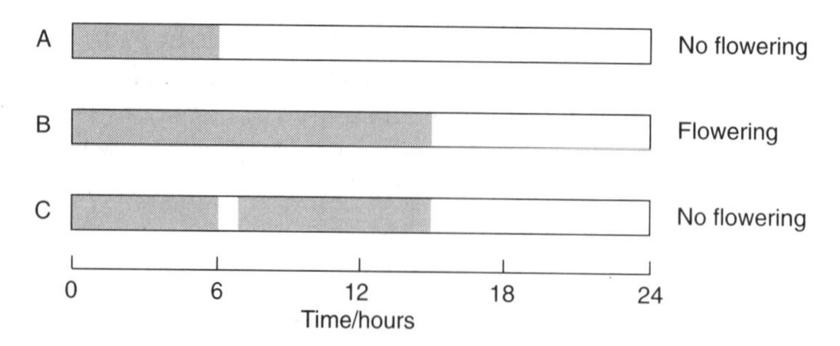

(c) In a further experiment a period of 15 hours darkness was interrupted by flashes of red light (R) and/or far-red light (F).

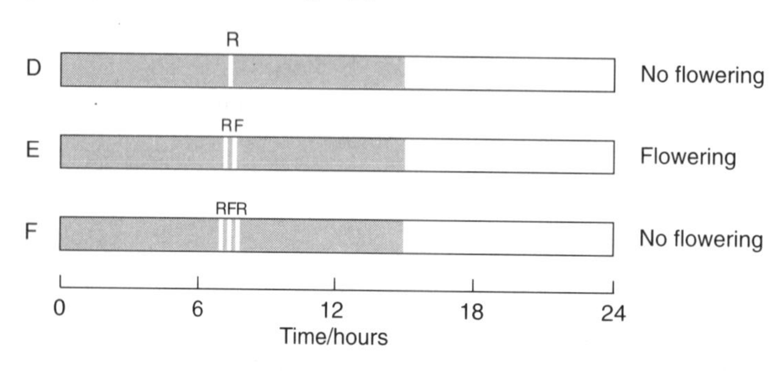

(i) Draw conclusions about the effects of these two types of light on flowering.
(ii) How might these conclusions be used commercially to enable flowers of this species to be produced out of season?

Time allowed 15 minutes

Pitfalls

The important experimental conclusions in part (a)(ii) must be general ones which apply to all experiments, e.g. the need to use a large sample of plants and/or repeat the experiment many times. Trivial or highly specific points will not bring marks, e.g. 'make sure the timing device is accurate'. In (a)(iii) the key words are 'with reasons' and you must therefore justify your choice of photoperiodic group. 'From the results' means that you must refer to the graph provided.

In part (b)(i) it is bar graph B (the one where flowering occurs) which is all important, whereas (b)(ii) requires you to compare B and C. Both have a short light period, but C has its dark period broken by a short light interval and this is sufficient to prevent flowering.

Part (c) requires you to appreciate that flashes of red light have a similar effect to a long period of daylight, while flashes of far-red light correspond to long periods of darkness. In experiment D, therefore, the red light, acting as a period of daylight, is sufficient to make a long dark period into two short ones and so prevent flowering. What is apparent from experiments E and F is that it is the last in a series of flashes which determines whether flowering occurs.

Points

(a) (i) They are plants in which flowering occurs irrespective of the length of the day.
(ii) Use large samples and/or repeat the experiment many times. Maintain all other factors, e.g. temperature, at a constant level throughout the experiment.
(iii) The number of plants flowering is very low when the length of day is less than 10 hours but increases when the length of day is longer. Duckweed is therefore a long-day plant.

(b) (i) Short-day plant
(ii) A long, continuous period of darkness induces flowering.

(c) (i) Red light has an effect similar to that of a long period of daylight, whereas the effect of far-red light is similar to that of a long period of darkness. When more than one type of light is used it is the last one which influences flowering.
(ii) By means of flashes of red or far-red light a grower can induce flowering out of season, e.g. far-red light can be used to make short-day species (normally flowering in spring and autumn) flower during the summer.

2 One theory of colour vision suggests there are three different types of cone cell in the human retina, each containing a different variety of the colour-sensitive pigment iodopsin. There are three varieties of iodopsin, one sensitive to red light, one to green and one to blue. The absorption of different wavelengths of light by the three types of cone is given below.

(a) From the data explain the following:
(i) Light of wavelength 430 nm appears blue.
(ii) Light of wavelength 550 nm appears yellow.
(iii) Light of wavelength 570 nm appears orange.

(b) From your knowledge of the retina explain why two small objects close together can be more easily distinguished by cones than rods.

Wavelength	Amount of light absorbed as a percentage of maximum		
(nm)	Red cones	Green cones	Blue cones
660	5	0	0
600	75	15	0
570	100	45	0
550	85	85	0
530	60	100	10
500	35	75	30
460	0	20	75
430	0	0	100
400	0	0	30

Time allowed 10 minutes

Pitfalls

You should analyse the table carefully and note the shift from absorption by red cones, through green, to blue cones as the wavelength of light decreases. Always read the column headings on tables and be clear in your own mind what the figures represent. In this case each figure is the percentage of light being absorbed by a particular cone compared to the total amount it could theoretically absorb. For example, with light of wavelength 530 nm, the red cones absorb 60% of the maximum they could absorb, the green cones absorb their maximum and the blue cones absorb a mere 10% of what they could theoretically take in. The answers to part (a) are derived from imagining the colours which result from mixing impulses from different cones, in the same way that different paints or coloured lights can be mixed, e.g. at wavelength 550 nm.

You will find part (b) easier to answer after a study of Fig. 8.18 and by recognising that groups of rod cells share a single nervous pathway to the brain so that the stimulation of more than one of a group still only produces a single impulse.

Points

(a) (i) Only blue cones absorb light at 430 nm and so the brain only receives nervous impulses from blue cones and therefore interprets the light entering the eye as being blue.
(ii) At a light wavelength of 550 nm, red and green cones are stimulated equally (85% of their maximum) while blue cones are not stimulated at all. Impulses

reaching the brain are received with equal frequency from red and green cones. The brain mixes the two equally to give the sensation of yellow light.

(iii) Light of wavelength 570 nm is absorbed in the approximate proportions of $\frac{2}{3}$ by red cones (100%), $\frac{1}{3}$ by green cones (45%) and nothing by blue cones. Orange results from mixing light (or paints) in these proportions.

(b)

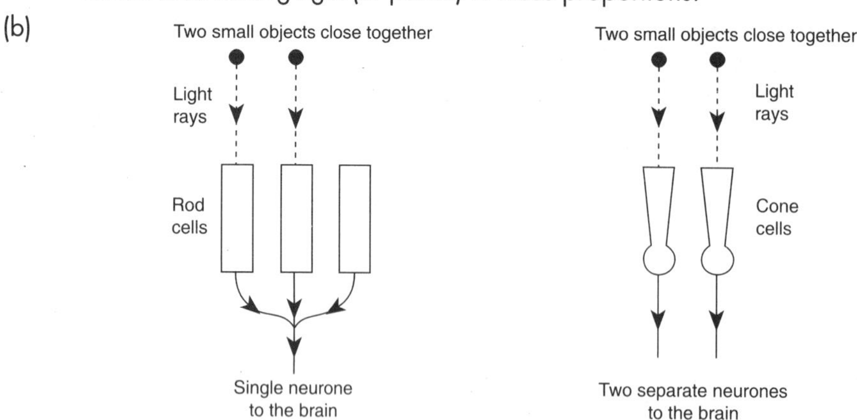

The figure illustrates that the cone cells of the retina have their own independent connections to the brain. Light rays from two objects close together stimulate two cones, the brain receives two impulses and interprets this as two separate objects. Rods, however, are connected in groups to a single neurone. The two objects will stimulate two rods but, if these are part of one group, only a single message is received by the brain which therefore interprets the objects as a single item. If the two stimulated rods belonged to different neurone groups the objects would be distinguished as separate items, but this would not occur in every case as it would if cones were stimulated.

3 Study the diagram below of a nerve synapse and answer the questions that follow.

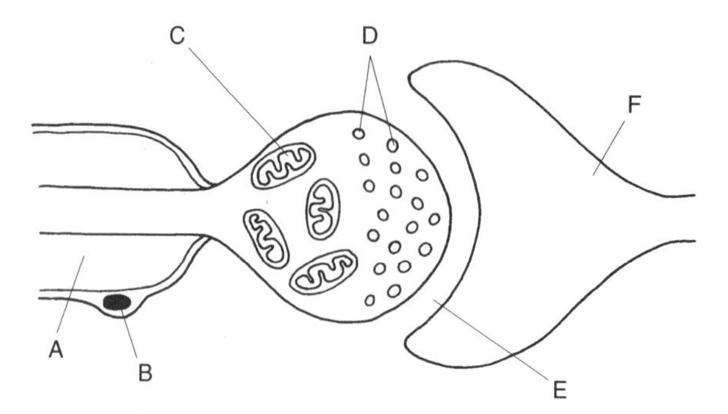

(a) Identify structures A–F.

(b) Why are structures C present in large numbers?

(c) Describe the methods of transmission of an impulse across the synapse.

(d) State three possible functions of a synapse.

Time allowed 15 minutes

Pitfalls

Be as precise as you can in answering part (a). For example, label B should be Schwann cell nucleus and F should be postsynaptic neurone. Similarly in part (b) it would not be adequate for you to simply state that mitochondria release energy; you should explain why the presynaptic neurone requires this energy. You will need to give a step-by-step account in part (c), again in some detail. The danger is that you will omit the sort of fine detail that brings credit at A level, so take care to include terms such as depolarisation and names like acetylcholine and acetylcholinesterase. To avoid confusion make it clear whether you are referring to the presynaptic or postsynaptic neurone or membrane. In part (d) make clear with a short explanation what you mean by 'valve', 'junction' and 'filter out'.

Points

(a) A myelin sheath
B Schwann cell nucleus
C mitochondrion
D synaptic vesicle
E synaptic cleft (gap)
F postsynaptic neurone

(b) Mitochondria are responsible for releasing energy in cells. During transmission of an impulse across a synaptic cleft the chemicals in structures D are released and need to be resynthesised before a new impulse can pass. The resynthesis requires energy and hence structures C are present. Vital activities frequently depend on a rapid succession of impulses and so resynthesis needs to take place as soon as possible. Large numbers of structure C ensure a greater release of energy and hence more rapid resynthesis of the chemicals in D.

(c) When an impulse travelling along the presynaptic neurone arrives at the synaptic knob, the synaptic vesicles (D) move towards the presynaptic membrane. These vesicles contain a transmitter substance (usually acetylcholine or noradrenaline) which diffuses across the synaptic cleft (E) to the postsynaptic membrane where special receptor sites receive the chemical. The transmitter substance creates a change in permeability of the postsynaptic membrane, thus allowing an influx of sodium ions. This causes depolarisation of the membrane, which passes along the postsynaptic neurone (i.e. a new impulse is initiated in the postsynaptic neurone). The transmitter substance is broken down by a specific enzyme, e.g. acetylcholine is broken down by acetylcholinesterase. The component parts of the transmitter substance diffuse back through the presynaptic membrane and are recombined using the energy provided by the mitochondria (C).

(d) ❶ They act as valves in as much as impulses may travel either way along a neurone but only in one direction across a synapse.

❷ They act as junctions. Impulses from many neurones may meet at a synapse but only a single neurone may leave. Similarly, a single presynaptic neurone may create impulses along many postsynaptic ones.

❸ Synapses may 'filter' impulses. Impulses of low frequency reaching a synapse may not initiate an impulse in the postsynaptic neurone. Only when a certain threshold frequency of incoming impulses is reached will a new postsynaptic impulse be initiated. Low-frequency impulses are thus 'filtered out'.

TEST RUN

In this section:

- This section should be tackled towards the end of your revision programme, when you have covered all your syllabus topics, and attempted the practice questions at the end of the relevant chapters.
- The Test Your Knowledge Quiz contains short-answer questions on a wide range of syllabus topics. You should attempt it without reference to the text.
- Check your answers against the Test Your Knowledge Quiz Answers. If you are not sure why you got an answer wrong, go back to the relevant unit in the text: you will find the reference next to our answer.
- Enter your marks in the Progress Analysis chart. The notes below will suggest a further revision strategy, based on your performance in the quiz. Only when you have done the extra work suggested should you go on to the final test.
- The Mock Exam is set out like a real exam paper. It contains a wide spread of topics and question styles, as used by all the examination boards. You should attempt this paper under examination conditions. Read the instructions on the front sheet carefully. Attempt the paper in the time allowed, and without reference to the text.
- Compare your answers to our Mock Exam Suggested Answers. We have provided tutorial notes to each, showing why we answered the question as we did and indicating where your answer may have differed from ours.

TEST YOUR KNOWLEDGE QUIZ

For each of the following, give the word or words which best answer the question.

Background questions

1 What is a micrometre (μm)?

2 What scale measures acidity and alkalinity?

3 If the objective lens of a microscope magnifies 10 times and the eyepiece lens magnifies 6 times what is the total magnification of an object viewed with this microscope?

4 What is the average of a group of values called?

5 What unit is energy measured in?

6 What unit is pressure measured in?

7 What is the study of animals called?

8 How many nanometres are there in a metre?

9 What name is given to a positively charged ion?

10 What prefix means 'many'?

Chapter 1

11 What is the type of bond formed when the negative region of an atom is attracted to the positive region of another?

12 What term do we apply to fats with double bonds?

13 Give the general name for the nonprotein part of a protein.

14 What is the process called by which a cell membrane invaginates to engulf particles too large to be absorbed by diffusion, osmosis or active transport?

15 Name the spherical body in the nucleus, which stores RNA.

16 What do we call the membrane surrounding the central vacuole in plants?

17 What is the stack of thylakoids within a chloroplast called?

18 Which two proteins make up the bulk of a voluntary muscle fibre?

19 What type of epithelium lines alveoli?

20 Name two types of sclerenchyma cells.

Chapter 2

21 What do we call features which have a similar origin, structure and position, regardless of their function in the adult?

22 Complete the last two words in the following series: phylum, class, order, family, ________________, ________________.

23 Name the protein coat which surrounds the nucleic acid core of a virus.

24 What is the important enzyme which retroviruses can produce and which accounts for their name?

25 Name the often unicellular group of eukaryotic organisms which are neither plants, animals nor fungi.

26 To which group do plants with flowers belong?

27 Name the two forms in which individuals belonging to the group Cnidaria exist.

28 Which is the largest phylum in the animal kingdom?

29 Which group of organisms is described as being unsegmented animals with a head, foot and visceral hump?

30 What do we call the vertebrate group which possesses dry scaly skins?

Chapter 3

31 What do the letters DNA stand for?

32 State the names of three types of RNA.

33 Which base pairs with adenine in DNA?

34 During which stage of cell division do chromosomes first appear as distinct structures?

35 In what type of cell division is the number of chromosomes halved?

36 What genetic term is used to describe the actual appearance of an organism?

37 What term is used to describe the alternative forms of a gene?

38 In a cross between a normal male and a female carrier for the disease, what percentage of their sons might be expected to be haemophiliac?

39 What word describes the type of mutation in which whole sets of chromosomes become duplicated?

40 What are agents which increase the rate of mutation called?

Chapter 4

41 What term describes a group of genetically identical individuals formed asexually from a single individual?

42 What is the name of the tubules which make up the testes?

43 What do we call the small, nonfunctional eggs produced during oogenesis?

44 Which hormone causes the mammary glands to produce milk?

45 A collection of sepals is called what?

46 Name the two nuclei found in a pollen grain.

47 What is the young shoot of a plant called?

48 In which type of germination are the cotyledons carried above the soil surface?

49 What is the type of growth in which different organs grow at different rates?

50 What is the general term used to describe those cells in plants which retain the ability to divide?

Chapter 5

51 Name the process by which simple chemicals are built up into complex ones.

52 How do we describe chemical reactions which liberate free energy?

53 From your knowledge of naming enzymes, devise a name for an enzyme which adds oxygen to polyphenols.

54 What is the process called by which ions can be separated from larger molecules by placing them in a membrane permeable only to the ions?

55 Name the five-carbon compound which combines with carbon dioxide in the light-independent stage of photosynthesis?

56 What two products of the light-dependent stage of photosynthesis are utilised in the light-independent stage?

57 Which inorganic ion, found in meat and green vegetables, is an important constituent of haemoglobin?

58 Name the enzyme which breaks down fat into fatty acids and glycerol.

59 What do we call organisms which obtain energy from the dead and decaying remains of other organisms?

60 What is the secondary host of the platyhelminth which causes bilharzia (schistosomiasis) in humans?

Chapter 6

61 Give the term for the point where the death and birth rates of a population are equal and its size becomes stable.

62 What term describes competition between individuals of one species?

63 Name the term which describes the physical position of a species within a habitat.

64 What type of organisms are primary consumers?

65 What percentage of the atmosphere is carbon dioxide?

66 Name a bacterium which derives its energy from oxidising nitrites into nitrates.

67 What term describes the condition in which biological activities, such as bacteria decomposing waste, deplete the oxygen content of a body of water?

68 What word describes resources which are not, for all practical purposes, replaceable?

69 What material have Brazilians used to produce alcohol (gasohol) which can be used to fuel motor vehicles?

70 What chemicals used in aerosol sprays, refrigerators and some plastic foams deplete the ozone layer and so contribute to global warming?

Chapter 7

71 What are the two specialised structures, one on leaves and the other on woody stems, which allow gaseous diffusion in plants?

72 Name the muscles found between the ribs in mammals which are important in ventilating the lungs.

73 Which particles found in the blood play an important role in blood clotting?

74 Would it be safe or dangerous to transfuse blood from a group O donor into a group A recipient?

75 What is the name of the group of specialised cells in the wall of the right atrium of myogenic hearts which initiate the heart beat?

76 What is a potometer used to measure?

77 What is the name of the pair of specialised cells which surround the stomatal pore and control its aperture size?

78 What name is given to the pathway of water movement which takes place along the cell walls of adjacent plant cells?

79 Name the long hairpin loop found in the nephrons of mammalian kidneys which helps to ensure greater reabsorption of water.

80 What is the hormone called which increases the permeability of the collecting ducts of the kidney to water and so reduces water loss?

Chapter 8

81 What term describes organisms which derive their body heat predominantly from external sources?

82 What do we call the processes by which arterioles narrow their internal diameter?

83 State the term used to describe the breakdown of glycogen.

84 How would you describe, in two words, the movement of a part of a plant away from gravity?

85 Apart from auxins, inhibitors and ethene, what are the other two groups of plant growth substances?

86 Name the hormone produced by the α-cells of the islets of Langerhans which raises the blood glucose level.

87 What are the gaps between adjacent Schwann cells of a neurone called?

88 Name the two main neurotransmitters.

89 What are the component parts of rhodopsin (visual purple) which are formed as a result of light splitting the rhodopsin molecule?

90 Name the filamentous protein muscle which is displaced by troponin.

General questions

91 State two situations where a countercurrent flow occurs in living organisms.

92 In drawing a graph on which axis should you plot the independent variable?

93 Name the enzyme which can make DNA from its mRNA code.

94 In which mammalian organ is urea made?

95 Where does a centromere occur?

96 Which group of organisms possesses nematoblasts?

97 Where would you find a Casparian strip?

98 What chemical acts as an immediate store of energy for a cell?

99 What process do phytochromes control?

100 What word describes a reaction in which a molecule is split into small parts by means of water?

TEST YOUR KNOWLEDGE QUIZ ANSWERS

The figures in brackets refer to the appropriate unit of the text in which the answer is defined or described.

1 A unit of length (equal to 10^{-6} m)
2 pH
3 60 times (10 × 6)
4 Mean
5 Joule (or kilojoule, kJ)
6 Pascal (or kilopascal, kPa)
7 Zoology
8 1 000 000 000 (1 nm = 10^{-9} m)
9 Cation
10 Poly
11 Hydrogen bond (1.1)
12 Unsaturated (1.1)
13 Prosthetic group (1.1)
14 Phagocytosis (1.2)
15 Nucleolus (1.2)
16 Tonoplast (1.2)
17 Granum (1.2)
18 Actin and myosin (1.3)
19 Squamous (pavement) epithelium (1.3)
20 Fibres and sclereids (1.3)
21 Homologous (2.1)
22 Genus, species (2.1)
23 Capsid (2.2)
24 Reverse transcriptase (2.2)
25 Protoctista (2.2)
26 Angiospermophyta (2.4)
27 Polyp and medusa (2.5)
28 Arthropoda (2.5)
29 Mollusca (2.5)
30 Reptilia (2.5)
31 Deoxyribonucleic acid (3.1)
32 Messenger, transfer and ribosomal (3.1)
33 Thymine (3.1)
34 Prophase (3.2)
35 Meiosis (3.2)
36 Phenotype (3.3)
37 Allele (3.3)
38 50% (3.3)
39 Polyploidy (3.4)
40 Mutagens (3.4)
41 Clone (4.1)
42 Seminiferous tubules (4.2)
43 Polar bodies (4.2)
44 Prolactin (4.2)
45 Calyx (4.3)
46 Generative nucleus and tube nucleus (4.3)
47 Plumule (4.3)
48 Epigeal (4.3)
49 Allometric growth (4.4)
50 Meristems (meristematic) (4.4)
51 Anabolism (5.1)

52 Exogonic (5.1)
53 Polyphenol oxidase (5.1)
54 Dialysis (5.1)
55 Ribulose bisphosphate (5.2)
56 ATP and $2NADPH_2$ (or 2H) (5.3)
57 Iron (Fe^{2+}) (5.3)
58 Lipase (5.3)
59 Saprobionts (5.4)
60 Freshwater snails (5.4)
61 Carrying capacity (6.1)
62 Intraspecific (competition) (6.1)
63 Ecological niche (6.2)
64 Herbivores (6.2)
65 0.04% (6.2)
66 Nitrobacter (6.2)
67 Biochemical oxygen demand (6.3)
68 Nonrenewable (6.3)
69 Sugar cane waste (6.3)
70 Chlorofluorocarbons (CFCs) (6.3)
71 Stomata (leaves), lenticels (woody stems) (7.1)
72 Intercostal muscles (7.1)
73 Platelets or thrombocytes (7.2)
74 Safe (7.2)
75 Sinoatrial node (7.2)
76 Water uptake by a plant (rate of transpiration) (7.3)
77 Guard cells (7.3)
78 Apoplast pathway (7.3)
79 Loop of Henle (7.4)
80 Antidiuretic hormone (ADH) (7.4)
81 Ectotherm (8.1)
82 Vasoconstriction (8.1)
83 Glycogenolysis (8.1)
84 Negative geotropism (8.2)
85 Gibberellins and cytokinins (8.2)
86 Glucagon (8.3)
87 Nodes of Ranvier (8.4)
88 Acetylcholine and noradrenaline (8.4)
89 Opsin and retinal (8.4)
90 Tropomyosin (8.5)
91 Fish gills and the nephron of a kidney (7.1 and 7.4)
92 Abscissa (*x*- or horizontal axis) (p.22)
93 Reverse transcriptase (3.1)
94 Liver (7.4)
95 In chromosomes (3.2)
96 Cnidaria (2.4)
97 Endodermis of a root of a dicotyledonous plant (7.3)
98 Adenosine triphosphate (5.5)
99 Flowering in plants (8.2)
100 Hydrolysis (1.1)

PROGRESS ANALYSIS

Place a tick next to those questions you got right.

Question	*Answer*	*Question*	*Answer*	*Question*	*Answer*	*Question*	*Answer*
1		26		51		76	
2		27		52		77	
3		28		53		78	
4		29		54		79	
5		30		55		80	
6		31		56		81	
7		32		57		82	
8		33		58		83	
9		34		59		84	
10		35		60		85	
11		36		61		86	
12		37		62		87	
13		38		63		88	
14		39		64		89	
15		40		65		90	
16		41		66		91	
17		42		67		92	
18		43		68		93	
19		44		69		94	
20		45		70		95	
21		46		71		96	
22		47		72		97	
23		48		73		98	
24		49		74		99	
25		50		75		100	

My total mark is: ______ out of 100

If you scored 1–40

You need to do some more work. You are not yet ready to take the Mock Exam because you do not have sufficient knowledge or understanding of the syllabus content. Starting at Section 2, Chapter 1, look at the list of units at the beginning of each chapter and revise those units on which you scored poorly in the Test. When you consider you have completed your revision get a friend to ask you questions (not necessarily those in the test) and if you are still weak on some units, look at them again. You should then attempt the Test Your Knowledge Quiz again.

If you scored 41–60

You are getting there, but you must do some more work. Go through the list of units at the beginning of each chapter and mark those which you could not answer questions about correctly in the test. In addition look through the practice questions at the end of each chapter and the points which accompany them. Go over some of your weak topics with a friend and then attempt the Test Your Knowledge Quiz again.

If you scored 61–80

You are nearly ready to attempt the Mock Exam, but to get the best out of it, brush up on those units which the test shows you have not fully understood. Also look at the practice questions at the end of each chapter and check those questions which relate to the subject areas you do not feel confident about. You should then be ready to go on to the Mock Exam.

If you scored 81–100

Well done! You can tackle the Mock Exam with confidence, although you will first need to revise some of the units which let you down in the Test Your Knowledge Quiz. Reassure yourself that there are no gaps in your knowledge and then set aside a time to do the Mock Exam.

LETTS SCHOOL EXAMINATIONS BOARD

General Certificate of Education Examination

ADVANCED LEVEL BIOLOGY

Paper 1

Time allowed: 2 hours

Answer all three sections.

You are advised to divide your time as follows:

Section A (Multiple choice) (15 marks) – 20 minutes
Section B (Structured questions) (50 marks) – 60 minutes
Section C (Essay) (35 marks) – 40 minutes

Answer all questions in Sections A and B, but *one* question only from Section C.

MULTIPLE-CHOICE ANSWER SHEET

When you have selected what you consider to be the most appropriate answer to a question, mark the response by making a *thick* pencil stroke under the appropriate letter. Use an HB pencil *not* ink or ballpoint pen. If you wish to change your answer, simply rub out your first mark completely and make your new response.

	A	B	C	D	E
1	( )	( )	( )	( )	( )
2	( )	( )	( )	( )	( )
3	( )	( )	( )	( )	( )
4	( )	( )	( )	( )	( )
5	( )	( )	( )	( )	( )
6	( )	( )	( )	( )	( )
7	( )	( )	( )	( )	( )
8	( )	( )	( )	( )	( )
9	( )	( )	( )	( )	( )
10	( )	( )	( )	( )	( )
11	( )	( )	( )	( )	( )
12	( )	( )	( )	( )	( )
13	( )	( )	( )	( )	( )
14	( )	( )	( )	( )	( )
15	( )	( )	( )	( )	( )

Section A

You are advised to spend about 20 minutes on this section.

Answer all 15 questions.

For each question there are five suggested answers; you should choose the most appropriate and indicate it on the answer sheet provided.

1 In which of groups A to E are all three organs not homologous?

A bat wing, bird wing, human forearm
B insect wing, bird wing, bat wing
C fish pectoral fin, whale flipper, penguin flipper
D human forearm, bird wing, penguin flipper
E whale flipper, bat wing, human forearm

2 The general formula for a disaccharide is:

A $C_{12}H_{22}O_{11}$
B $C_5H_{10}O_5$
C $C_3H_6O_3$
D $C_{12}H_{24}O_{12}$
E $C_6H_{12}O_6$

3 Which of the following cells would most probably contain the greatest number of Golgi bodies?

A muscle cell
B secretory cell
C nerve cell
D white blood cell
E epithelial cell

4 Which of the following components of an ecosystem has the greatest biomass?

A primary producers
B primary consumers
C secondary consumers
D tertiary consumers
E decomposers

5 Which of the following best represents the enzyme composition of pancreatic juice?

A amylase, peptidase, trypsinogen, rennin
B amylase, pepsin, trypsinogen, maltase
C lipase, amylase, pepsin, maltase
D lipase, amylase, trypsinogen, peptidase
E peptidase, amylase, pepsin, rennin

6 Twenty-four hours after removing the liver of a mammal the concentrations of urea and amino acids in the blood would be changed in which of the following ways?

A both urea and amino acid levels would have risen
B both urea and amino acid levels would have fallen
C the urea level would have risen while the amino acid level would have fallen
D the amino acid level would have risen while the urea level would have fallen
E there would be no change in the level of either

7 When formulating his theory of evolution Darwin considered each of the following except:

A the ecology of plants and animals
B genetic theory
C the morphology of living organisms
D the geographical distribution of organisms
E the structure of fossils

8 The initiation of the mammalian heart beat occurs in:

A left atrium
B atrio-ventricular node
C Purkinje fibres
D sino-atrial node
E right ventricle

9 If light were limiting the rate of photosynthesis in an aquatic plant in a laboratory experiment, which of the following would increase the rate of photosynthesis fourfold?

A increasing the carbon dioxide concentration four times
B increasing the temperature by 20 °C
C increasing the temperature by a factor of four
D halving the distance between the plant and the light source
E reducing the distance between the plant and the light source to one-quarter of its original distance

Matching pairs questions

Questions 10–13 refer to the diagram below, which represents the structure of a DNA molecule. For each question select one of the labels A, B, C, D or E which best fits the description given. Each letter may be used once, more than once or not at all.

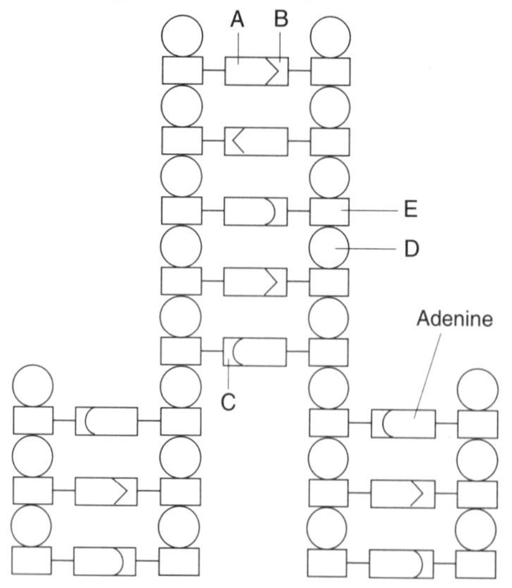

Which letter represents:

10 cytosine

11 guanine

12 an organic base not found in an RNA molecule

13 a molecule whose structure is modified in RNA by having an additional oxygen atom

Multiple-completion questions

For the following questions, determine which of the responses that follow are correct answers to the question. Give the answer A, B, C, D or E according to the key below.

- A 1, 2 and 3 are correct
- B 1 and 3 only are correct
- C 2 and 4 only are correct
- D 4 only is correct
- E 1 and 4 only are correct

14 Direct effects of follicle stimulating hormone (FSH) include

1. development of the corpus luteum
2. development of the Graafian follicle
3. ovulation
4. stimulation of sperm production

15 In constructing a natural classification the following should be used:

1. comparative embryology
2. individual pedigrees
3. habitat preferences between organisms
4. structural similarities between organisms

Section B

You are advised to spend 60 minutes on this section.
Answer *all* questions.
Write your answers in the spaces provided.

1 Exposed photographic film has black silver salts bonded to it by a thin layer of gelatin (a protein). In an investigation into the digestion of gelatin by the enzyme trypsin the end point is shown by the clearing of the film, as in the diagram.

Seven test tubes, each with a different buffered pH solution and 1 cm^3 of 0.5% trypsin solution, were placed in a water bath at 35 °C for 5 minutes. Small pieces of exposed film were simultaneously placed into each test tube and the time taken for the film to clear was noted.

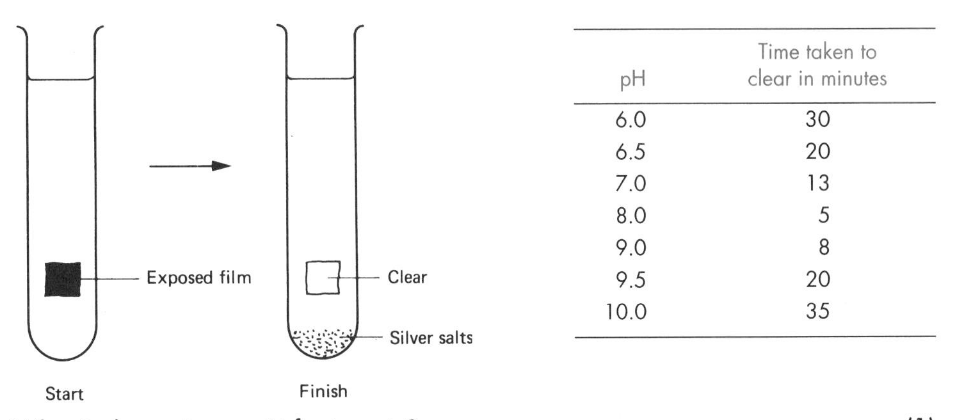

pH	Time taken to clear in minutes
6.0	30
6.5	20
7.0	13
8.0	5
9.0	8
9.5	20
10.0	35

(i) What is the optimum pH for trypsin? ________________ (1)

(ii) What is meant by 'buffered solution'? ________________

________________ (2)

(iii) What control will be required for this experiment? ________________

________________ (2)

(iv) Why was a control necessary? ______________________________

______________________________ (1)

(v) Why was a waterbath used? ______________________________

______________________________ (1)

(vi) Why were the seven test tubes placed in the water bath for 5 minutes before inserting the film into each? ______________________________

______________________________ (1)

(vii) What would have happened to the 'time to clear' had the experiment been carried out at 70 °C? ______________________________

______________________________ (1)

Total marks 9

2 In a paternity suit, a mother of blood group A has a child of blood group O. The man she claims to be the father is blood group B. Explain whether the information can settle the issue one way or another.

Total marks 6

3 Study the diagram below and then answer the questions that follow.

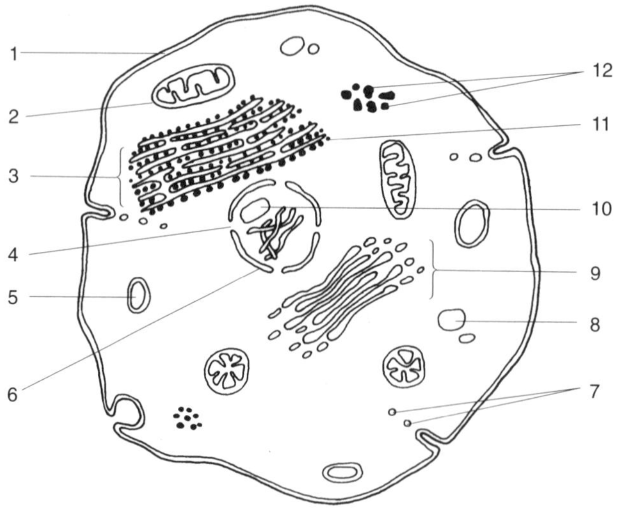

(a) Label parts 1–12.

1 ______________________________

2 ______________________________

3 ______________________________

4 __________________________________

5 __________________________________

6 __________________________________

7 __________________________________

8 __________________________________

9 __________________________________

10 __________________________________

11 __________________________________

12 __________________________________ (6)

(b) State the function of the part labelled 2. ____________________ (1)

(c) Describe the events which link structures 10, 11 and 12. __________

____

____ (3)

(d) What is the approximate magnification of this drawing? __________

____ (2)

Total marks 12

4 The following table gives the amount of DNA in a cell at various stages of cell division. The least amount of DNA present at any stage is taken as 1.0 and this is used as a basis for comparison of the other stages.

DNA content of cell	Examples of stages of cell division
1.0	meiosis late telophase II
2.0	mitosis early interphase mitosis late telophase meiosis metaphase II
4.0	mitosis prophase meiosis anaphase I

Explain the differences in DNA content between:

(i) mitosis early interphase and mitosis prophase ________________

____ (3)

(ii) mitosis prophase and mitosis late telophase ________________

____ (3)

(iii) meiosis anaphase I and meiosis metaphase II ____________________

____ (3)

(iv) meiosis metaphase II and meiosis late telophase II ______________

____ (3)

Total marks 12

5 (a) Study the following statements and then relate each one to one of the five terms A, B, C, D and E.

1 a natural community of plants and animals
2 the study of interrelationships between living organisms and their environment
3 the wise management and use of natural resources
4 that part of the earth and its atmosphere inhabited by living things
5 a naturally occurring group of organisms inhabiting a common environment

A ecology = statement __________

B conservation = statement __________

C community = statement __________

D ecosystem = statement __________

E biosphere = statement __________ (5)

(b) A simple food web of five organisms A–E is shown below.

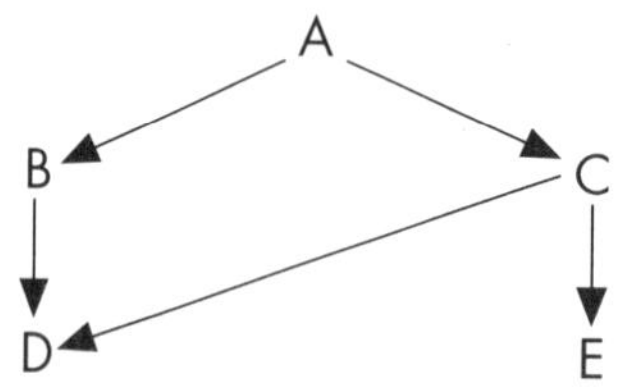

If organism C were suddenly to be removed from the food web, how would the populations of organisms A, D and E be affected? Explain your answers.

Organism A ____

____ (2)

Organism D ____

____ (2)

Organism E ____

____ (2)

Total marks 11

Section C

You are advised to spend about 40 minutes on this section.
Answer only *one* question.

Either

1 (a) Explain how a flowering plant obtains
  (i) water
  (ii) ions

(b) Describe the pathways and mechanisms of water transport in the plant.

or

2 With reference to a mammal, describe the means by which blood circulation is maintained and controlled. How is blood maintained at a constant temperature? Review the part played by the blood in conferring immunity to infectious disease.

MOCK EXAM SUGGESTED ANSWERS

Section A

1 In order to be homologous, structures must have evolved from a common ancestral structure. The wing of the insect has very different origins from the vertebrate limbs.
Answer = B

2 Disaccharides are formed from two monosaccharides by the loss of a water molecule, e.g. two hexoses ($C_6H_{12}O_6$) combined give $C_{12}H_{24}O_{12}$ which, on losing water (H_2O), gives $C_{12}H_{22}O_{11}$.
Answer = A

3 One major function of the Golgi body is to add carbohydrate to the proteins produced by the endoplasmic reticulum and to sort out and package them into secretory vesicles.
Answer = B

4 Energy is lost at each stage of a food chain or web and hence the biomass is always greater at the beginning of any chain.
Answer = A

5 Rennin and pepsin are produced by the gastric glands of the stomach wall (eliminating options A, B and E). Maltase is produced by the wall of the small intestine (eliminating B and C). The composition of D does correspond with that of the pancreatic juice.
Answer = D

6 One major function of the liver is to deaminate amino acids and so produce urea. In the absence of the liver the level of amino acids will rise (assuming amino acids are still absorbed from the diet) because it is no longer removing them. The level of urea will fall as it is no longer being produced by the liver but is still being removed by the kidneys.
Answer = D

7 Genetic theory was not formulated until after the publication of Darwin's theory of evolution.
Answer = B

8 The key word is 'initiation'. While the other structures all have a role to play in the beating of the heart, only the sino-atrial node *initiates* the beat.
Answer = D

9 If light is limiting, only an increase in its intensity can increase the rate of photosynthesis. This leaves only options D and E. Light intensity is inversely proportional to the square of the distance from the source. Halving the distance

therefore increases light intensity by 2^2, i.e. four times.
Answer = D

10–13 On the diagram the circles (D) and squares (E) represent the alternating phosphate and deoxyribose molecules which form the structural 'uprights' of the DNA 'ladder'. The organic bases (adenine, A, B and C) form the 'rungs'. Adenine is labelled. It is a purine and as such has a longer molecular structure than its partner thymine (label C) which is a pyrimidine. Guanine is also a purine and it is therefore logical to assume that label A represents guanine and label B its partner cytosine. Thymine does not exist in RNA (it is replaced by uracil) and the deoxyribose molecule (label E) is modified in RNA to ribose.
Answers 10 = B 11 = A 12 = C 13 = E

14 Ovulation (option 3) and development of the corpus luteum (option 1) are stimulated by luteinising hormone (LH) and not follicle stimulating hormone (FSH). Option 4 is the distractor because many candidates fail to appreciate that FSH is produced in males where it stimulates sperm production.
Answer = C

15 The pedigree of individuals is not used in classification because it varies widely between individuals; habitat differences also vary between individuals. Natural classifications are based upon the more permanent genetic differences of comparative embryology and structural similarities.
Answer = E

SECTION B

1 Tutorial note

In part (i) you must remember that the optimum pH is where the trypsin breaks down the gelatin in the shortest time. In part (iii) remember that a control should exactly mimic the experimental conditions in all respects except the factor under test. In part (vii) you need to be aware that at temperatures around 60 °C many enzymes become denatured.

Suggested answer

(i) pH 8.0
(ii) One in which the pH remains constant despite the addition of small amounts of solutions of different pH.
(iii) Boiling the enzyme beforehand (to denature it) or replacing the enzyme with an equal quantity of water.
(iv) To show that any change was due to trypsin acting enzymatically and not to some other factor (e.g. the pH buffer).
(v) To maintain a constant temperature – water has a high thermal capacity and therefore alters temperature very slowly.
(vi) To bring the tubes to the required experimental temperature before commencing the experiment.
(vii) It would have been infinity, i.e. the film would not clear because the trypsin would have been denatured.

2 Tutorial note

You need to be aware that alleles A and B are equally dominant and that allele O is recessive. Begin by ascertaining the child's genotype and then work on the basis that one of each of its alleles must have come from each parent.

Suggested answer

The child must have the genotype I^OI^O as this is the only one possible for group O. One I^O must have come from each parent. The mother as blood group A has two possible genotypes

– I^AI^A and I^AI^O. Likewise the father (group B) could have genotypes I^BI^B or I^BI^O. If the mother is I^AI^O and the father I^BI^O, the following offspring are possible:

	Father's gametes I^B	I^O
Mother's gametes I^A	I^AI^B	I^AI^O
I^O	I^BI^O	I^OI^O

All possible blood groups, including O, could result. The man could be the father, but then so could all fertile males carrying the I^O allele – this includes about half the male population of the world. The issue cannot therefore be resolved.

3 Tutorial note

This question is largely one testing recall and only parts (c) and (d) may present problems to a knowledgeable candidate. It is detail of protein synthesis needed in (c) and in (d) you will need to make an appropriate measurement rather than use memory of organelle size. Choose an organelle with a fairly constant dimension, rather than one which varies considerably.

Suggested answer

(a)
1 plasma membrane
2 mitochondrion
3 rough endoplasmic reticulum
4 nuclear pore
5 lysosome
6 nuclear envelope
7 pinocytic vesicles
8 oil droplet
9 Golgi apparatus
10 nucleolus
11 ribosome
12 glycogen granules

(b) This is a mitochondrion, which is the centre of energy production within the cell. The invaginations of the inner membrane (cristae) have attached to them the enzymes involved in the Krebs (citric acid) cycle and the electron (hydrogen) carrier system.

(c) Structure 10 is the nucleolus, which contains RNA. One type of RNA called messenger RNA leaves the nucleus and enters the cytoplasm of the cell. It comprises a sequence of organic bases that have been determined by the sequence of such bases on an appropriate section of DNA within the nucleus. The messenger RNA wraps itself around some ribosomes (structure 11). The sequence of organic bases ultimately determines the sequence of amino acids in a protein which is synthesised. This protein could be an enzyme that is involved in the production of glycogen (structure 12), the main carbohydrate store of an animal cell.

(d) The best means of calculating this is to take an organelle whose actual size is known. The best organelle is usually a mitochondrion because its size, although variable in length, is reasonably constant in diameter. The diameter is usually 0.75 µm. If the diameter of a few mitochondria are measured an average can be found. In this case it is near enough 5.0 mm (it is a good idea to approximate to a figure that will make the calculation reasonably easy).

Actual size = 0.75 µm
Size drawn = 5.0 mm
Therefore magnification = 5.0 mm/0.75 µm
= 5000 µm/0.75 µm (1 mm = 1000 µm)
= 6666 times approximately

4 Tutorial note

You need to remember that DNA replicates in interphase and that the cell actually divides into two during telophase (effectively halving the DNA content). Take care in reading the question and be sure about whether you are considering mitosis or meiosis, and if it is meiosis, whether it is meiosis I or meiosis II.

Suggested answer

(i) In mitosis early interphase, the chromatids disappear having completed a cell division. Towards the end of this division the chromosomes separate into their chromatids. During interphase these chromatids, which comprise the DNA of the cell, replicate. By the first stage of the next mitotic cell division (mitosis prophase) the DNA content has therefore doubled from 2.0 to 4.0.

(ii) In mitosis prophase the chromosomes of a cell comprise two chromatids, one of each pair moving to an opposite pole. During late telophase the cell divides into two, thus halving the DNA content *per cell* compared with mitosis prophase, from 4.0 to 2.0.

(iii) Meiosis comprises a double division of the cell. Between the first meiotic division and the second meiotic division the chromosome, and hence DNA, content is halved. Anaphase I is followed by telophase I, where the actual division of the cell, and hence the halving, occurs. No further division of the cell takes place before metaphase II and so the DNA content is simply halved, from 4.0 to 2.0.

(iv) Between these two stages the second meiotic division of the cell has occurred and so the DNA content has been further halved, in this case from 2.0 to 1.0.

5 Tutorial note

It is reasonable to assume from the wording of the question that in part (a) each of the terms relates to a different statement. In part (b) you should assume that the direction of the arrows indicates the flow of energy. The principles to apply are that the removal of a food source will reduce an organism's population, while the removal of a predator will increase it. Always bear in mind (and make it clear to the examiner) that organisms may compensate for food losses by using more of an alternative source (if they have one). The words 'suddenly removed' are important – you must assume an *immediate* and complete loss of the species.

Suggested answer

(a) A = 2 B = 3 C = 5 D = 1 E = 4

(b) Organism A: If C were to suddenly disappear, one means by which A is consumed would be removed and the population of A would increase in size. The increase would not be indefinite because A is also eaten by organism B and, in the event of the population of A increasing, B would have a larger food supply and its population would also increase. More A would be consumed until a new equilibrium between the two populations was established.

Organism D: The arrow indicates that C is eaten by D. The sudden removal of C would therefore remove one source of food for D whose population would decrease. To what extent it would decrease would depend on how much the shortfall in food could be compensated for by consuming more of organism B. However, D is unlikely to disappear altogether as it has B as an alternative food source.

Organism E: Assuming that there is no other food source for organism E than organism C, then the complete removal of C would result in the extinction of E in due course. The time lag between the removal of C and the extinction of E would depend on the internal and external food stores of E.

Section C

1 Tutorial note

In part (a) (i) you should begin your explanation by mentioning the sources of water available to the plant. The major one is the available water, i.e. the water between the soil particles and to some extent the film of water over the particles. Another source, likely to be forgotten, is metabolic water originating as a by-product of biochemical processes, such as respiration, within the plant. While water uptake is part of the answer there are other processes involved before the plant can actually absorb the water. One of these is hydrotropism, a growth movement of plant roots towards the wettest area of the soil. Another point to mention before discussing the actual absorption of water is the large surface area provided by the finely divided lateral roots and the root hairs. In addition, the main roots branch widely to absorb water over a large region of the soil. The root hairs are the cells which absorb the water and you should refer to how they are adapted for this function. The description of the mechanism of absorption should include reference to the more negative water potential ψ in the root hair cell compared with the soil solution. This more negative ψ is due to salts and sugars in the sap of the root hair cell. Mention should also be made of absorption into and through cell walls.

In (a) (ii) you should again be aware of all the processes involved. It is a common error to give only one correct answer, which typically comprises an account of diffusion with no mention of the equally important process of active transport. Discussion of the latter should include the need for a source of energy, the fact that absorption occurs against a concentration gradient and the use of a carrier. The point that the ions are obtained from solution in the soil water and are absorbed through the root hairs should not be forgotten.

For part (b) the water potential gradient across the cortex and water movement via the endodermis and into the xylem should be included, followed by the transpiration pull created in the leaves by the evaporation of water through the stomata. In all, part (b) is extensive and care should be taken to give detail and yet be concise. Annotated diagrams may help, but detailed drawings of leaf, stem and root structure will only waste time. This could be better spent explaining words such as adhesion, cohesion, capillarity and transpiration pull as examinations at A level set out to test understanding rather than your ability to simply recall words and phrases.

Answer plan

(a) (i) Water

1. Sources – available soil water, metabolic by-product
2. Hydrotropism
3. Root hair adaptations – large surface area, thin-walled, no cuticle
4. Uptake – root hair cell sap has a more negative ψ than soil solution; osmosis carries water in; movement through cell walls alone

(ii) Ions

1. Taken in by root hair cells from soil solution
2. Diffusion – along concentration gradient
3. Active transport – against concentration gradient using energy and carrier

(b) Water transport in plant

1. Water potential gradient across cortex
2. Root pressure
3. Into xylem via endodermis using active transport
4. Along xylem: (i) continuous from root to leaves; (ii) no end walls in xylem to restrict flow; (iii) cohesion, adhesion, capillarity
5. Across leaf – osmotic gradient
6. Into atmosphere – through stomata by evaporation (transpiration pull)
7. Movement across cortex and leaf may also be through cell walls alone and plasmodesmata

2 Tutorial note

In the first part the key words are 'maintained and controlled'. Although the two processes are often connected it may be easier to deal with them separately. Remember to limit the answer to 'a mammal'. The maintenance of circulation is achieved mainly by the pumping action of the heart. Any details of the heart structure should be brief and limited to aspects which relate to its function. A short account of features of cardiac muscle such as its moderately powerful action and the fact that it is not fatigued would be better than an elaborate diagram of the heart. The remaining factors are recoil of the elastic walls of the arteries, and the veins' residual heart pressure, muscular contraction (explain this fully), inspiratory movements and gravity (in some situations). The 'control' of circulation can be conveniently divided into control of heartbeat and control of peripheral flow (blood vessels). The control of peripheral flow involves the autonomic nervous system controlling vasoconstriction and vasodilation of blood vessels.

The second part of the question deals with thermoregulation. The maintenance of constant blood temperature is achieved via mechanisms for losing heat when the blood temperature is above normal and conserving heat when it is below it. The amount of information required by the question as a whole does not allow, in the time given, for a full account of each mechanism. The major mechanisms are listed in the answer plan and a small amount of detail should be given to show how each affects blood temperature. For example, to describe shivering the candidate should state that the involuntary muscular contractions of the body generate heat which is then transferred to the blood.

All factors should relate to the particular mammal chosen. Hair erection, for example, has little effect, if any, in conserving heat in humans or whales. A classic error, even at A level, is to refer to blood capillaries moving nearer to or further from the skin surface when discussing vasodilation or vasoconstriction. The widening of the superficial arterioles (vasodilation) increases the flow of blood near to the body surface and allows more heat to be lost to the environment. The blood flows either in vessels close to the surface or in ones deeper in the skin. The vessels themselves do not move. Finally, it is important to mention control of these processes, in particular the role of the hypothalamus in detecting changes in blood temperature and the thermoregulatory centres of the brain.

In the last part of the question the key word is 'review'. The question as a whole involves many processes and details and one danger is that you might overrun the time allowance. It is therefore important to isolate the essential points. The other important word is 'immunity'; only those processes that confer immunity are required. The answer should include a review of antigen–antibody action. It should then include points such as the continued production of antibodies even after the initial infection has been overcome and how these immediately destroy the same infective agents if they enter the body on a subsequent occasion. More specific knowledge of the exact role and mechanism of the antigen–antibody reaction would be useful.

Answer plan

Maintenance of circulation

Arteries

Mechanism of heartbeat (SA node, AV node, Purkinje fibres, cardiac muscle features)
Recoil action of arterial wall

Veins

Residual heart pressure
Muscular action and role of pocket valves
Inspiratory movements
Gravity

Control of circulation

Control of heartbeat

Chemical including hormones
Nervous

Control of blood flow

Vasoconstriction
Vasodilation
Role of autonomic nervous system

Regulation of blood temperature

1. Role of hypothalamus in detecting blood temperature changes

Mechanisms to lower blood temperature

Sweating
Vasodilation
Lowering of hairs (some mammals)
Behavioural mechanisms, e.g. being nocturnal, avoiding direct sunlight

Mechanisms to raise blood temperature

Increased metabolic activity
Shivering
Vasoconstriction
Hair erection (some mammals)
Behavioural mechanisms, e.g. lying in sunlight

2. Control by thermoregulatory centres of the brain

Immunity

1. Antibody production by lymphocytes in response to foreign antigen
2. Production continues for some time and any subsequent infective agent is immediately destroyed

INDEX